ADVANCES IN X-RAY ANALYSIS

Volume 26

A Continuation Order Plan is available for this series. A continuation order will bring delivery of each new volume immediately upon publication. Volumes are billed only upon actual shipment. For further information please contact the publisher.

ADVANCES IN X-RAY ANALYSIS

Volume 26

Edited by

Camden R. Hubbard

National Bureau of Standards
Washington, D.C.

Charles S. Barrett
and Paul K. Predecki

University of Denver
Denver, Colorado

and

Donald E. Leyden

Colorado State University
Fort Collins, Colorado

Sponsored by
University of Denver Research Institute
and
JCPDS-International Centre for Diffraction Data

PLENUM PRESS • NEW YORK AND LONDON

The Library of Congress cataloged the first volume of this title as follows:

Conference on Application of X-ray Analysis.
Proceedings 6th- 1957- [Denver]

v. illus. 24-28 cm. annual.
No proceedings published for the first 5 conferences.
Vols. for 1958- called also: Advances in X-ray analysis, v. 2-
Proceedings for 1957 issued by the conference under an earlier name: Conference on Industrial Applications of X-ray Analysis. Other slight variations in name of conference.
Vol. for 1957 published by the University of Denver, Denver Research Institute, Metallurgy Division.
Vols. for 1958- distributed by Plenum Press, New York.
Conferences sponsored by University of Denver, Denver Research Institute.
1. X-rays—Industrial applications—Congresses. I. Denver. University. Denver Research Institute II. Title: Advances in X-ray analysis.
TA406.5.C6 58-35928

Library of Congress Catalog Card Number 58-35928
ISBN-13: 978-1-4613-3729-4 e-ISBN-13: 978-1-4613-3727-0
DOI: 10.1007/978-1-4613-3727-0

Proceedings of the 1982 Denver Conference on the Applications of X-Ray Analysis, held August 1–6, 1982, at the University of Denver, in Denver, Colorado

Plenum Press is a division of Plenum Publishing Corporation
233 Spring Street, New York, N.Y. 10013
Softcover reprint of the hardcover 1st edition 1982

FOREWORD

At the Denver X-Ray Conference, the topic for the plenary lectures alternates annually between x-ray diffraction and x-ray fluorescence. This year is a "diffraction" year, and the theme is accuracy in powder diffraction. Instead of comprehensive coverage, such as was attempted at the Accuracy in Powder Diffraction Meeting held at the National Bureau of Standards in 1978, this meeting focuses on recent developments in measurement accuracy of two-theta and intensity.

The focus on accuracy, from the practical point of view, is important in a wide range of x-ray diffraction measurements. Accurate data improve our ability to identify phases in a mixture using the Powder Diffraction File. Improved accuracy is essential for better characterization of the lattice, crystallite size, strain and structure. Finally, the accuracy of quantitative analysis is of great concern in many laboratories. The five invited papers of the plenary session give a broad perspective of recent activity throughout the world on uses of more accurate data, on methods to achieve greater accuracy, and on fundamental factors affecting the accuracy.

The scope of the conference, however, is much broader than that of the plenary session. The workshops lead off with many practical aspects of x-ray analysis. Many of the contributed papers expand on the theme of accuracy in x-ray powder diffraction. In particular, the session on XRD quantitative phase analysis provides an exceptional coverage of the limitations in quantitative analysis and of the techniques being employed to improve the results.

Assisting in the organization of the diffraction sessions of the conference and reviewing the papers has been a great pleasure. I wish to extend my appreciation to all of the workshop instructors, contributing speakers, and invited speakers who really made the meeting a success.

Camden R. Hubbard
Gaithersburg, MD
January, 1983

PREFACE

This volume constitutes the proceedings of the 1982 Denver Conference on the Applications of X-ray Analysis and is the 26th in the series. The conference was held August 2-6, 1982 at the University of Denver. The general chairmen were D. E. Leyden of Colorado State University and P. K. Predecki of the University of Denver, with C. S. Barrett of the University of Denver as honorary chairman. The conference advisory committee this year consisted of: C. S. Barrett-University of Denver, C. R. Hubbard-NBS, R. Jenkins-Philips Electronic Instruments, D. E. Leyden-Colorado State University and J. Russ-North Carolina State University. The continuing and vital contributions of this committee to the planning and execution of the conference are indispensable. We take this opportunity to thank the committee members for their willingness to serve in this capacity.

The invited conference chairman, C. R. Hubbard of the National Bureau of Standards, organized and chaired the conference plenary session entitled "Accuracy in X-Ray Powder Diffraction."

The names of the invited speakers on the program and the titles of their papers are listed below.

Robert L. Snyder, "Accuracy in Angle and Intensity Measurements in X-Ray Powder Diffraction"

Allan Brown, "Precision Lattice Parameter Measurements with Guinier Camera and Counter Diffractometer: Comparison and Reconciliation of Results"

Ron Jenkins, "Effects of Diffractometer Alignment and Aberrations on Peak Positions and Intensities"

W. Parrish and T. C. Huang, "Accuracy and Precision of Intensities in X-Ray Polycrystalline Diffraction"

C. R. Hubbard, "New Standard Reference Materials for X-Ray Powder Diffraction"

L. D. Calvert, G. J. Gainsford, A. F. Sirianni, and C. R. Hubbard, "A Comparison of Methods for Reducing Preferred Orientation"

Tutorial workshops on various topics in diffraction and fluorescence were held during the first two days of the conference. These are listed below with the names of the workshop organizers and instructors.

1D "Treatment of X-Ray Spectra - XRD and XRF." J. C. Russ-North Carolina State U. (chair), F. Schamber-Tracor Northern and D. Gedcke-EG&G Ortec

2D "Quantitative XRD." C. R. Hubbard-NBS (chair), R. L. Snyder-Alfred Univ. (chair), and R. P. Goehner-General Electric Co.

3D "Commercial Minicomputer Search/Match Methods." W. Parrish-IBM (chair), G. J. McCarthy-North Dakota State Univ., J. Padur-Bausch & Lomb, W. D. Stewart-Dapple Systems, G. L. Ayers-IBM, T. C. Huang-IBM, R. A. Sparks - Nicolet XRD Corp., R. Jenkins-Philips Electronic Instruments, W. N. Schreiner-Philips Laboratories, M. Fornoff-Rigaku/USA, and R. L. Snyder-Alfred Univ.

4D "XRD Sample Preparation." V. E. Buhrke-The Buhrke Co. (chair), M. C. Morris-NBS, D. K. Smith-The Pennsylvania State Univ., A. J. Gude-USGS, and P. L. Hauff-USGS.

5D "JCPDS Workshop on Manual Search Methods." C. O. Ruud-The Pennsylvania State Univ. (chair), G. J. McCarthy-North Dakota State Univ., and D. K. Smith-The Pennsylvania State Univ.

6D "Guinier Camera Alignment and Trouble Shooting." A. Brown-Studsvik Energiteknik AB (chair), and C. M. Foris-DuPont Co. (chair).

1F "Treatment of X-Ray Spectra-XRD and XRF." J. C. Russ-North Carolina State Univ. (chair), F. Schamber-Tracor Northern, and D. Gedcke-EG&G Ortec.

2F "Energy Dispersive Spectrometry." W. D. Stewart-Dapple Systems (Chair), H. Acree-IBM, R. Vane-Kevex Corp., and H. E. Marr-UPA Technology.

3F "Isotope Source XRF." J. R. Rhodes-Columbia Scientific Industries (chair), D. A. Coppell-Amersham Corp., D. Kalnicky-Princeton Gamma-Tech, and A. S. Klein-Woodside, CA.

4F "Wavelength Dispersive XRF." D. W. Beard-Siemens Corp. (chair), J. F. Croke-Philips Electronic Instruments, and M. F. Garbauskas-General Electric Co.

5F "XRF Sample Preparation." V. E. Buhrke-The Buhrke Co. (chair), J. F. Croke-Philips Electronic Instruments, J. Renault-New Mexico Bureau of Mines, and J. E. Taggart, Jr.-USGS.

6F "Fundamental Parameter, XRF." J. W. Criss-Criss Software (chair), D. Gedcke-EG&G Ortec, J. C. Russ-North Carolina State Univ., and M. F. Garbauskas-General Electric Co.

The workshop attendance this year was 230 out of a total of 316 conference attendees. We are particularly indebted to the workshop organizers and instructors who gave unselfishly of their time and talent to make the workshops a success.

We are grateful to C. R. Hubbard, the conference chairman, for organizing and chairing a plenary session of high quality and to all those who co-chaired the various contributed sessions. These were: C. S. Barrett, D. W. Beard, L. D. Calvert, C. M. Foris, J. V. Gilfrich, R. P. Goehner, R. Jenkins, J. J. LaBrecque, W. H. Lemons, H. E. Marr III, G. J. McCarthy, C. O. Ruud, E. Sabino, M. A. Short, D. K. Smith and R. Vane.

We are also grateful to the conference aids who worked long and unusual hours to make sure the conference ran smoothly. These were: Dorothy Barrett, Barb Cain, Penny Hudson, Cynthia King, Rich Miller, Dorothy Predecki, Stewart Presnall and Yvonne Shinton.

A special word of thanks and appreciation to Mildred Cain, the conference secretary, for running the whole show.

Paul Predecki
For the Conference Committee

Unpublished Papers

The following papers were presented orally only and are not published here for a variety of reasons.

"Influence Coefficients for Cement: A Comparison of Calculation Methods," John A. Anzelmo and Harley C. Renton, Bausch & Lomb, Dearborn, Michigan; and George C. Kistler, Louisville Cement Co., Logansport, Indiana

"Assessment of Precision and Accuracy of X-Ray Fluorescence Determination of Sulfur in Petroleum Products," Leif H. Christensen and Kaj Heydorn, Risø National Laboratory, Roskilde, Denmark; and Stig Røen, Haldor Topsøe A/S, Lyngby, Denmark

"Data Sets for Evaluation of Powder Diffraction Search/Match Algorithms: A Progress Report," Wilson H. De Camp, Food and Drug Administration, Washington, D.C., and Gregory P. Hamill, General Telephone and Electronics Laboratories, Inc., Waltham, Massachusetts.

"A Flexible, RSX-11M, PDP-11/34 Based Multiple Spectrometer Control System," G. J. DeMott, Corning Glass Works, Corning, New York

"A Standardless Method of Quantitative XRD Using Factor Analysis," J. H. Fang, University of Alabama, University, Alabama; T. H. Starks, Southern Illinois Univ., Carbondale, Illinois; and L. Zevin, Ben-Gurion Univ. of the Negev, Israel

"Application of Search/Match Techniques to Special Purpose Data Bases," Phoebe L. Hauff and Richard E. Phillips, USGS, Denver, Colorado; and Robert A. Sparks, Nicolet XRD Corp., Fremont, Calif.

"Procedures for Obtaining High Quality Reference Patterns," Marlene C. Morris, National Bureau of Standards, Washington, D.C.

"Automation of an X-Ray Diffractometer with an Electron Column Automation System," Mark E. Palmer, Rockwell Hanford Operations, Richland, Washington

"A Comparison of the Design Parameters and Use of Curved Crystal and Flat Crystal Goniometers and Scanners," D. F. Sermin, B. J. Price and W. Wittwer, Bausch & Lomb, Sunland, California

"Powder Pattern Simulation, Models for Testing Computer Algorithms," Deane K. Smith and M. E. Zolensky, The Pennsylvania State Univ., University Park, Pennsylvania; and Monte C. Nichols, Sandia National Laboratories, Livermore, California

"Pitfalls in Using Filter Paper Disks as Standard Supports in X-ray Fluorescence Spectrometry," G. K. H. Tam and G. Lacroix, Health and Welfare Canada, Ottawa, Ontario.

"A New Lixiscope Fluoroscopic System," Lo I Yin, Jacob I. Trombka, and Arthur P. Ruitberg, NASA Goddard Space Flight Center, Greenbelt, Maryland

ACKNOWLEDGEMENTS

We acknowledge with gratitude sponsorship of the Conference by the University of Denver through its Research Institute, by the JCPDS International Centre for Diffraction Data and by Colorado State University Department of Chemistry through the participation of D. E. Leyden.

CONTENTS

I. ACCURACY IN X-RAY POWDER DIFFRACTION

Accuracy in Angle and Intensity Measurements in X-Ray Powder Diffraction. 1
Robert L. Snyder

Precision Lattice Parameter Measurements with Guinier Camera and Counter Diffractometer: Comparison and Reconciliation of Results 11
Allan Brown

Effects of Diffractometer Alignment and Aberrations on Peak Positions and Intensities 25
Ron Jenkins

Accuracy and Precision of Intensities in X-Ray Polycrystalline Diffraction 35
W. Parrish and T. C. Huang

New Standard Reference Materials for X-Ray Powder Diffraction 45
Camden R. Hubbard

Precision and Reproducibility of Lattice Parameters from Guinier Powder Patterns: Follow-up and Assessment . . 53
A. Brown and C. M. Foris

II. SEARCH/MATCH PROCEDURES, POWDER DIFFRACTION FILE

POWDER-PATTERN: A System of Programs for Processing and Interpreting Powder Diffraction Data. 63
Nikos P. Pyrros and Camden R. Hubbard

An Evaluation of Some Profile Models and the Optimization Procedures Used in Profile Fitting 73
Scott A. Howard and Robert L. Snyder

Computer-Aided Qualitative X-Ray Powder Diffraction Phase Analysis. 81
Raymond P. Goehner and Mary F. Garbauskas

The JCPDS Data Base - Present and Future 87
Winnie Wong-Ng, Mark Holomany, W. Frank McClune and Camden R. Hubbard

Search/Match Implications of the Frequency Distribution of "d" Values in the JCPDS Powder Data File 89
Shozo Toyohisa, Iwao Fujiwara, Takuji Ui and Eiichi Asada

Computer Search/Match of Standards Containing a Small Number of Reflections. 93
T. C. Huang, W. Parrish and B. Post

Comparison of the Hanawalt and Johnson-Vand Computer Search/Match Strategies 99
Satyam C. Cherukuri, Robert L. Snyder and Donald W. Beard

III. QUANTITATIVE XRD ANALYSIS

A Comparison of Methods for Reducing Preferred Orientation . 105
L. D. Calvert, A. F. Sirianni, G. J. Gainsford and C. R. Hubbard

The Dramatic Effect of Crystallite Size on X-Ray Intensities. 111
James P. Cline and Robert L. Snyder

X-Ray Diffraction Intensity of Oxide Solid Solutions: Application to Qualitative and Quantitative Phase Analysis. 119
Ronald C. Gehringer, Gregory J. McCarthy, R. G. Garvey and Deane K. Smith

Quantitative Analysis of Platelike Pigments by X-Ray Diffraction. 129
P. Kamarchik and J. Ratliff

Preparation and Certification of Standard Reference Materials to be Used in the Determination of Retained Austenite in Steels 137
George E. Hicho and Earl E. Eaton

Profile Fitting for Quantitative Analysis in X-Ray Powder Diffraction. 141
Walter N. Schreiner and Ron Jenkins

XRD Quantitative Phase Analysis Using the NBS*QUANT82 System. 149
Camden R. Hubbard, Carl R. Robbins and Robert L. Snyder

SCRIP - Fortran IV Software for Quantitative XRD. . . . 157
Edward R. Wong, John Yeko, Philip Engler and Richard A. Gerron

IV. XRD APPLICATIONS AND AUTOMATION

An X-Ray Diffraction Study of $CaNi_5$ Hydrides Using In Situ Hydriding and Profile Fitting Methods 163
G. J. Gainsford, L. D. Calvert, J. J. Murray and J. B. Taylor

The Measurement of Thermally Induced Structural Changes by High Temperature (900°C) Guinier X-Ray Powder Diffraction Techniques 171
T. G. Fawcett, P. Moore Kirchhoff and R. A. Newman

The Use of X-Ray Diffraction and Infrared Spectroscopy to Characterize Hazardous Wastes 181
Douglas S. Kendall

Comparison of X-Ray Powder Diffraction Techniques . . . 185
K. Das Gupta

The Use of Multi-Scan Diffraction in Phase Identification. 189
Gordon S. Smith and M. C. Nichols

An Automated X-Ray Diffractometer for Detection and Identification of Minor Phases. 197
John C. Russ and Thomas M. Hare

Time Share Computer Capability for Phase Identification by X-Ray Diffraction 205
Andrew M. Wims

V. X-RAY STRESS DETERMINATION, POSITION SENSITIVE DETECTORS, FATIGUE AND FRACTURE CHARACTERIZATION

The Use of Mn-Kα X-Rays and a New Model of PSPC in Stress Analysis of Stainless Steel. 209
Yasuo Yoshioka, Ken-ichi Hasegawa and Koh-ichi Mochiki

Measurement of Stress Gradients by X-Ray Diffraction . . 217
J. M. Sprauel, M. Barral and S. Torbaty

A Method for X-Ray Stress Analysis of Thermochemically Treated Materials. 225
Rolf A. Prümmer and H. W. Pfeiffer-Vollmar

Application of a Position Sensitive Scintillation Detector for Nondestructive Residual Stress Measurements Inside Stainless Steel Piping. 233
C. O. Ruud, P. S. DiMascio and D. M. Melcher

On the X-Ray Diffraction Method of Measurement of Triaxial Stresses with Particular Reference to the Angle $2\theta_o$. 245
S. Torbaty, J. M. Sprauel, G. Maeder and P. H. Markho

Direct Determination of Stress in a Thin Film Deposited on a Single-Crystal Substrate from an X-Ray Topographic Image. 255
Wayne S. Berry

The Determination of Elastic Constants Using a Combination of X-Ray Stress Techniques. 259
Charles Goldsmith and George A. Walker

One-Dimensional, Curved, Position-Sensitive Detector for X-Ray Diffractometry. 269
B. Sleaford, V. Perez-Mendez and C. N. J. Wagner

A Phi-Psi-Diffractometer for Residual Stress Measurements . 275
C. N. J. Wagner, M. S. Boldrick and V. Perez-Mendez

X-Ray Fractography on Fatigue Fractured Surface. 283
Shotaro Kodama, Hiroshi Misawa and Yuji Sekita

X-Ray Diffraction Observation of Fracture Surfaces of Ductile Cast Iron. 291
Zenjiro Yajima, Yukio Hirose and Keisuke Tanaka

Analytical and Experimental Investigation of Flow and Fracture Mechanisms Induced by Indentation in Single Crystal MgO. 299
T. Larchuk, T. Kato, R. N. Pangborn and J. C. Conway, Jr.

X-Ray Diffraction Study of Shape Memory in Uranium-Niobium Alloys 307
D. A. Carpenter and R. A. Vandermeer

VI. NEW XRF INSTRUMENTATION AND TECHNIQUES

X-Ray Fluorescence Analysis Using Synchrotron Radiation. 313
J. V. Gilfrich, E. F. Skelton, D. J. Nagel, A. W. Webb, S. B. Qadri and J. P. Kirkland

Energy Resolution Measurements of Mercuric Iodide Detectors Using a Cooled FET Preamplifier. 325
Lawrence Ames, William Drummond, Jan Iwanczyk and Andrzej Dabrowski

X-Ray Polarization: Bragg Diffraction and X-Ray Fluorescence 331
John D. Zahrt

A New Technique for Radioisotope-Excited X-Ray Fluorescence 337
J. J. LaBrecque and W. C. Parker

Bragg-Borrmann X-Ray Spectroscopy from a Line Source. . . 341
K. Das Gupta

VII. XRF COMPUTER SYSTEMS AND MATHEMATICAL CORRECTIONS

Automated Qualitative X-Ray Fluorescence Elemental Analysis 345
Mary F. Garbauskas and Raymond P. Goehner

A New Method for Quantitative X-Ray Fluorescence Analysis of Mixtures of Oxides or Other Compounds by Empirical Parameter Methods 351
Michael Mantler

FPT: An Integrated Fundamental Parameters Program for Broadband EDXRF Analysis Without a Set of Similar Standards 355
D. A. Gedcke, L. G. Byars and N. C. Jacobus

A Comparison of the XRF11 and EXACT Fundamental Parameters Programs When Using Filtered Direct and Secondary Target Excitation in EDXRF 369
Ronald A. Vane

A Generalized Matrix Correction Approach for Energy-Dispersive X-Ray Fluorescence Analysis of Paint Using Fundamental Parameters and Scattered Silver Kα Peaks 377
Leif Højslet Christensen and Iver Drabæk

Multielement Analysis of Unweighed Biological and Geological Samples Using Backscatter and Fundamental Parameters. . 385
K. K. Nielson and R. W. Sanders

A Correction Method for Absorption in the Analysis of Aerosols by EDX Spectrometry. 391
A. S. M. de Jesus, D. J. van der Bank and E. S. Wesolinski

XRF Analysis by Combining the Standard Addition Method with Matrix-Correction Models 395
Peter B. De Groot

Accurate Geochemical Analysis of Samples of Unknown Composition 401
J. Kikkert

VIII. XRF GENERAL APPLICATIONS

XRF Analysis of Vegetation Samples and Its Application to Mineral Exploration. 409
T. K. Smith and T. K. Ball

Application of XRF to Measure Strontium in Human Bone _In Vivo_ . 415
L. Wielopolski, D. Vartsky, S. Yasumura, and S. H. Cohn

Determination of Boron Oxide in Glass by X-Ray Fluorescence Analysis 423
T. Arai, T. Sohmura and H. Tamenori

Measurement of Composition and Thickness for Single Layer Coating with Energy Dispersive XRF Analysis. 431
Robert Shen and Alan Sandborg

Determination of Light Elements on the Chem-X Multichannel Spectrometer 437
Y. M. Gurvich, A. Buman and I. Lokshin

Simultaneous Determination of 36 Elements by X-Ray Fluorescence Spectrometry as a Prospecting Tool . . . 443
Clive E. Feather and Fritz C. Baumgartner

Elemental Analysis of Geological Samples Using a Multichannel, Simultaneous X-Ray Spectrometer. 451
J. B. Cross and L. V. Wilson

Chemical Analysis of Coal by Energy Dispersive X-Ray Fluorescence Utilizing Artificial Standards. 457
Bradner D. Wheeler

Author Index. 467

Subject Index 469

ACCURACY IN ANGLE AND INTENSITY MEASUREMENTS IN X-RAY POWDER DIFFRACTION

Robert L. Snyder

N.Y.S. College of Ceramics
Alfred University
Alfred, N.Y.

INTRODUCTION

The advent of computer automation and profile fitting techniques in powder diffraction, along with a general solution to the problem of preferred orientation, has opened a series of new horizons for this method. The new levels of accuracy attainable have brought us to the threshold of routine reliable qualitative phase identification, high precision quantitative analysis and the ability to perform crystal structure analysis on some of the most important technological materials. It has been primarily the question of accuracy which has held up these developments until now.

A large percentage of the 38,000 standards in the current powder diffraction file (PDF)[1] were determined with highly error prone Debye-Scherrer and uncalibrated, and often misaligned, diffractometers. The average $\Delta 2\theta$ error in the 1638 indexable cubic patterns in sets 1 through 24 of the PDF is 0.091 degrees[2]. Modern automated techniques using profile fitting can produce patterns with an average $\Delta 2\theta$ 1000 times smaller. This paper will survey the developments leading to the current levels of accuracy and attempt to look into the near future for new applications.

THE ASSESSMENT OF DIFFRACTION INTENSITIES

The evaluation of the absolute error in diffraction intensity measurements still appears to be a Platonian ideal. The closest we can get to a "true" intensity is to calculate it, when the crystal structure is well known. However even calculated intensities are subject to a large number of errors and uncompensated distortions. Examples of these are: temperature parameters, when known at all, are subject to large error. The atomic scattering factors only approximate the true oxidation state in most inorganic compounds. The relationship between integrated and peak intensities is not well known and the profile models most commonly used to derive peak intensities are poor approximations. Even well known factors like the polarization correction give rise to errors when applied to the highly mosaic graphite monochrometer. However, all of these problems typically result in only a few percent error. A more serious problem is the assumption of crystalline and stoichiometric perfection which is seldom the case in most materials. The most reasonable function in use today is:

$$R = \frac{\Sigma |I_{obs} - I_{calc}|}{\Sigma I_{calc}}$$

This function has been used by a number of authors, incorporated into NBS*AIDS80[3] and adopted by the JCPDS. Its many deficiencies are only offset by the lack of a better alternate. Frevel[4] was the first to publish a function to measure intensity error but it contains a qualitative term and is of limited usefulness.

The JCPDS associateship at NBS has for many years adopted the practical approach of evaluating intensity reproducibility from multiple mountings and measurements. However these relative intensities will only relate to samples of similar crystallite size, mosaisity and stoichiometry measured on a diffractometer with Cu radiation and side drifted into the sample holder. In even the best of cases we find relative intensity discrepancies, from those on the PDF card, on the order of three to five percent.

The establishment of the fact that spherical agglomerates of crystallites will eliminate preferred orientation[5] has led to the first general method for eliminating this most severe source of error. The spray drying of crystallites under ten microns in size when properly done will produce agglomerates in the range of 50 microns. As long as the agglomerate size is kept less than 100 microns, reliable, reproducible orientation free intensity measurements can be made. The only remaining factors which can affect the background corrected relative intensities are the choice of a different wavelength and the choice of the diffractometer divergence slit or the use of a theta compensating slit. Thus spray drying of finely ground crystallites puts all factors affecting relative intensities for a particular material under the control of the experimentalist. The solution to the classical problem of large relative intensity errors is to remove them at the time of the experiment, by forming spherical agglomerates, rather than to face the unsolved problem of assessing them a posteriori.

The measurement and evaluation of the accuracy of absolute intensities in multi-phase mixtures is an even more complex problem. Factors affecting crystallinity and stoichiometry[6] will strongly affect the relative intensities from one sample to another and the absolute intensities for any particular sample. The removal of preferred orientation has exposed the serious affect of particle size on absolute intensities[7] and indicates deficiencies in our current models for the phenomenon known as microabsorption. Since these affects will strongly influence quantitative analysis and also limit the development of highly reliable second generation computer search match algorithms, solutions are needed before the next major advances in powder diffraction applications can occur.

THE ASSESSMENT OF DIFFRACTION ANGLES

DeWolff[8] published the first quantitative figure of merit for the metric aspects of a powder diffraction pattern. This function, called M_{20}, was intended for the evaluation of the correctness of a computer indexing. Frevel[4] also published a function to assess the error in d values but this, like his intensity function, contains a qualitative term and therefore cannot be evaluated in general. An extensive study[2] of possible metric figures of merit led Smith and Snyder[9] to propose the function F_N for general use:

$$F_N = \frac{N_{observed}}{|\overline{\Delta 2\theta}|\, N_{possible}}$$

F_N serves the dual application of evaluating the error in two theta angles and indicating the validity of a computer indexing. It has gained wide acceptance by authors, the JCPDS, the IUCr and the crystallographic journals. It should be noted however that this function does not really measure accuracy. Since the calculated two theta values, used in determining the average delta two thetas, are derived from the lattice parameters which in turn are derived from a least squares refinement of the observed angles. F_N is really a measure of internal consistancy. Absolute accuracy in the measurement of angles can only be attained by the use of an internal standard with accurately known lattice parameters. The greatest advance in attaining the goal of highly accurate published powder patterns has been the development and general distribution of the computer program NBS*AIDS80[3]. This program not only evaluates the metric figures of merit but evaluates the unit cell, computes the reduced cell, checks the chemical formula and molecular weight, the observed and calculated densities and performs a large number of other consistency checks. It should be used by all potential authors of powder patterns. This program along with the adoption of a set of publication standards[10] should, in time, produce a much more reliable data base.

IMPROVED ACCURACY VIA COMPUTER AUTOMATION

The development and widespread use of computer automated powder X-ray diffraction during the 1970's has lead to a dramatic improvement in accuracy. This has resulted from the introduction of programmed intelligence into both data collection and data reduction. Most of the data collection algorithms now in use fall into the "dumb", fixed angle increment move and fixed time count, category. Limited extensions toward a "smart" algorithm were made by Jenkins[11], Mallory and Snyder[12] and Goehner[13]. The first general extension of the computer's ability to make decisions and optimize into the data collection process, came with the development of AUTO[14]. The AUTO algorithm optimizes all aspects of powder data collection to produce minimum error, except for the conventional pattern scan for qualitative analysis. The world still awaits the development of an intelligent pattern scanning algorithm which will minimize scan time and/or maximize peak detection limits.

The principal effect of computer automation on accuracy and precision to date, has come from the development of highly intelligent data reduction procedures. Most of the research in this area has been on the development of relatively rapid first or second derivative techniques for finding peaks and on slower but more precise profile fitting methods to achieve high accuracy. All current peak finding methods use a data file of 2θ and intensity values collected in the "dumb" manner described above. The procedures typically involve five steps: 1. Background determination, 2. Data Smoothing, 3. Spectral stripping, 4. Peak location and 5. Profile fitting.

1. Background Determination. The principal problem with all peak finding algorithms is that they rely on heuristic methods for determining background. Of the nine published procedures for determining the background level in a powder pattern, none are based on a theoretically sound foundation and none will work for complex patterns with extensive line overlap. The problem in complex patterns is that at high angles we reach a continuum of peaks and the pattern never returns to true background. Since all background finding techniques rely on the sensing of true background points across the scan, and the fitting of various types of polynomials through overlapping 2θ segments, they are doomed to failure. The most commonly used techniques[15-19] all work

reasonably well for simple and moderately complex patterns. However the location of true background remains the most important unsolved problem limiting the development of a "turn key" pattern analysis system.

2. Data Smoothing. Statistical fluctuations and the possible presence of noise spikes in the intensity measurements can lead to the detection of false peaks. Three basic approaches have been taken to minimize these problems. The first and oldest approach is to improve the signal to noise ratio (S/N)[20] by summing the results of multiple scans. This approach is called box car averaging or CAT scanning. Since identical results will be obtained by simply increasing the count time at each point when the data is collected, this technique is no longer used. Another approach has been developed by Kane and Fisher[21]. They use the fast Fourier transform algorithm to transform the entire pattern into frequency space. A Gaussian filter is then applied and the pattern is transformed back into 2θ space. This procedure is being successfully applied, however the lack of a theoretical basis for the parameters used in the filter do not permit the computation of the affect on the S/N ratio or the subsequent loss in resolution. Future work in this area could produce a superior method.

Most of the current algorithms for data smoothing have adopted a least squares polynomial fit to small 2θ regions. This method is referred to as digital filtering. Savitzky and Golay[22] developed a very fast computational approach for this method using a set of convoluting integers and a scaling factor. In addition to smoothing this method has the advantage of producing a polynomial expression for each peak region so that first and second derivatives are readily computed. Although this technique has won the widest acceptance it should be emphasized that it, like any other smoothing method, will distort the data and reduce the resolution. Thus, unless a calibration step is incorporated somewhere in the procedure, accuracy will be degraded.

3. Spectral Stripping. Diffractometer algorithms often incorporate a routine for the removal of $K_{\alpha 2}$ peaks. Two approaches to this have been pursued. The first is to remove the $K_{\alpha 2}$ peaks by a Fourier deconvolution[23], while the Rachinger method[24] subtracts half of the intensity due to the $K_{\alpha 1}$ from the calculated $K_{\alpha 2}$ position. Both of these procedures assume that the profiles of the two peaks are identical. However, since the $K_{\alpha 2}$ is more wavelength broadened than the $K_{\alpha 1}$, both procedures fail to accurately remove the $K_{\alpha 2}$ peaks. Ladell et al.[25] modified the Rachinger model to allow for the difference in peak shapes, producing an algorithm which gives a minimum amount of high angle "ringing". Since this ringing seldom produces peaks above background, the procedure is effective.

4. Peak Finding. Jenkins[11], Goehner[13] and Gobel[26] have used first derivative peak maxima finding procedures. However, since these methods cannot detect shoulders indicating peak overlap, most algorithms[27] use the second derivative procedure proposed by Savitsky and Golay[22]. The number of peaks found in any procedure will depend on the amount of data smoothing and the S/N ratio which is a direct function of the count time. The number of false peaks found can be minimized by correlating the smoothing parameters with the count time.

The accuracy of the peak locations will depend to some extent on each of the above four steps. The peaks produced by any algorithm must be corrected for both instrumental and sample abberations. A factor of two improvement in accuracy can routinely be achieved[12] by applying an external standard calibration curve. This curve is obtained by determining $\Delta 2\theta$ vs. 2θ for a set of pure standards of well known lattice parameter. Another factor of two improvement in accuracy of peak location can be achieved by the use of an internal standard[28]. Since sample displacement is one of the worst sources of peak shifts it is recommended that an internal standard be used whenever high accuracy is desired.

5. Profile Fitting. The asymmetric shape of an X-ray diffraction peak is the result of the convolution of eight different functions. A very large increase in accuracy can be achieved by optimizing the match between the observed intensities and an analytical profile model. Brown and Edmonds[29] have reported average $\Delta 2\theta$ values in the $.001^{\circ}$ range while Parrish[30] has achieved values in the range of $.0001^{\circ}$. The approaches various researchers have taken, both in selecting the profile model and the optimization procedure, have been recently evaluated by Howard and Snyder[31]. The best profile fits published to date, have been obtained with the multiple Lorentzian approach of Parrish and co-workers and the split-Pearson function of Brown and Edmonds.

Profile fitting not only achieves extremely high precision but since a large number of intensity measurments go into the location of each peak there is a commensurate improvement in the statistical measures of both peak location and intensity. The problem of tracking the original counting statistics through the optimization procedures used in profile fitting in order to be able to predict the uncertainties in the refined parameters, has yet to be solved. However it is clear that the uncertainties decrease dramatically.

FUTURE APPLICATIONS OF POWDER DIFFRACTION

A number of very recent developments promise to help extend the applications of X-ray powder diffraction beyond the horizons opened by the techniques reviewed above. The most important of these is the development of synchrotron radiation sources. The growing number of these sources, and in particular the new National Synchrotron Light Source at Brookhaven National Laboratory, have made this new tool available to most researchers. A synchrotron source has intensities many orders of magnitude greater than the conventional laboratory generator. In addition, it has a continuous wavelength distribution from very low wavelengths out to well beyond the air absorption limits. These sources should bring new life to the area of energy dispersive diffraction.

The development of extremely efficient position sensitive detectors (PSD) has already led Gobel[32] to some exciting applications for the study of short time duration phenomena. These detectors have been receiving considerable attention and it seems reasonable to predict a number of new applications from them. Two recent designs[33-34] for Guinier geometry diffractometers, one of them using a PSD, also portend new applications.

The principal areas where I expect to see major developments in the near future are:

1. Micro-phase Analysis. The conventional limits of detectability for X-ray powder diffraction are usually quoted in the range of 1%. However, since the S/N ratio is a function only of the number of counts collected at each point, there is in fact no theoretical lower limit to the concentration of detectable phases. There is of course a practical limit to the amount of time which can be spent at each point in a scan. We have recently shown that 1 ppm of Si_2O in Al_2O_3 can be observed at a >3 sigma significance level by counting for 3000 seconds per point[35]. The future extension of routine phase analysis even to grain boundary phases, I believe, can be achieved with some combination of synchrothon or rotating anode sources, PSD's, long count times and automated S/N optimization.

2. Reliable Computer Phase Identification. The developments reviewed above and the current work on the remaining problems will reduce most phase identification problems to a fully automated status. The current approach[36] to a second generation computer search algorithm is only the first step. As the currently attainable accuracy in 2θ measurements becomes more widely achieved, a purely metric search will become feasible. The current computer indexing methods coupled with the cell reduction procedures in AIDS80 will allow phase identification purely from lattice parameters using the Crystal Data, data base. The techniques described by Parrish for isolating the sample broadening parameter S, should, in principle, lead to the separation of the lines due to different phases in mixtures so that they may be independently indexed and identified.

3. Highly Accurate Quantitative Analysis. This most desirable goal is still somewhat out of reach. The solution to the problem of preferred orientation coupled with an optimized data

collection algorithm such as AUTO and the sophisticated data reduction in NBS*QUANT82[37], are major steps toward this goal. However the effects of crystallite size and perfection require more research. I think we shall see routine amorphous phase analysis, from the area under the amorphous scattering peak, in the near future.

4. Analysis of Rapid Events. We have already seen developments in this area using PSD's. I believe that synchrotron radiation and energy dispersive diffraction will produce other applications.

5. Routine Crystal Structure Analysis. Every one of the developments reviewed in this paper have a bearing on this topic. The increasing levels of accuracy in both 2θ and intensity, the new developments in experimental procedures and the progress currently being made[38] on Reitveld refinement techniques all lead to the most exciting potential application of all, routine structure analysis from powders. Since one of the most serious impediments in the development of materials science is our inability to readily characterize the structures of real materials, which may never be obtained as single crystals, this application holds the promise of having the most far-reaching effects on both science and technology.

REFERENCES

1. JCPDS International Center for Diffraction Data, 1601 Park Lane, Swarthmore, PA 19081.
2. Snyder, R. L., Johnson, Q. C., Kahara, E., Smith, G. S. and Nichols, M. C., "An Analysis of the Powder Diffraction File," Lawrence Livermore Laboratory (UCRL-52505), 61 pages (June 1978).
3. Hubbard, C. R., Stalick, J. K. and Mighell, A. D., "NBS*AIDS80: A FORTRAN Program to Evaluate Crystallographic Data," Adv. X-ray Anal. 24, 99-109 (1981).
4. Frevel, L. K. and Adams, C. E., "Quantitative Comparison of Powder Diffraction Patterns," Anal. Chem. 40 [8] 1335-1340 (1968).
5. Smith, S. T., Snyder, R. L. and Brownell, W. E., "Minimization Preferred Orientation in Powders by Spray Drying", Adv. X-ray Anal., 22, 77-88 (1979).
6. McCarthy, G. J., Gehringer, R. C., Smith, D. K., Pfoertsch, D. E. and Kobel, R. L., "Internal Standards for Quantitative X-ray Phase Analyses: Crystallinity and Solid Solution," Adv. X-ray Anal. 24, 253-264 (1981).
7. Cline, J. and Snyder, R. L., "The Dramatic Effect of Crystallite Size on X-ray Intensities," Adv. X-ray Anal., this volume.
8. DeWolff, P. M., "A Simplified Criterion for the Reliability of Powder Pattern Indexing," J. Appl. Cryst. 1, 108 (1968)

9. Smith, G. S. and Snyder, R. L., "FN: A Criterion for Rating Powder Diffraction Patterns and Evaluating the Reliability of Powder Pattern Indexing," J. Appl. Cryst., 12, 60-65 (1979).
10. Calvert, L. D., Flippen-Anderson, J. L., Hubbard, C. R., Johnson, Q. C., Lenhert, P. G., Nichols, M. C., Parrish, W., Smith, D. K., Smith, G. S., Snyder, R. L. and Young, R. A., "Standards for the Publication of Powder Patterns: The American Crystallographic Association Subcommittee Final Report," Accuracy in Powder Diffraction, National Bureau of Standards Special Publication 567, p. 513-536 (1980).
11. Jenkins, R., "Quantitative Analysis with the Automatic Powder Diffractometer," Norelco Rep. 22 [1] 7-12 (1975).
12. Mallory, C. L. and Snyder, R. L., "The Alfred University X-ray Powder Diffraction Automation System," N.Y.S. College of Ceramics Technical Publication No. 144, 172 pages (1979).
13. Goehner, R. P. and Hatfield, W. T., "PEAKSEARCH: A Program to Find Diffraction Peaks," private communication (1981).
14. Snyder, R. L., Hubbard, C. R. and Panagiotopoulos, N. C., "Auto: A Real Time Diffractometer Control System," National Bureau of Standards Publication NBSIR 81-2229, 102 pages (1981).
15. Sonneveld, E. J. and Visser, J. W., "Automatic Collection of Powder Data from Photographs," J. Appl. Cryst., 8 [1] (1975).
16. Segmuller, A. and Cole, H., "Procedures to Run an Automated Micro-densitometer on a Shared Computer System," Adv. X-ray Anal., 14 338-351 (1970).
17. Huang, T. C. and Parrish, W., "Accurate and Rapid Reduction of Experimental X-ray Data," Appl. Phys. Lett., 27 [3] 123-4 (1975).
18. Goehner, R., "Background Subtract Subroutine for Spectral Data," Anal. Chem., 50 [8] 1223-4 (1978).
19. Mallory, C. L. and Snyder, R. L., "Threshold Level Determination from Digital X-ray Powder Diffraction Patterns," Accuracy in Powder Diffraction, National Bureau of Standards Special Publication 567, p. 93 (1980).
20. Rex, R. W., "Numerical Control X-ray Powder Diffractometry," Adv. X-ray Anal., 10 366-73 (1966).
21. Kane, W. T. and Fisher, G. R., "Conditioning of Powder Diffractometer Data Using Fourier Digital Filtering Techniques," Abstract D5, Am. Cyrst. Assoc. Meeting, University of Hawaii, Honolulu, Hawaii, March 28, 1979.
22. Savitzky, A. and Golay, M., "Smoothing and Differentiation of Data by Simplified Least Squares Procedures," Anal. Chem., 36 [8] 1627-39 (1964).
23. Gangulee, A., "Separation of the α_1-α_2 Doublet in X-ray Diffraction Profiles," J. Appl. Crystallogr., 3 272-7 (1970).

24. Rachinger, W. A., "A Correction for the 112 Doublet in the Measurement of Widths of X-ray Diffraction Lines," J. Sci. Instrum., 25 [7] 254-5 (1948).
25. Ladell, J., Zagofsky, A. and Pearlman, S., "CuK$_2$ Elimination Algorithm,: J. Appl. Cyrstallogr., 8 499-506 (1970).
26. Jobst, B. A. and Gobel, H. E., "IDENT - A Versitile Microfile-based System for Fast Interactive XRPD Analysis," 25, 273-282 (1982).
27. Mallory, C. L. and Snyder, R. L., "The Control and Processing of Data from an Automated X-ray Powder Diffractometer," Adv. X-ray Anal., 22, 121-132 (1979).
28. Snyder, R. L., Hubbard, C. R. and Panagiotopoulos, N. C., "A Second Generation Automated Powder Diffractometer Control System," Adv. X-ray Anal. 25, 245-260 (1982).
29. Brown, A. and Edmonds, J. W., "The Fitting of Powder Diffraction Profiles to an Analytical Expression and the Influence of Line Broadening Factors," Adv. X-ray Anal., 23, 361-374 (1980).
30. Parrish, W. and Huang, T. C., "Accuracy of the Profile Fitting Method for X-ray Polycrystalline Diffraction," in Accuracy in Powder Diffraction, NBS Special Pub. 567, 95-110 (1980).
31. Howard, S. A. and Snyder, R. L., "An Evaluation of Some Profile Models and the Optimization Procedures Used in Profile Fitting," Adv. X-ray Anal., this volume.
32. Gobel, H. E., "The Use and Accuracy of Continuously Scanning Position Sensitive Detector Data in X-ray Powder Diffraction," Adv. X-ray Anal., 24, 123-138 (1981).
33. Visser, J. W., private communication (1981).
34. Gobel, H. E., "A Guinier Diffractometer with a Scanning Position Sensitive Detector," Adv. X-ray Anal., 25, 315-324 (1982).
35. Bliss, M., "The Application of X-ray Powder Diffraction to Micro-phase Analysis," N.Y.S. College of Ceramics, Senior Thesis (1981).
36. Snyder, R. L., "A Hanawalt Type Phase Identification Procedure for a Minicomputer," Adv. X-ray Anal. 24, 83-90 (1981).
37. Snyder, R. L. and Hubbard, C. R., "NBS*QUANT82: A System for Quantitative Analysis by Automated X-ray Powder Diffraction," NBS Special Publication, in press, (1982).
38. Young, R. A. and Wiles, D. B., "Application of the Rietveld Method for Structure Refinement with Powder Diffraction Data," Adv. X-ray Anal. 24, 1-23 (1981).

PRECISION LATTICE PARAMETER MEASUREMENTS WITH GUINIER CAMERA AND COUNTER DIFFRACTOMETER: COMPARISON AND RECONCILIATION OF RESULTS

Allan Brown

Studsvik Energiteknik AB
S-611 82 Nyköping, Sweden

1. INTRODUCTION

Different procedures used in precision measurements of lattice parameters are, strictly, only valid if they can be shown to give results that are mutually reproducible. For this purpose reproducibility is defined in terms of the parameters a_i and standard deviations σ_i obtained for X-ray specimens of one or more reference materials. The requirement is that all systematic errors should be minimized to a level below that of the random measurement errors. Where these have a Gaussian distribution the significance of the difference, Δa_o, between two measurements can then be tested by evaluating $K \stackrel{\text{D}}{=} \Delta a_o/(\sigma_1^2 + \sigma_2^2)^{\frac{1}{2}}$. Thus, if $K < 2$ the difference, Δa_o, cannot be distinguished from the effects of random measurement errors. This condition should be met for specimens of the same sample if reproducibility is good. For $K \geq 3$ the value of Δa_o is then taken to reflect real differences in the crystalline lattice of two X-ray specimens of a given compound. A basis is thus created for the study of solid solubility and for the precise characterization of crystalline compounds.

Where the results are influenced by systematic errors, the same sample is likely to yield $K \geq 3$ when the measurement procedures are altered. Since the evaluation of K cannot indicate which procedure is affected by non-random errors, the techniques in use should be tested against one or more standard materials whose lattice parameters have been established beyond the σ's used in the test.

An interlaboratory comparison of this type has been made with three different Guinier-type focusing cameras[1,2]. A number of well defined compounds were made up into X-ray specimens together with SRM 640, a silicon powder employed as an internal calibrant. The measured parameters are accordingly based on the published diffraction angles for this material[3]. With standardized specimen preparation, film measurement and calibration procedures these parameters comply with the requirement $K < 2$. Precision is of the order of 0.002 % for the simplest measurement methods. It attains 0.0005 % when low background film, such as CEA 15, is used and the powder pattern is measured instrumentally to $< 0.003°(2\theta)$.

This comparison has now been extended to the powder diffractometer, in order to examine in what way the results from these two types of instrument can be made compatible. Thus in the camera, calibration is essential because of Seeman-Bohlin focusing geometry and an upper 2θ limit of 100°, imposed by the transmission of the monochromatized X-ray beam through the specimen. Specimen displacement, film shrinkage on development, and uncertainty as to the position of the zero-2θ limit, are known sources of systematic error in angular measurements.

For the diffractometer the physical conditions are more favourable for automatic angular calibration. The upper 2θ limit is 150 - 160°, where the influence of small, random measurement errors is minimal. Moreover, Bragg-Brentano focusing geometry is associated with known systematic errors that extrapolate to zero at 180°. With the methods described over the years for applying systematic error corrections[4,5,6] precision parameters are frequently quoted without reference to a standard material.

A figure of merit serves to indicate the relative quality of the measurements given by the different techniques. The quantity used in this comparison is F_N[7]. This takes into account the number of peaks actually measured relative to the number theoretically allowed and the average measurement error $|\Delta 2\theta|$. The definition of F_N is†

$$F_N = \overline{|\Delta 2\theta|} \cdot N/N_{poss}$$

2. COLLECTION AND TREATMENT OF 2θ DATA

The conditions used to determine the precision and reproducibility of calibrated Guinier data are published separately[1,2]. This section will therefore describe only the diffractometer techniques used in the present comparison.

---

†Editorial note: N is usually limited to the first 30 observed lines and N, N_{poss} and $|\Delta 2\theta|$ are usually reported along with F_N.

2.1 Goniometers and 2θ Measurements: Powder specimens (< 30 µm particle size) were scanned in two Philips PW 1050 goniometers mounted on opposite sides of a copper fine-focus tube, operated at 40 kV and 30 mA. Each goniometer is adapted for computer steering, with a stepping motor to drive the 2θ shaft and the detector angle read at intervals of 0.002° by an absolute encoder coupled to this shaft. Peak search and measurement are performed by an on-line, time sharing computer. A first derivative strategy is used to detect and estimate the angular range of the peak, after which five points across the upper 30 % of the intensity distribution are measured with an optimizing strategy, e.g. 10^5 cts or 50 s. The five 2θ,I data pairs are then fitted to a parabola, the reproducibility of the peak angle being 0.001° for maxima above 100 cps and P/B > 5.

Using this technique the Kα doublet of the silicon 111 peak is repeatedly resolved with a separation of 0.071 - 0.075° compared with the theoretical value of 0.072°(2θ).

Below the count rate and P/B limits quoted above, reproducibility is rapidly lost and below 50 cps detectability is poor. This has serious consequences above 80° where structure factors are low and diffracted intensities tend to be weak. Failure to detect, and misreading of, peak positions accordingly increases with 2θ. A change of divergence slit at 80° would compensate for the loss of intensity to some extent, but this is not compatible with automated operation. In place of fixed divergence slits, a θ-compensating divergence slit can be used to optimize the diffracted intensity and P/B level over the whole angular range[8]. The length of irradiated specimen is thereby maintained at 12.5 mm, the length irradiated by a 1° divergence slit at 28°.

Goniometer 1 is fitted with a θ-compensating slit, PW 1386/50, a reflected beam monochromator of pyrolitic graphite, PW 1152, and a 0.1 mm receiving slit. Post-specimen Soller slits are absent. Gonio 2 operates with a fixed 1° divergence slit. Arrangement 2A uses a 0.1 mm receiving slit, post-specimen Soller slits and a nickel foil, 18 µm thick, as a β-filter. In arrangement 2B the goniometer is fitted with a reflected beam monochromator and a 0.13 mm receiving slit (1/4° anti-scatter slit). Gonio 1 has a take-off angle, Ψ, of 4.5° and a resulting angular range of 159°(2θ). For Gonio 2, Ψ was increased to 6° which limited the maximum angle to 158° where the detector met the X-ray tube.

A single knife-edge was used to determine the zero 2θ limit to within ± 0.001° with tube settings of 20 kV and 5 mA. For Gonio 2 this limit was re-established after installing the monochromator.

Specimens were made from sieved powders by side filling an aluminium holder under vibration[9]. Reproducible packing is ensured by maintaining a constant level of powder in the hopper during two minutes' vibration.

2.2 Correction of the Data for Systematic Errors: Lattice parameters are calculated from measured 2θ data using the method of least squares. Input values of $\sin^2\theta$ are corrected for known systematic errors that extrapolate to zero at $2\theta = 180°$ using, essentially, the method proposed by Mueller et al.[5]. Up to three trigonometric functions are combined to correct for specimen displacement, absorption/divergence, and flat specimen error. A choice of different functions can be combined for successive runs with each set of input data. The $\sin^2\theta$ data are then weighted according to the scheme proposed by Hess on the assumption that systematic errors have been accounted for[10]. This gives $\omega = \lambda^4/\sin^2 2\theta$.

A cross correlation matrix, based on the variances of all of the parameters and error coefficients, is also calculated. This is used to determine the level at which the specified systematic errors influence the estimated parameter values. Repeated calculation for a range of samples gave the following information.

- All systematic error functions, however they are formulated and combined, are always highly correlated ($r^2 > 0.95$).

- With 2θ data limited to the range below 140°, the estimated parameters are correlated with the systematic errors ($r^2 > 0.7$), as described in 3.2.

- A single set of trigonometric functions serves to correct for systematic errors that extrapolate to zero at 180°. For errors in $\sin^2\theta$ these are

  specimen displacement: $\cos\theta.\cot\theta$
  absorption/divergence: $\sin^2 2\theta(1/\sin\theta + 1/\theta)$
  flat specimen error: $\cos^2\theta$

Other functions than the above give correlations that are either higher or only marginally lower. The function $\cos^2\theta$ relates to flat specimen errors produced by a constant divergence slit. Nevertheless this gave a lower correlation for data collected with the θ-compensating slit than the function $\sin^2\theta$, which is relevant to constant specimen length.

For the majority of peaks above 28°, 2θ values were obtained for both α_1 and α_2 components of the $K\alpha$ doublet. Only the $K\alpha_1$ values were used, however, to keep the number of coefficients in the least squares equation to a minimum. Similarly, the expression used to correct for specimen displacement is also used to account for the shift in the $K\alpha_1$ peak position produced by overlap with the $K\alpha_2$ component of the doublet.

2.3 Zero Setting Error: The path of the incident beam is located with a single knife-edge using low tube loadings. There are indications that the zero-2θ setting thus obtained shifts by 0.02 to 0.03 on raising the tube power to the operating level. Reasons for this change have been advanced in terms of a shift in the position of the focal line as a result of a change in focusing conditions and/or stability of the incident electron beam within the X-ray tube[10]. The possibility that the shift might be negative on one side of the line focus and positive on the other has also been taken into account. On the other hand, measurements of 2θ for the 111, 200 and 311 peak positions of gold give reproducibilities better than 0.002° for all tube settings between 40 kV, 30 mA and 20 kV, 5 mA. Clearly then a shift in zero 2θ between these tube loadings would introduce a constant error in the measured peak position over the whole angular range.

A further systematic error that does not extrapolate to zero at 180°(2θ) is the read-out error between the driving shaft and the toothed wheel that carries the detector. Quality control during manufacture should limit this to ± 0.01° over one half cycle but a value of 0.02° might arise with wear. Following zero setting of the detector, the read-out error can be expected to accumulate and reach a maximum at 140° - 180°(2θ).

The systematic error functions described in 2.2 will minimize discrepancies between observed and calculated $\sin^2\theta$ in the low and middle angular range. Because they are so highly correlated these functions are, moreover, able to accomodate errors of 0.05°(2θ) in this region by a process of redistribution. Systematic errors that do not extrapolate to zero or are otherwise unspecified will, accordingly, tend to accumulate in the high-angle region and so constitute an effective zero setting error.

2.4 Correction for Residual Errors: A consistent difference was observed between lattice parameters obtained from calibrated Guinier patterns and values for the same samples as given by diffractometer measurements. Thus for Gonio 1 these parameters were greater than the Guinier values such that $K \simeq 4$, while for Gonio 2 the discrepancy was somewhat larger but negative.

With the Guinier results for ThO_2 and α-quartz (Arkansas stone) as reference values, correspondence between the two sets of parameters was sought by correcting the 2θ data in steps of 0.005°. For Gonio 1 the best agreement for these two materials was obtained with $\Delta 2\theta = -0.02^\circ$. The absolute encoder was reset by this amount, and the specimens were made up again and re-run to check the effectiveness of the correction. Subsequent treatment of the Gonio 2 data gave a best fit at $\Delta 2\theta = +0.03^\circ$ for the 2A and $+0.035^\circ$ for the 2B configurations.

An attempt was subsequently made to correct for this residual error as part of the least squares calculation by introducing an additional term into the equation for $\sin^2\theta$. Trial calculations with two separate data sets for Mo_3Ge_7 (Gonio 1 and 2B) and one for As_2O_3 (Gonio 1) demonstrate the difficulties of this approach.

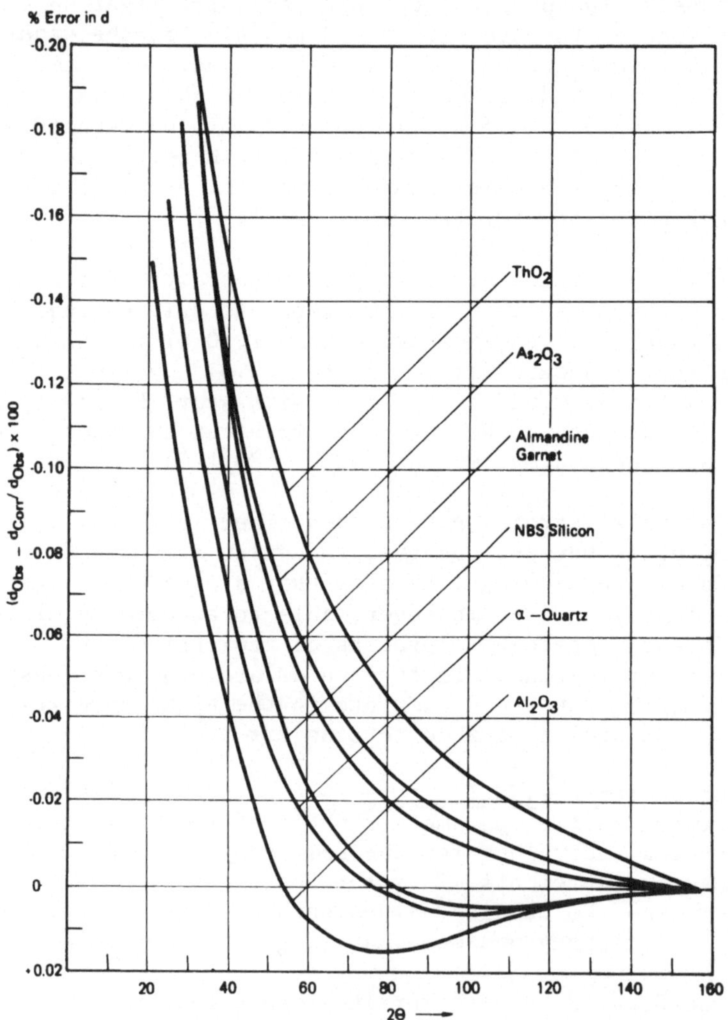

Fig. 1. Precision of interplanar spacing, d, before systematic error correction.

The parameter of Mo_3Ge_7 exhibited a difference, Δa, of 2.10^{-4} Å for $\sigma = 4.10^{-5}$ Å (K = 3.5) while the computed $\Delta 2\theta$ for the three calculations differed by 0.02 and 0.04°(2θ).

A similar lack of consistency emerged when $\Delta 2\theta$ was varied to obtain a minimum either in σ for the lattice parameter or in F_N.

3. RESULTS AND DISCUSSION

Lattice parameters for a number of well crystallized compounds with cubic, hexagonal and tetragonal symmetry have been measured. Tables 1 - 5 list values obtained from calibrated Guinier patterns and with the two goniometers described in 2.1. Goniometer results refer to data corrected for systematic and residual errors as described in 2.2 and 2.3. The results are given for $\lambda = 1.54051$ Å in the form: a_o, c_o, F_N $\sigma(a)$, $\sigma(c)$.

3.1 Comparison of Goniometer Data: Considerable variation is present in the raw 2θ measurements for the same sample recorded with the three goniometer configurations. Discrepancies of 0.01° are exhibited, even with the same instrument after repacking and repositioning the specimen holder between consecutive runs. Although these discrepancies tend to occur in the low and middle 2θ range, this is by no means the rule. For the 444 peak of ThO_2 at 145°, for example, differences are of the order of 0.01 - 0.02°. Similarly, there is no clearly defined pattern of discrepancies as between specimens of strongly absorbing ThO_2 and α-Fe_2O_3 and those of silicon, quartz and α-Al_2O_3. The overall corrections, on the other hand, display clear family similarities according to how strongly absorbing the sample is. This is seen in the curves for $\Delta d/d$ versus 2θ shown in Fig. 1.

All three correction functions described in 2.3 are active in the parameter calculations. Although the contribution made by one or other may be insignificant in a few instances, the coefficients display no clear pattern and no particular value can be assigned previous to calculation. This lack of evident system and the inability to establish characteristics for a particular goniometer configuration or diffracting sample is probably caused by the high level of correlation between the trigonometric functions mentioned earlier. Thus, these functions appear to operate collectively, serving as a sink for those parts of the 2θ measurement that would otherwise produce inconsistency in the overall calculation. In this context inconsistency is defined with reference to the measurements beyond 120 - 130°, on which increasing reliance is placed by the nature of the correction functions and the diminishing influence of random measurement errors on $\sin^2\theta$.

Table 1. Cubic Parameters

	CeO_2	ThO_2	Mo_3Sb_7
Guinier	5.412 14 (256) 11	5.597 28 (332) 9	9.568 88 (221) 9
Gonio 1	1 94 (707) 4	28 (399) 6	9 06 (284) 5
Gonio 2(A)	1 87 (263) 10	21 (261) 12	9 20 (66) 19
2(B)	2 12 (446) 6	28 (142) 24	8 95 (135) 9

Despite notable discrepancies between the sets of raw data, the goniometer scans give lattice parameters at a high level of mutual agreement after introducing the zero-2θ correction described in 2.4. With $K \leq 1.6$ the discrepancies in a_o are not statistically significant.

3.2 Correlation of Lattice Parameters and Systematic Errors: The validity of the approach described in 2.2 and 2.3 to correct for systematic errors was tested by scanning specimens of SRM 640 silicon. With Gonio 1 all 12 peaks up to 159° were measured and for λ = 1.54051 Å the statistics are

$$a_o = 5.43058,\ \sigma = 4\cdot 10^{-5},\ F_N = 368,\ r_i^2 = 0.43,\ 0.56,\ 0.48$$

Table 2. α-Quartz (Arkansas Stone)

Source	2θ	a_o	c_o	F_N
BFB	99	4.913 04 5	5.404 60 8	389
Gonio 1	157	07 5	63 7	121
2A	151	14 8	71 21	73
2B	157	19 11	65 13	68

Table 3. α-Al_2O_3 (Edmonds)

Source	2θ	a_o	c_o	F_N
BFB	98	4.759 05 5	12.991 32 19	478
Gonio 1	152	8 99 5	06 20	178
2A	150	9 14 8	03 28	100
2B	152	9 09 7	10 22	139
Edmonds, Gaithersburg '79 (ref. 12)		9 11 5	11 10	199

Table 4. α-Fe_2O_3 (LME)

Source	2θ	a_o	c_o	F_N
BFB	94	5.034 53 20	12.747 22 72	143
Gonio 1	148	78 11	80 52	104
2B	148	37 17	04 77	70

Table 5. SnO_2 (Fisher Scientific)

Source	2θ	a_o	c_o	F_N
BFB	96	4.737 50 10	3.186 32 10	215
Gonio 1	147	45 5	41 5	422
2A		66 15	51 16	134
2B		61 10	49 9	182

This yields $a_o/\lambda = 3.525183 \pm 30$ as compared with the value of 3.525176 ± 23 quoted by Hubbard et al.[3]

With Gonio 2A, only 11 reflections up to 136.9 could be measured and the corresponding statistics were

$$a_o = 5.43083,\ \sigma = 23 \cdot 10^{-5},\ F_N = 173,\ r_i^2 = 0.68,\ 0.81,\ 0.73$$

With the addition of the 444 peak value from the Gonio 1 scan, however, the values improve to

$$a_o = 5.43067,\ \sigma = 8 \cdot 10^{-5},\ F_N = 193,\ r_i^2 = 0.43,\ 0.56,\ 0.48$$

These results emphasize the importance of making peak measurements beyond 140° in order to effectively separate systematic errors from lattice parameters.

Fig. 2 was obtained from As_2O_3 data collected up to 154° with Gonio 1. Values of a_o, $\sigma(a)$ and r_i^2 were calculated at successively lower 2θ limits to investigate the influence of this limit on reliability. The tight coupling that exists between the three systematic error functions is evident in the closely similar development of the curves of r_i^2. The curve of σ as a function of 2θ is interpreted as follows. With the cut-off imposed below 120° the calculation is dominated by random measurement errors according to $\Delta d/d = |\cot\theta\ \Delta\theta|$. Above this limit $\cot\theta$ is low and falls less rapidly. Systematic errors now play a more prominent role, as shown by the conformity of the σ and r_i^2 curves between 120° and 154°(2θ). In the range up to 135°, moreover, the discrepancy between the parameter calculated with all the 2θ data and that at successive, lower limits is such that $K \simeq 4$. This reflects the basic unreliability of calculations based on insufficient high-angle information.

Fig. 2 indicates the positions of the last measurable pair of reflections of silicon, CeO_2 and ThO_2. The influence of the 444 peak on the parameter calculation of silicon has been described. It further appears that the reliability of the CeO_2 parameter is marginal when the limit is set at the 622 peak. If the detector could reach 162° to measure the 444 peak reliability should be improved.

3.3 Significance of the θ-Compensating Slit and Diffracted Beam Monochromator: A general observation is that Gonio 1 gives smaller σ's and higher F_N's than Gonio 2 in either its A or B configurations. This is attributable to the higher count rate made possible by the θ-compensating slit used with Gonio 1. Thus, the number of reflections measured with Gonio 2 in the statistically important high-angle range is fewer by 30 - 50 % than that given by Gonio 1. The improvement obtained by adding a monochromator to increase the P/B ratio is seen from a comparison of the 2A and 2B data. These results emphasize the importance of count rate and background level for the quality of the diffractometer data and the reliability of the parameter measurements.

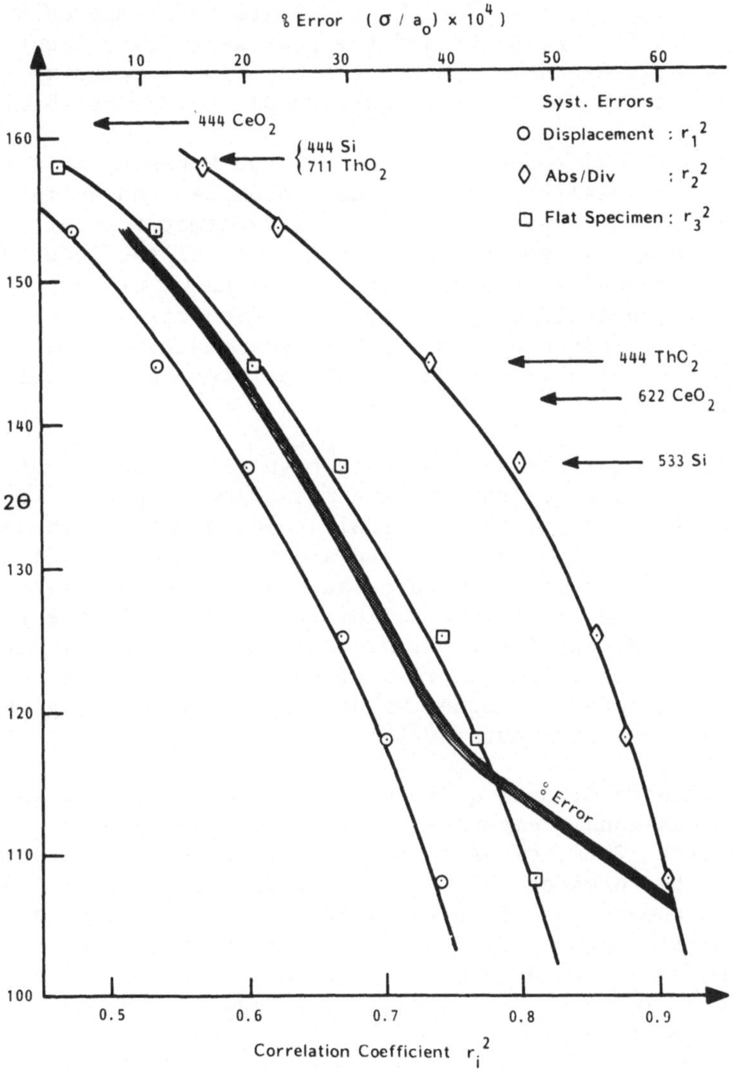

Fig. 2. Influence of upper 2θ limit on precision of lattice parameter and correlation coefficient between parameter and systematic errors.

3.4 Comparison of Guinier and Diffractometer Results: Agreement at the level K<2 is observed in Tables 1 - 5 between lattice parameters measured with the Guinier camera (BFB) and those obtained with Gonio 1 for the conditions employed in this study. Precision lies in the range 10^{-3} to $5 \cdot 10^{-4}$ % for these two instruments and $|\Delta 2\theta|$ is of the order 0.002 - 0.003°. This corresponds to the individual measurement error in both cases.

Differences in F_N between the two sets of parameters generally arise through failure of the goniometer to detect reflections in the range 80 - 130° with the limitations imposed by the level of diffracted intensity and the peak search and location routine in use at Studsvik. In the common angular range up to 100° the two instrument types exhibit comparable detectability.

An essential difference between the two types of measurement is the use of autocalibration for the goniometer and an internal standard for the Guinier camera. The diffractometer results are thereby made heavily dependent on the quality of the input data as a whole. Successful reduction in the influence of systematic errors can be jeopardized by misreading a proportion of the detectable 2θ's. This can occur either as a result of low count rates or because of shifts produced by the overlap of adjacent intensity profiles.

For the Guinier camera, the only significant source of systematic error associated with the measurement appears to be the lattice parameter of the internal standard. Errors in measuring the pattern of the internal standard are traceable by back calculating the 2θ's as long as the measurements are correlated in an equation that describes the calibration curve[2]. Such errors should not exceed that for the individual measurement by more than a factor of two. Local errors of measurement are then essentially reflected in higher values of σ, but do not give rise to a shift in parameter outside this limit.

The influence of data quality on F_N is most obvious in the case of the hexagonal compounds, α-Al_2O_3, α-Fe_2O_3 and, in particular, α-quartz. Thus, owing to an appreciably greater instrument function and the presence of $K\alpha_2$ radiation in the diffracted beam, diffractometer profiles run a greater risk of overlapping in the line-rich region beyond 80°(2θ). In the instance of α-quartz it proved necessary to delete six pairs of 2θ's with a separation less than 0.8° in order to avoid serious loss of reproducibility and overall precision.

For SnO_2 only one pair of 2θ's need be deleted because of overlap, and this compound is to be recommended as a potential non-cubic standard. A similar consideration applies to α-Al_2O_3, for which only two pairs of 2θ's were deleted. The relatively low F_N values for the diffractometer measurement of both α-Al_2O_3 and α-Fe_2O_3 are otherwise attributable to inability to detect reflections in the high-angle range that give a low count rate. The agreement obtained with the earlier measurement[12] for the same sample of α-Al_2O_3 is particularly satisfactory since this refers to yet another set of measuring conditions.

Sintered plates of high grade α-Al_2O_3 are readily available owing to their use in the manufacture of integrated circuits. One such plate is now used by the writer to check the zero setting of the two goniometers after realignment. The requirement is that the lattice parameters established for this specimen at a precision of 10^{-3} % during the current study should be reproducible to K<2.

4. FINAL COMMENTS

An essential feature of the good agreement between the Guinier and Gonio 1 results is the correction made to the zero setting of the goniometer. This step might be interpreted as a subjective and unwarranted adjustment of otherwise valid diffractometer measurements to fit those obtained with Guinier films. Further consideration of the overall situation, as it developed in this study, will show that this is not the case.

In the first instance discrepancies between diffractometer and Guinier parameters, particularly when they are as widespread and consistent as those observed here, point to the need to bring the two sets of results into line. This is seen to be possible by a single resetting of the goniometer zero 2θ and thus indicates the operation of systematic errors not included in the autocalibration process. In the absence of this adjustment to meet the Guinier results the additional agreement obtained between the different sets of diffractometer results would not have been possible.

In the final analysis the Guinier data, on which the zero correction is based, are themselves based on SRM 640, used as an internal standard. The correction is therefore an indirect calibration of the angular read out of the goniometer against the lattice parameter of this material. Since, for purely mechanical reasons, the vital high-angle peak of SRM 640 may lie beyond the upper limit of the goniometer use must be made of secondary standards that provide measurable reflections up to 150 - 156° (2θ). The Guinier measurements employed in this comparison serve as secondary reference data of this type.

REFERENCES

1. A. Brown, J. W. Edmonds and C. M. Foris, "Reproducibility and Precision of Guinier Patterns Using Powdered Silicon Calibrant", Adv. X-ray Anal. 24 111 (Edited by Deane K. Smith, Plenum Press, 1981).
2. A. Brown and C. M. Foris, "Precision and Reproducibility of Lattice Parameters from Guinier Powder Patterns: Follow-up and Assessment". (For publication in Adv. X-ray Anal. 1983).
3. C. R. Hubbard, H. E. Swanson and F. A. Mauer, "A Silicon Powder Diffraction Standard Reference Material", J. Appl. Crystallogr. 8 45 (1975).
4. W. Parrish and A. J. C. Wilson, "Precision Measurements of Lattice Parameters of Polycrystalline Specimens", Int. Tables X-ray Cryst., Vol. II, 216 (Edited by J. S. Kasper and Kathleen Lonsdale, Kynoch Press, 1959).
5. M. H. Mueller and L. Heaton, "Determination of Lattice Parameters with the Aid of a Computer", AEC Research and Development Report, ANL 6176 (1961).
6. K. E. Beu and D. R. Whitney, "Further Developments in the Likelihood Ratio Method for the Precise and Accurate Determination of Lattice Parameters", AEC Research and Development Report, GAT-T-1289/Rev 1 (1965).
7. G. S. Smith and R. L. Snyder, "F_N: A Criterion for Rating Powder Diffraction Patterns and Evaluating the Reliability of Powder-Pattern Indexing", J. Appl. Cryst. 12 60 (1979).
8. R. Jenkins and F. R. Paolini, "An Automatic Divergence Slit for the Powder Diffractometer", Norelco Reporter 21 9 (1974).
9. A. M. Byström-Asklund, "Sample Cups and a Technique for Sideward Packing of X-ray Diffractometer Specimens", Am. Mineral 51 1233 (1966).
10. J. B. Hess, "A Modification of the Cohen Procedure for Computing Precision Lattice Constants from Powder Data", Acta Cryst. 4 209 (1951).
11. Discussions with D. W. Beard, Siemens Corporation and J. Pichert, Amperex Electronic Corp.
12. J. W. Edmonds, "Precision Guinier X-ray Powder Diffraction Data", National Bureau of Standards Special Publication 567. Proceedings of Symposium on Accuracy in Powder Diffraction, Gaithersburg, MD, 1979 (Edited by S. Block and C. R. Hubbard, NBS, 1980).

EFFECTS OF DIFFRACTOMETER ALIGNMENT AND ABERRATIONS ON PEAK POSITIONS AND INTENSITIES

Ron Jenkins

Philips Electronic Instruments Inc.

Mahwah, N.J. 07430

INTRODUCTION

The two basic parameters traditionally employed in qualitative phase identification using the powder diffractometer are the diffracted peak maximum intensity and the interplanar spacing "d". Most routine powder diffractometry carried out today is performed with the parafocussing geometry, and this arrangement gives a diffraction pattern in terms of diffraction angle (2-theta) vs. intensity. During the course of obtaining a set of experimental 2θ and intensity values from a specimen to be analysed and in the subsequent conversion of these data to a d/I list for qualitative phase identification, errors will of course accrue. Most qualitative search methods in use today give high credence to the d-value, since this is by far the most accurately known of the two search parameters, and d-values of a few tenths of a percent are reasonably easily obtained with a well aligned and properly calibrated diffractometer. Intensities, on the other hand, can be subject to errors of the order of tens of percent, and problems arise in qualitative analysis where the need arises to subtract an identified phase from a multi-phase pattern to allow identification of further phases.

In most quantitative analyses the situation is reversed and great accuracy in the d-value is generally not required, except in those cases where one is trying to correlate changes in unit cell dimension with compositional variations. An accurate intensity measurement is generally a critical first step in any quantitative analytical scheme, and the final accuracy in calculated phase concentration obtained is invariably limited by the accuracy of the measured intensity. It is thus clear that for powder work in general, the analyst must be able to control the accuracies of measured 2θ and intensity values and the calculated d-values, and the subject of this paper is to discuss the effect of alignment of the diffractometer on these parameters.

Fig. 1 reviews the basic features of the parafocussing geometry and it will be recalled that two circles define the optical arrangement. The focussing circle has a radius equal to r and the source, receiving slit and specimen all lie on the circumference of this circle. As the specimen moves through an angle theta, the receiving slit moves through an angle 2-theta and the value of r varies accordingly. A second circle also exists, the goniometer circle, which has the specimen at the center and the source and receiving slit on the circumference. The radius of the goniometer circle is R and a relationship exists between R and r, namely $R = r.2(\mathrm{Sin}\theta)$. As θ varies from zero to 90 degrees, r varies from infinity to R/2. In effect the focussing circle becomes very large at low angles of diffraction and, in general, so too do the systematic errors in the measured 2θ value, making it very difficult to obtain large d-values with any great accuracy.

Various slits and/or apertures are employed in a typical diffractometer, and the two most important of these are the divergence slit and the receiving slit. The receiving slit RS is placed on the goniometer circle at a distance r from the center of the specimen, and its major function is to maintain the focussing condition and hence the shape of the diffracted line profile. The divergence slit DS is placed between the source and the specimen and has the function of controlling the length s of the specimen irradiated. Parallel plate Soller collimators are always placed between source and

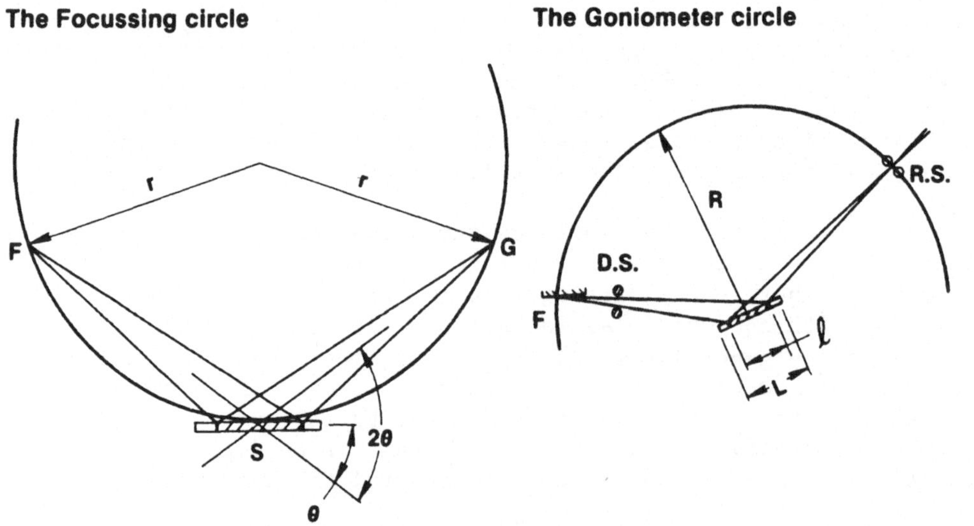

Fig. 1. Parafocussing geometry showing focussing and goniometer circle.

specimen and generally between receiving slit and detector, also. The function of these collimators is to limit the axial divergence of the beam as it passes from source to specimen, to receiving slit, and finally to the detector. The axial divergence of the beam is one of the greatest sources of instrumental influence on the diffracted line profile. The implementation of this geometry varies from manufacturer to manufacturer with the goniometer in either the horizontal or vertical configuration; nevertheless, most instruments will include at least the basic components described. Typical arrangements might include a diffracted beam monochromator and a variable divergence slit. Additional attachments might typically include automated specimen loaders, spinners, controlled atmosphere chambers, high and low temperature stages and so on.

ALIGNMENT OF THE DIFFRACTOMETER

The alignment of the powder diffractometer involves a series of complex, synergistic, three dimensional adjustments of the various component parts to achieve the focussing conditions already described. Fig. 2 shows the four critical motions required for the alignment process. While each of these steps is critical, by far the most important and by far the most difficult to achieve is the setting of the mechanical zero. There are many critical tolerances that the manufacturer must hold in designing a goniometer, and these would include such things as maintaining accurate circular motion of θ and 2θ gears to ensure good $\theta/2\theta$ tracking, maintaining slits exactly parallel and in line with the goniometer movement, ensuring that the specimen reference surface is at the true focussing circle, and so on. These criteria are made even more difficult to achieve because of the requirement for interchangeability of parts, particularly at the specimen mounting region. There is a need for interchange not just of specimen mounts, but also to replace fixed holders with rotating holders, to add specimen changers, high temperature stages, plus a host of other possibilities. In practical terms this means that the θ shaft must be removable and replaceable with a tolerance of around 10 microns--clearly a difficult task and one which directly impacts the accuracy of the setting of the mechanical zero.

The mechanical zero represents the angle at which the center line from the source bisects divergence and receiving slits and the axis of rotation of the goniometer. When a specimen is placed with its center line at the axis of rotation of the goniometer it will form a tangent to the focussing circle. The setting of the zero is typically performed by use of a pinhole or single knife edge at the sample position, with the $\theta/2\theta$ movement decoupled. Scanning of the 2θ will give an intensity curve of sharply changing slope as the direct beam is crossed. The slit is rotated through 180 degrees

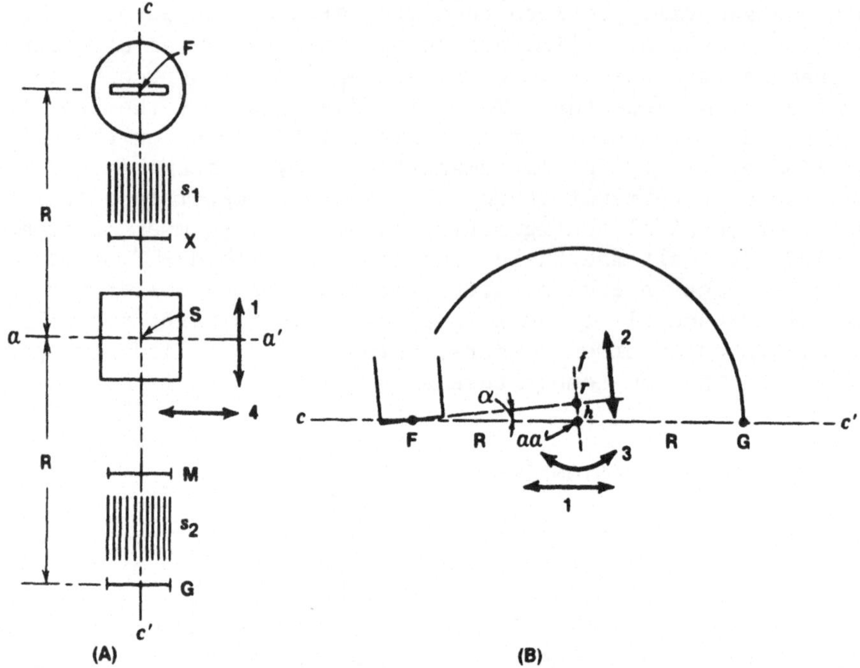

Fig. 2. The principal motions required in aligning an x-ray diffractometer.

and the process repeated. The point at which the two curves intersect is taken as the mechanical zero. The major problem in obtaining an accurate zero is buildup of tolerance, including the θ shaft hole in the θ gear wheel, the θ shaft itself, the accuracy of the sample reference surface on the shaft, and finally the flatness of the knife edge tool. Great care is required to reduce this error to the level of 0.01 degrees or so and a separate calibration curve should always be taken following the completed alignment to establish the magnitude of any residual zero error.

In addition to the introduction of alignment errors such as zero offset and 2:1 errors, any given diffractometer arrangement will have associated with it certain inherent aberrations.[1] In order to obtain high d-spacing accuracy these errors will almost certainly have to be calibrated out of the system and it is necessary that the type and form of the errors be known before corrections can be routinely applied. A problem will almost certainly arise for the diffractionist in separating inherent aberrations from misalignment errors. The situation is further complicated by the fact that the diffractionist himself may introduce experimental errors during the course of obtaining diffraction data.

TYPES OF ERROR

Errors break down into three main categories: inherent aberrations, alignment errors and experimental errors. Included within the category of inherent aberrations would be errors due to transparency, axial divergence, flat specimen, etc., and in the range of alignment problems would be included mis-setting of $\theta/2\theta$, zero line offset and so on. Experimental errors would include specimen displacement, introduction of orientation and general problems due to mis-reading of data in the form presented by the diffractogram. These errors are included in Table 1, and it is useful to differentiate between the theoretical 2θ value, i.e. those values which are dependent only upon the size and distribution of atoms in the unit cell of the phase; the practical value which is fixed for a given diffractometer with its aberrations and thirdly the experimental value, which is also dependent on alignment errors. Finally, errors may arise in the conversion of the experimental 2θ value to the "d" spacing because of uncertainties in defining peak maxima, because of non-monochromaticity of the source and aberrations in the shape of the diffracted line profile. One must clearly be careful in defining what a given set of d/I data really means. Especially since the whole basis of our qualitative analytical procedures is the comparison of our experimental data with somebody else's single phase data taken on a different system.

By far the most critical experimental error is the displacement error which is due to mechanical displacement of the sample above or below the focussing circle due to incorrect sample mount alignment, poor sample holder design or poor technique in loading the sample. Additionally, axial divergence occurs because of divergence of the source in an axial plane across the specimen leading in turn to deterioration of the focussing condition.

Table 1. Commonly Employed Systematic Error Correction Functions

1. Specimen Displacement:	$\Delta 2\theta = -2s\,(\cos\theta/R)$
2. Flat Specimen:	$\Delta 2\theta = -1/6\,\alpha^2 \cot\theta$
3. Transparency:	$\Delta 2\theta = \sin 2\theta/2\mu R$
4. Axial Divergence:	$\Delta 2\theta = -h^2 (K_1 \cot 2\theta + K_2$ cosec $2\theta/3R^2$

s = Specimen Displacement
α = Angular Aperture
R = Radius of the Goniometer Circle
μ = Linear Absorption Coefficient
h = Axial Extension of Specimen
K_1 & K_2 = Constants

Correction functions for these errors can, in principle at least, be applied relatively easily with the modern computer-controlled diffractometer since they can all be described in terms of known parameters.[2] The only term really in any doubt is the axial divergence term which was first described by Pike.[3] It seems, at least intuitively, that there must be a problem with the Pike axial divergence term at low angles since it involves cot and csc functions which become infinite as 2θ approaches zero. Infinite peak shifts are physically not possible. The infinities probably result from approximations made in the derivations which are not valid as 2θ approaches zero.

A calibration curve can be obtained for a given standard reference material and the experimentally determined data used as a measure of how well a given diffractometer is aligned. In our own laboratory we use a Novaculite (alpha-quartz) specimen which has been surface ground to a flatness of better than 10 μm as a calibration check standard. Since it can be assumed that there is no significant displacement error and since our practical 2θ value curve includes the appropriate correction for flat specimen, transparency and axial divergence errors, any deviation of our experimental points from the practical curve can be attributed to misalignment. Also, the form of the deviation may provide a clue as to the type of misalignment. As an example, a systematic error in the zero setting of the goniometer will manifest itself as a constant displacement of our data points above or below the practical curve, whereas a 2:1 error would show as a change in the curvation of the fitted line.

In practice, even a well aligned diffractometer can give a shift of the order of 0.05-0.1 degrees 2θ where systematic error corrections are not applied. This is borne out by a detailed analysis of the data in the JCPDS powder data file by Snyder, et al.[4] who clearly showed that, whereas the average 2θ error in NBS patterns was a few hundredths of a degree, other patterns had errors three to four times larger. This is directly attributable to the use of internal standards by NBS to calibrate out systematic errors. The systematic error correction is applied in terms of the measured parameter 2θ but for qualitative work this must be converted to a "d" spacing. This makes the alignment of the diffractometer particularly critical at low angles, which is unfortunate since two additional problems arise in this low angle region.

LOW ANGLE PROBLEMS

The first of these errors is associated with the lack of good low angle reference standards making both the checking of the integrity of alignment and the establishment of a correction curve rather difficult. We have recently described[5] the use of fatty acid monolayer crystals for this purpose. As an example, lead stearate has a d spacing of about 50Å and its first reflection comes at about 1.8 2θ. We then observe about thirty harmonics out to around 60 degrees. This problem is also under investigation by NBS and a later paper will discuss the use of fluorophlogopite mica as a low angle reference standard.[6]

The second problem involves the appearance of low angle artifact due to poor choice of the divergence slit aperture. As was previously stated, the function of the divergence slit is to limit the irradiation length of the specimen. For a fixed slit the irradiation length will increase with decrease of goniometer angle. At high angles the beam is completely intercepted by the sample, but as the angle is decreased the beam is wider than the sample but still smaller than the maximum width of the sample support. At low angles the beam is wider than the sample support. As we pass from the first to the third condition, i.e., from a fully intercepted and scattered primary beam to a partially intercepted and scattered primary beam, two effects will occur. First as the scattering (diffraction) angle gets smaller the intensity will increase. Second, as less of the sample support intercepts the primary beam, the total amount of scatter will get less. The overall effect is to produce a scatter curve with a peak. The actual position of the scatter intensity maximum may vary from sample to sample, but it will typically occur at about 3-4 degrees 2θ.

A typical low angle diffraction diagram is made up of four basic components as illustrated in Fig. 3. The dotted curve shows a typical diffractogram over the range 0 to 18 degrees as it might occur if recorded with a fixed 1 degree divergence slit. The background between about 4-18 degrees is made up of sample scatter and sample support scatter. Below 4 degrees the major contribution to the background is direct radiation which has passed over the top of the sample support. This background is a very high intensity, so high in fact, that if the dead time of the detector system is more than a couple of microseconds the counter may saturate, giving rise to another "peak" at about 1 degree. Superimposed on top of the background are the diffraction peaks. From the foregoing it will be apparent that when too wide a divergence slit aperture is employed in the recording of a diffractogram one or more broad lines can occur at low angles which may complicate the interpretation of a diffractogram.[7]

MAGNITUDE OF ERRORS

It is useful to look at the relative sizes of the various errors, and as an example fig. 4 shows the effect of axial divergence, flat specimen, transparency (for quartz), a +0.02 zero angle offset, a

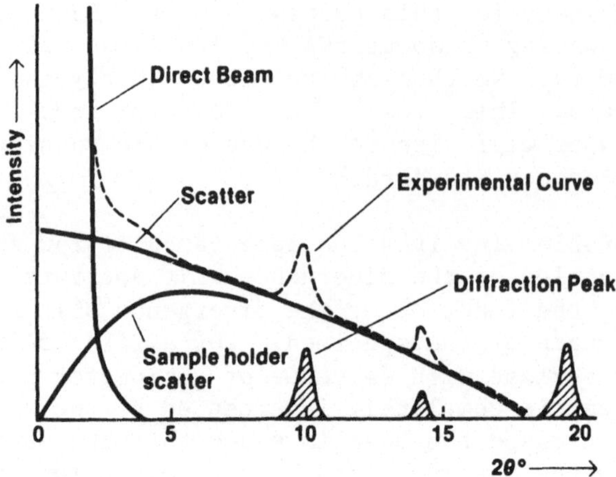

Fig. 3. Composition of a typical low-angle diffractogram.

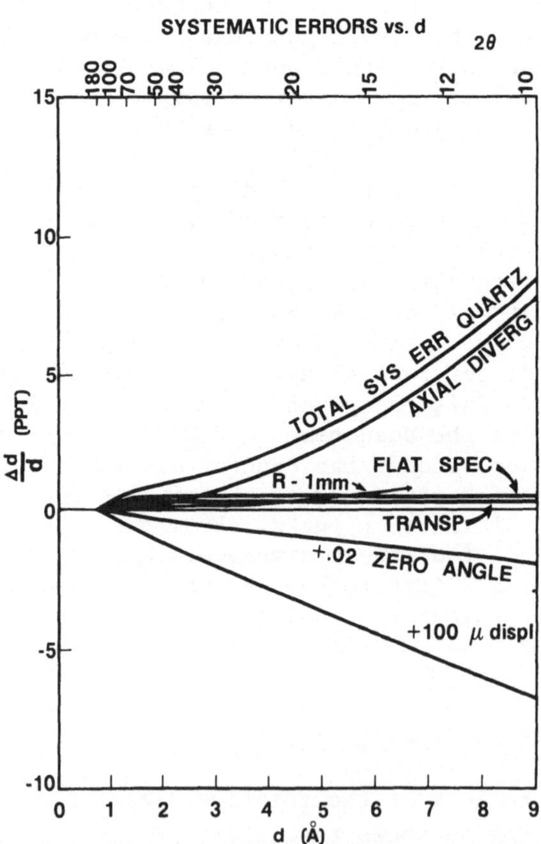

Fig. 4. Effect of various diffractometer errors on Δd/d.

-1mm error in specimen to receiving slit distance and a 100μm displacement error, on Δd/d. It is seen that displacement error is by far the largest source of error. The only other possibly significant errors are axial divergence and zero angle offset. However, as we noted before, it seems unlikely that axial divergence error is correct at low angles, i.e., large d. Furthermore, zero angle error behaves almost the same as specimen displacement error in the important region of 0-60 2θ.

When one makes a similar study for the effects of these various errors on x-ray intensity, it is found that most misalignment errors manifest themselves as distortions of the diffracted line profile. The effect on the peak intensity maximum is generally small, but what can be important is the shift in the centroid of the peak.

CONCLUSIONS

In conclusion one can say that a well aligned diffractometer with a correctly established error function curve should give Δd/d values of better than one part per thousand. A trend appears to be developing, however, in which more reliance is being placed on the computer correction of poor data from badly aligned diffractometers. The rationale is that since the computer can easily generate an error curve, the alignment for the diffractometer is relatively non-existent. My personal view is that this is a dangerous assumption, especially because the errors tend to couple in rather complex ways and irreproducibility of data may result. In the obtaining of accurate d-spacing a well designed diffractometer is a vital first step.

REFERENCES

1. W. Parrish and A. J. C. Wilson, Acta Cryst. 7:622 (1954).
2. W. N. Schreiner, C. Surdukowski, R. Jenkins, and C. Villamizar, Norelco Reporter 29:42 (1982).
3. E. R. Pike, J. Sci. Instrum., 36:52 (1959).
4. R. L. Snyder, et al., Lawrence Livermore Laboratory Report UCRL52505, June 1978.
5. R. Jenkins, T. Hom, K. Villamizar and W. N. Schreiner, Adv. in X-Ray Anal. 26, this issue (1983).
6. C. R. Hubbard, Adv. in X-Ray Anal. 26, this issue (1983).
7. R. Jenkins and B. Squires, Norelco Reporter 29:20 (1982).

ACCURACY AND PRECISION OF INTENSITIES IN X-RAY POLYCRYSTALLINE DIFFRACTION

W. Parrish and T. C. Huang

IBM Research Laboratory
5600 Cottle Road
San Jose, California 95193

ABSTRACT

The integrated and peak intensities of a series of silicon powder samples of various crystallite sizes were measured with a computer automated diffractometer and a profile fitting method (PFM). The accuracy of the PFM was better than 0.003% in computing the integrated intensities. The PFM gave more precise values than would be expected from counting statistics of the peak intensity. The average difference between each measurement and the average intensity was 0.5% with little dependence on the absolute intensity. Crystallite sizes have a large effect and it is essential to rotate the specimen around the diffraction vector. The best results were obtained with <10 μm particles. Larger sizes decrease the absolute intensities and change the relative intensities. Structure refinement using the POWLS (powder least squares refinement) program showed the presence of (111) preferred orientation even in the <10 μm specimens. R(Bragg) decreased from 4.3% to 0.7% by including the preferred orientation correction in the refinement.

INTRODUCTION

The large majority of publications on accuracy in powder diffraction deal with diffraction angle and lattice parameter determinations. This paper describes a preliminary study of a number of factors that contribute to the accuracy of the intensities. Accurate intensity data are required in a number of applications such as quantitative phase analysis, crystal structure refinement, line profile determination and similar studies. The introduction of the counter tube diffractometer 35 years ago greatly improved the accuracy over film methods but it remains surprisingly difficult to make accurate measurements even now. Much of the problem resides in the specimen preparation which is a dominant factor limiting the accuracy.

We will confine the description to measurements of silicon powder which is widely used for angular calibration. It is ideal for these studies because it has much smaller preferred orientation than micas or organics for example, the pattern is simple with no overlaps, it is inexpensive and can be prepared in various particle size fractions. The accuracy of the intensity measurements can be determined from the structure refinement because the atomic positions are fixed by the symmetry.

EXPERIMENTAL METHOD

The data were collected by step scanning with a computer controlled diffractometer (1,2) and the intensities determined by the profile fitting method (2,3). The experimental conditions were: long fine focus Cu target X-ray tube, 12° take-off angle, 50 kV, 20 mA, vertical scan reflection specimen diffractometer geometry (4), R=185 mm, angular aperture 1°, anti-scatter slits, receiving slit 0.11° (0.35 mm), incident beam parallel slit aperture 4.5° (none in diffracted beam), curved graphite monochromator after receiving slit, vacuum path, and scintillation counter with pulse amplitude discrimination.

SPECIMEN PREPARATION

The preparation of satisfactory specimens is a very important factor in determining the accuracy of the intensities. It is a difficult art and needs considerably more development. Ideally there should be a sufficient number of randomly oriented crystallites to provide the correct intensities for all reflections. The effective number having the correct orientations is determined by the particle size, multiplicity factor, and instrument geometry (5). Because most of the diffraction occurs from the surface layers, the final surface finish and interparticle microabsorption can exert a major effect on the results.

We used silicon powders prepared from a number of vials of NBS Standard Reference Material 640 (6). The powders were sifted through 5, 10, 20 and 30 μm micromesh (Buckbee-Meers) using an acoustic sifter to provide <5, 5-10, 10-20, 20-30 and >30 μm size fractions. Scanning electron microscope photos indicated the size separations were satisfactory, Fig. 1.

A number of specimens were prepared from each size fraction. Two types of specimen holders were used: a 1 inch diameter cylinder with 0.5 or 1 mm deep recess, and a flat silicon single crystal plate cut parallel to (510) cemented on top of the cylinder. The former requires considerably more powder. The binder, collodion/amyl acetate, was spread on the holder and some powder added and mixed with a flat toothpick or spatula. More binder and then powder were added and the process repeated until a sufficient amount of the wet powder covered the top of the holder or crystal. The specimen was tightly packed and the surface made approximately flat by inserting the long edge of the toothpick into the mixture and repeating this across the surface in several directions. After overnight drying the surface was scraped with the edge of a microscope slide. The cylinder was then placed in a doughnut-shaped lapping holder to make certain the surface was normal to the axis of rotation and lapped lightly on a dry glass plate. The excess powder was

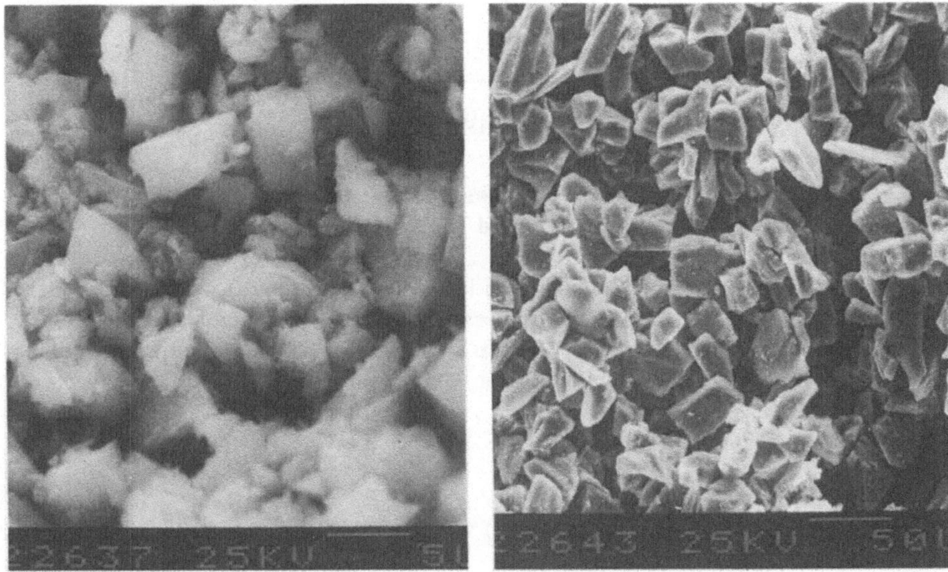

Fig. 1. Scanning electron microscope photos of silicon powder. <5 μm left, 20-30 μm right.

removed by tapping and light rubbing. A square was used to see if the surface was flat and normal to the axis. Once the specimen was prepared the particles were tightly bonded and the surfaces stable. All specimens were examined with a binocular microscope and given the X-ray test described below.

PRECISION OF METHOD

The inherent precision of the data collection and reduction methods was determined with carefully controlled conditions. Four reflections (111), (220), (311) and (400) of a rotating 5-10 μm silicon powder were measured ten times by step scanning over an angular range of 1.6° to 1.8° with 0.04° steps and 1 second count time per step. The specimen was not moved between runs and the experimental time was about three minutes for the four reflections. The same experiment was repeated a few months later using the same experimental conditions except the step increment was reduced to 0.02° and the specimen surface was relapped. The relapping increased the intensity about 12% and the relative intensities were virtually unchanged.

The integrated and peak intensities and reflection angles were computed by the profile fitting method. The previously determined instrument function (W*G) was convolved with the true diffraction effects S of the specimen, i.e., (W*G)*S. The program computes the R-factor relating the accuracy of the fitted profile to the experimental data. The accuracy of the integrated intensity is given by

$$RI(PF) = \left| \frac{\sum_{i=1}^{n} [Y_i(obs) - Y_i(calc)]}{\sum_{i=1}^{n} Y_i(obs)} \right| \times 100\%$$

where Y_i is the counts at each step. There were a large number of counts contained in the profiles and all the runs in this paper had RI(PF)<0.003%.

The results of this study, summarized in Table 1, are typical of a large number of method checks conducted over several years. The first row lists the averages of the peak intensities of the ten runs; the units are counts for the one second count time. The standard deviation σ was calculated from

$$\left\{ \sum_{i=1}^{10} [I_i - I(avg)]^2 / (10 - 1) \right\}^{1/2} ,$$

Δ is the difference between the I_i and I(avg) values, and Δ/I is in percent.

The expected precision of a peak intensity is usually estimated to be one standard deviation, $\sqrt{I(avg)}$. The values of σ were lower than $\sqrt{I(avg)}$ showing that profile fitting gives more precise values than those based only on counting statistics of the peak. The Δ/I values average 0.5% for all the runs and are about the same for the integrated and peak intensities. If we omit the (400) 0.04° run the remaining values have little dependence on intensity. In practice it will be necessary to measure weaker peaks than those used here and our experience is the method works equally as well over a very large intensity range. For example, in the structure refinements of other materials, peaks as small as 10 counts were measured and although RI(PF) was somewhat larger the method was used without intensity weighting factors over a range of 5000 in intensity.

The reproducibility of the peak reflection angles of the 40 reflections in the 0.04° runs was σ=0.0016° and in the 0.02° runs σ reduced to 0.0003°. The same precision was obtained with our Peak Search program (2).

CRYSTALLITE SIZE EFFECTS

The absolute and relative intensities have a high dependence on the crystallite sizes. Two of the most important factors required to achieve high accuracy are control of the particle sizes in the specimen preparation, and rotating the sample around the diffraction vector (4, p. 287, 5). The NBS silicon powder used was prepared from a large single crystal and each particle is a single crystal fragment so that the particle and crystallite sizes are the same. The study was made with a series of specimens prepared from the five particle size ranges described above.

The crystallite size effect can be observed by rotating the specimen around the axis normal to the center of the specimen surface while the diffractometer is set at the peak position, and recording the intensity on a strip chart, Fig. 2. The short intensity

Table 1. Precision of Intensities from Profile Fitting

(hkl)	(400)		(311)		(220)		(111)	
$\Delta 2\theta°$	0.04	0.02	0.04	0.02	0.04	0.02	0.04	0.02
				Peak Intensities				
I(avg)	2500.0	2832.0	10366.0	11922.0	19365.0	22255.0	36035.0	41194.0
$\sqrt{I(avg)}$	50.0	53.2	101.8	109.2	139.2	149.2	189.8	203.0
σ	41.1	23.7	74.2	65.7	145.4	67.1	144.7	83.8
Δ	30.0	19.7	58.2	51.6	116.3	51.5	120.9	69.8
Δ/I%	1.2	0.7	0.6	0.4	0.6	0.2	0.3	0.2
				Integrated Intensities				
I(avg)	591	684	2287	2670	4013	4700	6979	8130
σ	9.8	2.6	19.9	8.7	18.7	18.8	36.5	28.3
Δ	8.0	2.1	16.2	7.0	13.2	14.1	27.3	22.2
Δ/I%	1.4	0.3	0.7	0.3	0.3	0.3	0.4	0.3

lines to the right of each recording are the averages of the intensities at all azimuths with fast (60 rpm) rotation. The small wiggles result only from counting statistics. The numbers are the chart amplitudes and scaling to the highest=100, the (111) relative peak intensities were <5 μm=95, 5-10=100, 10-20=94, 20-30=88 and >30=59. The decrease is probably due mainly to increasing interparticle microabsorption, and the lower particle packing density of the specimen surface.

The variation of peak intensity with azimuth angle ϕ was recorded by substituting a 1/7 rpm motor. The diffractometer was stationary at the previously determined average intensity (horizontal dash line); the horizontal solid lines are ±10% of the average. The intensity fluctuations increase rapidly with increasing particle size. Moreover, there is no correlation between fluctuations of different reflections from the same specimen as shown by the 10-20 μm specimen for which the incident beam intensity was adjusted to give the same chart amplitude for the three reflections. It is clear that a stationary specimen could cause large errors in the relative intensities of the larger particle size specimens.

The method was used to test the sifting and specimen preparation. If large particles are embedded in a small particle matrix they may cause large intensity spikes. Another advantage of specimen rotation is that it averages the in-plane preferred orientation, although it has virtually no effect on preferred orientation parallel to the specimen surface.

Quantitative determinations of particle size effects were made by step scanning and profile fitting four noncoplanar reflections of a series of different particle size specimens. The results are summarized in Table 2. Two packed samples were made

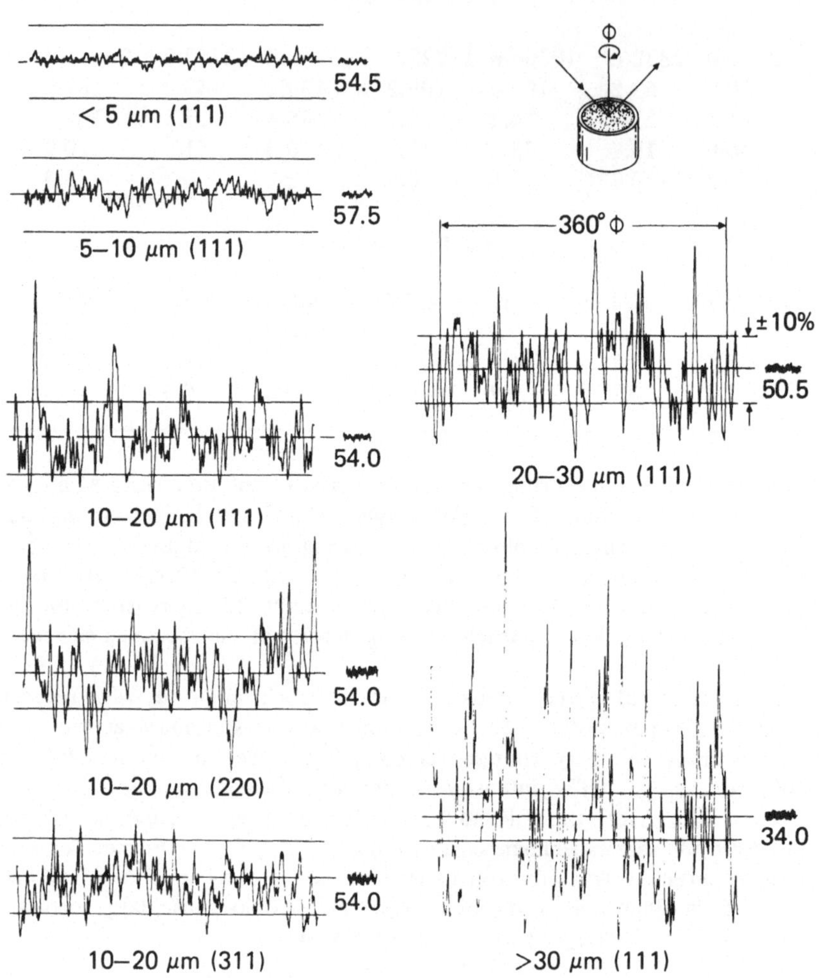

Fig. 2. Recordings of intensity variations of rotating silicon powder samples of various particle sizes.

Table 2. Effect of Crystallite Sizes on Silicon Intensities

Particle Size μm		Relative (111)=100						Absolute	
		Integrated			Peak			Integ	Peak
		400	311	220	400	311	220	111	111
<5	a	8.4	32.2	57.7	6.5	27.5	52.8	6991	34327
	b	8.4	32.2	57.7	6.4	27.3	52.1	8189	39934
	d	8.1	31.8	57.2	5.3	26.2	52.0	7051	34681
5-10	a	8.4	32.2	57.6	6.9	28.5	53.8	7012	36088
	c	8.4	32.8	57.8	6.9	28.9	54.0	8130	41194
10-20	a	8.3	35.1	58.0	7.2	31.8	54.5	6418	34677
	b	8.6	34.1	58.3	7.3	30.8	55.3	7616	39913
	d	8.0	31.2	55.2	7.3	28.6	52.5	6448	35305
20-30	a	8.9	34.0	59.1	7.9	31.3	57.7	5924	31782
	b	8.5	34.6	61.2	7.2	31.1	58.6	5890	36388
	d	7.8	36.9	60.2	6.9	33.9	58.1	6681	35776
>30	b	8.8	39.6	56.4	7.7	35.7	52.8	4829	23908
Unsifted	a	9.0	35.6	60.9	7.7	32.6	58.7	5765	28170

[a] $\Delta 2\theta = 0.04°$; t=18 sec (400), 4.5 (311), 2.5 (220), 1.5 (111)
[b] $\Delta 2\theta = 0.02°$; t=10, 4.5, 2.5, 1.5 sec
[c] $\Delta 2\theta = 0.02°$, t=1 sec all refl., avg. 10 runs.
[d] same as a

for each particle size range and the averages of the two runs are listed. The step scanning conditions are in the footnote. The count times were increased for the weaker reflections to obtain a more uniform counting statistical accuracy for the four reflections. The integrated and peak intensities of (400), (311) and (220) are the values relative to (111)=100 for each set. The last two columns are the absolute values of the integrated intensities in units of counts/degrees/seconds and the peaks in counts for one second.

The packed samples (letter "a") were later relapped by an improved method and rerun ("b" and "c"). This increased the absolute intensities and the relative intensities remained about the same in each particle size group. The relative integrated intensities are nearly the same for the <5 and 5-10 μm samples but they generally increase and the absolute intensity of (111) decreases in the >10 μm samples. The degree of randomness of the particle size distributions are thus size dependent. This effect could be magnified in materials having greater preferred orientation, and the intensities must be corrected as described in the following section.

The samples prepared on single crystal plates ("d") have about the same absolute intensities as the packed specimens but the relative values are generally lower indicating more (111) preferred orientation. It was difficult to prepare these samples with large particles.

SILICON STRUCTURE REFINEMENT

The crystal structure residual R(Bragg) is a good measure of the accuracy of the integrated intensities because silicon does not require determining atomic positions. Refinements were made of several sets of profile fitted data which were entered in the POWLS (Powder Least Squares) program (7,8). The parameters refined were: the isotropic temperature factor B, a scale factor and a preferred orientation factor p.

The best results were obtained with $<10\ \mu m$ particle samples and the refinements improved by including the quantity p in the list of variables. The correction (9) for the preferred orientation is

$$I(cal) = I(random) \times \exp[\, p\, (\pi/2 - \alpha)^2\,]$$

where α is the acute angle (in radians) between the diffracting plane (hkl) and the preferred orientation plane.

The results for one of the analyses are summarized in Table 3. Using the first eight reflections (111) to (440) measured with a 1° incident beam aperture, R(Bragg)=4.3%, which deceased to 0.7% by correcting for the preferred orientation. A similar improvement was obtained using all 12 reflections of Si (the first six with a 1° and the last six with a 4° aperture). Refinements were also made using (110) and (100) as preferred orientation planes but negative p-values indicated these preferred orientations were incorrect. Therefore, the actual preferred orientation was confined to (111).

The intensities calculated in the POWLS refinement were used with the W*G functions to generate a computer powder pattern. Figure 3 shows this pattern as an overlay on the the experiment data points but the match is too close to distinguish them. The small differences Δ can be seen in the upper horizontal lines.

Table 3. Silicon Structure Refinement

No. of Reflections	Pref. Orient. Plane	B	P	R(Bragg)%
8	None	0.55	0.0	4.3
	(111)	0.16	0.07	0.7
12	None	0.28	0.0	3.5
	(111)	0.24	0.06	1.5

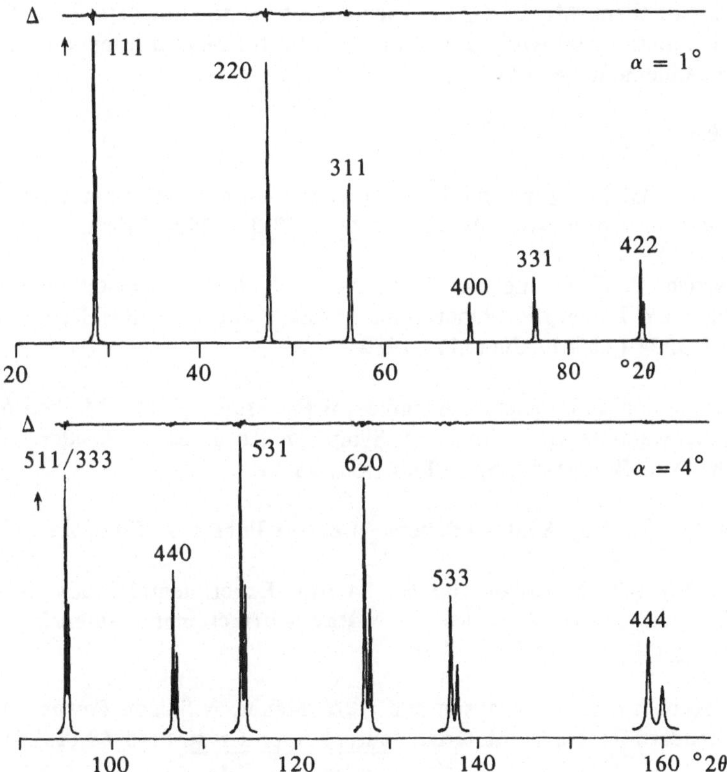

Fig. 3. Silicon powder pattern from calculated intensities of least squares refinement convolved with the instrument function. The matching of the experimental points can be seen in the upper Δ lines.

CONCLUSIONS

The profile fitting method is shown to have high precision in determining the integrated and peak intensities as well as the diffraction angles. The crystallite sizes have a strong effect on the absolute and relative intensities. It is essential to rotate the specimen around an axis normal to the surface and it is advisable to use $<10\ \mu m$ particles. Preferred orientation corrections are required to achieve accurate values.

ACKNOWLEDGMENTS

We wish to thank Mr. C. G. Erickson for sifting the powders, Miss G. S. Lim for aid in the experimental work, and Prof. G. Will for helpful discussions relating to the structure refinement using POWLS.

REFERENCES

1. W. Parrish, G. L. Ayers and T. C. Huang, A Minocomputer and Methodology for X-Ray Analysis, Adv. X-Ray Anal. 23:313-316 (1980).

2. W. Parrish, T. C. Huang and G. L. Ayers, X-Ray Polycrystalline Diffraction: Program Description-Operations Manual, 2nd ed., IBM Research Report RJ 3524 (41681), June 12, 1982

3. W. Parrish and T. C. Huang, Accuracy of the Profile Fitting Method for X-Ray Polycrystallite Diffractometry, Symp. on Accuracy in Powder Diffraction, in: Nat. Bur. Stand. Spec. Pub. 457, 95-110.

4. W. Parrish, "X-Ray Analysis Papers," Centrex Publ. Co., Eindhoven (1965).

5. P. M. DeWolff, J. Taylor and W. Parrish, Experimental Study of Effect of Crystallite Size Statistics on X-Ray Diffractometer Intensities, Journal Appl. Phys. 30:63-69 (1959).

6. C. R. Hubbard, H. E. Swanson and F. A. Mauer, A Silicon Powder Diffraction Standard Reference Material, Journal Appl. Cryst. 8:45-48 (1975).

7. G. Will, POWLS: A Powder Least-Squares Program, Journal Appl. Cryst. 12:483-485 (1979).

8. G. Will, W. Parrish and T. C. Huang, Crystal Structure Refinement by Profile Fitting and Least Squares Analysis of Powder Diffractometer Data, in preparation.

9. H. M. Rietveld, A Profile Refinement Method for Nuclear and Magnetic Structures, Journal Appl. Cryst. 2:65-71 (1969).

NEW STANDARD REFERENCE MATERIALS FOR X-RAY POWDER DIFFRACTION*

Camden R. Hubbard

Center for Materials Science
National Bureau of Standards
Washington, D.C. 20234

INTRODUCTION

Standard Reference Materials (SRMs) from the National Bureau of Standards are samples or artifacts certified for one or more particular parameters. The NBS has produced SRMs since 1905 to aid commerce, to improve measurement technology and to assist in the enforcement of regulations. Today nearly 900 different SRMs are available to serve major segments of industry such as ferrous metals, nonferrous metals, mining, glass, primary chemicals, computer, nuclear power and electronics. In addition to the industrial customers, major SRM users include both federal and state governments, universities and nonprofit research organizations.

For the x-ray powder diffraction (XRD) measurement method there are a large number of standards that can be developed. For example, to check instrument alignment standards for 2Θ, intensity, resolution and inherent instrumental line shape are potentially useful. Internal and external 2Θ standards have been shown to have significant value in obtaining more accurate line positions (Snyder, Hubbard and Panagiotopoulos, 1982). Relatedly, phase identification based on the Powder Diffraction File has been shown to be more reliable and more efficient when an internal standard is employed (Jenkins and Hubbard, 1979). To perform quantitative phase analysis by XRD a wide variety of standards are needed which fall into three general categories: internal standards, pure phase standards, and solid solution standards.

To meet these needs the National Bureau of Standards (NBS) initiated a program to develop XRD standards. The first standard, SRM640 silicon powder, was certified by NBS in 1975 (Hubbard, Swanson and Mauer). Since then over 850 units have been sold through the Office of Standard Reference Materials and the stock has been depleted. To replenish the stock and to improve the particle size SRM640a silicon powder has been certified. Certification of low 2Θ standard, SRM 675 fluorophlogopite, has been completed to complement SRM640a. A set of five intensity standards is nearing completion, and work is under way on a respirable quartz standard. In this paper some details of the certification methods and results will be presented for these four SRMs. Development of SRMs for quantitative analysis of retained austentite in steel is reported separately (Hicho and Eaton, 1983).

SRM640a, SILICON POWDER

The recertification of a <10 µm particle size silicon powder involved use of digital data collection, profile fitting, internal standard correction, and least-squares lattice parameter refinement. The profile model used was the asymmetric rational polynomial f_4 (Pyrros and Hubbard, 1982). Two internal standards, W and Ag, (Swanson et al., 1966) were used for calibration of the 2Θ scale. An example of the calibration data is given in Table 1. Corrections for the effect of thermal expansion were applied to each peak based on the measured sample temperature. The root mean square deviation of the observed Δ2Θ values from the fourth order polynomial calibration curve is 0.0047 deg. 2Θ. In Table 2 the corrected silicon 2Θ values for the same sample are given along with the calculated values based on the least-squares refined lattice parameter. In all refinements the Si 111 line was assigned a weight of 0.0 due to the uncertainty of extrapolation of the calibration curve below the first line of the internal standard. Altogether twelve samples from the bulk were studied. The certified lattice parameter and estimated standard deviation for SRM640a are $\langle a \rangle = 5.430825$Å $\sigma = 0.000011$Å for $\lambda(CuK\alpha_1) = 1.5405981$Å and T = 25.0°C (Hubbard, 1982a).

SRM 675, FLUOROPHLOGOPITE POWDER

For CuKα radiation the lowest 2Θ line for Si occurs at 28.443 degrees. Fluorophlogopite, $KMg_3Si_3AlO_{10}F_2$ (abbreviated as FP), was chosen in order to provide a calibration point at lower 2Θ This material is a synthetic mica with exceptional stability to

TABLE 1. Internal Standard Calibration (Si sample #115)

Std	hkl	$2\Theta_{obs}$	$2\Theta^{*}_{true}$	$\Delta 2\Theta_{obs}$	$\Delta 2\Theta^{\neq}_{calc}$
Ag	111	38.117	38.111	-0.006	-0.001
W	110	40.254	40.262	0.008	0.003
Ag	200	44.285	44.294	0.009	0.008
W	200	58.225	58.251	0.023	0.021
Ag	220	64.416	64.436	0.020	0.024
W	211	73.153	73.184	0.031	0.029
Ag	311	77.355	77.388	0.033	0.031
Ag	222	81.505	81.531	0.026	0.034
W	220	86.953	86.995	0.042	0.038
Ag	400	97.825	97.872	0.047	0.048
W	310	100.574	100.631	0.057	0.051
Ag	331	110.436	110.495	0.059	0.063
W	321	131.085	131.170	0.085	0.083
Ag	422	134.782	134.866	0.083	0.085
W	400	153.461	153.533	0.072	0.072

*Corrected for thermal expansion. Temperature of measurement varied from 25.9° to 26.2 °C.

$^{\neq}\Delta 2\Theta_{calc} = a_0 + a_1(2\Theta-75) + a_2(2\Theta-75)^2 + a_3(2\Theta-75)^3 + a_4(2\Theta-75)^4$ with $a_0 = 0.02990$, $a_1 = 0.58272 \times 10^{-3}$, $a_2 = 0.67635 \times 10^{-5}$, $a_3 = 0.22210 \times 10^{-6}$, and $a_4 = -0.40345 \times 10^{-8}$

Table 2. Observed and Calculated 2Θ (Si sample #115, a_i = 5.43081A)

hkl	$2\Theta_{obs}$	$2\Theta_{calc}$	weight	$\Delta 2\Theta$
111	28.425	28.443	0.0	0.018
220	47.299	47.304	1.0	0.005
311	56.124	56.124	1.0	0.000
400	69.128	69.132	1.0	0.004
331	76.382	76.378	1.0	-0.004
422	88.030	88.033	1.0	0.003
511	94.951	94.956	1.0	0.005
440	106.710	106.712	1.0	0.002
531	114.098	114.097	1.0	-0.001
620	127.551	127.550	1.0	-0.001
533	136.904	136.900	1.0	-0.004

x-ray exposure, humidity and heat. The powder is purposely of large crystallite size (<200 mesh) to enhance preferrential orientation in packed mounts. A well oriented sample can be prepared by front loading and pressing with a microscope slide. Parafocusing diffractometer measurements from such a sample will yield only 00ℓ diffraction peaks. A segment of a pattern of 10% FP and 90% Si by weight is shown in Figure 1. Note that the FP diffracted intensity is quite strong relative to that of silicon and that the line shapes are exceptionally sharp. Procedures similar to those for certification of SRM640a were used to determine d(001) except for the addition of a $k \cdot \sin\Theta$ factor to the least-squares refinement. This necessary correction factor probably results from sample transparency effects due to the large size of the FP particles and the small particle size of the heavily absorbing standards Ag and W. This correction factor, which is dependent on differences in the mass absorption coefficients of FP and the sample, may lead to errors in calibration using FP as large as 0.02° 2Θ at 90 degrees, 0.01° at 28 degrees and 0.003° at 10 degrees 2Θ. Smaller errors are expected when the mass absorption coefficients of the sample and FP are similar. Further details of the certification are to be published (Hubbard, 1982b).

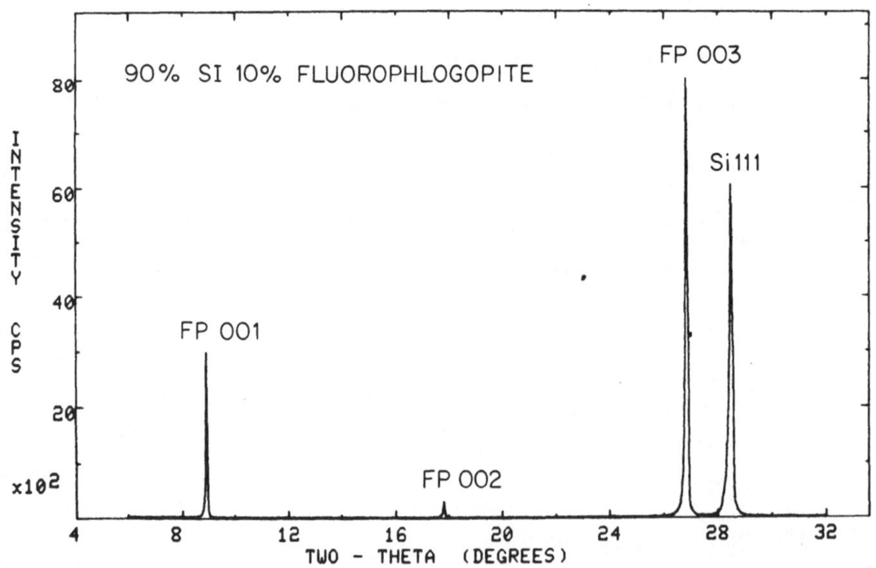

Figure 1. Oriented XRD Pattern of 10% Fluorophlogopite with 90% Silicon.

SRM 674, INTENSITY STANDARDS

Five crystalline phases have been choosen to be certified as XRD intensity standards. Both the relative intensities and reference intensity ratios (RIR) will be certified for α-Al_2O_3, ZnO, TiO_2(rutile), Cr_2O_3 and CeO_2. Both parameters will be given for a constant irradiated volume of sample (reflection geometry, "infinitly" thick sample and fixed divergent slit). Several mounting techniques were employed to test for elimination of preferred orientation. Included were the side drifting, back loading, front loading and spherical agglomeration (Smith and Barrett, 1979). Only the front loading method exhibited preferred orientation effects. The side drifting technique more closely represents what can be achieved in any laboratory and also has been shown at NBS to be relatively free of orientation effects. This technique was choosen for both the relative intensity and reference intensity ratio measurements. Calculated powder patterns were used to check the experimental results (Hubbard, Evans and Smith, 1976). Table 3 lists the average relative intensities for CuKα radiation for ZnO. The agreement between the calculated and experimental relative intensities is considered quite good considering the possible uncertainties arising in the calculated values (Hubbard and Smith, 1977). Similarly, good agreement was found for the other phases except for α-Al_2O_3. The somewhat larger differences for α-Al_2O_3 are possibly due to greater errors in the single crystal model. Table 4 presents the preliminary values for the reference intensity ratios for the ten possible binary mixtures. Multiple lines of the sample and of the reference phase were used to the greatest extent possible. Multiple mountings were always used. In Table 4 the sample phase is given in the left hand column while the reference phase is listed across the table. When α-Al_2O_3, corundum, is the reference phase the reference intensity ratio is known as I/Ic. The calculated values of I/Ic are given in the last column. The CeO_2 - TiO_2 mixture was quite

Table 3. ZnO Relative Intensities (CuKα, constant sample volume)

hkl	2θ	I_{calc}	I_{obs}	RMS DEV.
100	31.75	55	57.6	1.1
002	34.44	38	40.2	1.4
101	36.25	100	100	---
102	47.54	22	22.8	0.5
110	56.55	34	34.4	0.8
103	62.87	31	31.0	0.9
200	66.40	5	4.7	0.2
112	67.91	25	25.9	0.6

Table 4. Experimental and Calculated Reference Intensity Ratios (CuKα, constant sample volume)

	experimental				calculated
Sample Phase	Cr_2O_3	TiO_2	ZnO	Al_2O_3	I/Ic
ZnO				5.37(13)*	5.43
TiO_2			0.66	3.21(8)	3.44
Cr_2O_3		0.60	0.41	2.16(5)	2.16
CeO_2	3.61	1.69	1.49	7.37(18)	14.1

* Estimated standard deviation in the least significant digit(s) given in parentheses.

difficult to blend homogeneously. That RIR value, which is inconsistent with other values in the table, probably reflects the resulting phase segregation in that mixture. The calculated and experimental values agree within the expected uncertainties for ZnO, TiO_2 and Cr_2O_3. The large difference for CeO_2 possibly can be attributed to microabsorption effects.

FUTURE

The certification of a respirable quartz powder is well underway. The starting material, Minusil-5, was acid washed to remove trace metal contamination and hopefully any residual amorphous SiO_2 content. The powder will be certified for the percent amorphous silica by a modified spiking method of quantitative analysis. Particle size distribution, lattice parameters, relative intensities and I/Ic will be provided as supplemental data. A similar project leading to a cristobalite standard is just begining. SRMs of all the polymorphs of silica are needed for the calibration of XRD quantitative analyses of respirable dust samples. The cristobalite and tridymite standards will also be useful in the ceramics industries where the concentration of these polymorphs must be controlled to avoid such problems as thermally induced stress and cracking.

The possibility of certifying a crystallite size standard is also being studied. Such a standard would be used for testing and calibrating XRD procedures used for crystallite size analyses in the 100Å to 1000Å range. MgO has been choosen for several reasons: (1) individual well shaped microcrystals whose size can be measured by electron microscopy are formed in a MgO smoke; (2) the (200)-(400) and (111)-(222) reflection pairs, which are free of overlap, can be used in the Warren-Averback method of analysis; (3) MgO can be prepared in bulk quantities by calcination of a wide variety of precursors.

USER INPUT

Development of future x-ray powder (or single crystal) diffraction SRMs is dependent in part on the needs of the industrial and government laboratories. Suggestions for future SRMs should be sent to the author.

REFERENCES

Hicho, G.E. and Eaton, E.E. (1983). Preparation and Characterization of Standard Reference Materials to be Used in the Determination of Retained Austenite in Hardened Steels. Adv. X-Ray Analy. 26.

Hubbard, C.R. (1982a). Certification of Si Powder Diffraction Standard Reference Material 640a. J. Appl. Cryst. (to be published)

Hubbard, C.R. (1982b). Fluorophlogopite - Low 2Θ/Large d-spacing Powder Diffraction Standard Refence Material 675. (Submitted for publication)

Hubbard, C.R., Evans, E.H. and Smith, D.K. (1976). The Reference Intensity Ratio, I/Ic, for Computer Simulated Powder Patterns, J. Appl. Cryst. 9: 169-174.

Hubbard, C.R. and Smith, D. K. (1977). Experimental and Calculated Standards for Quantitative Analysis by Powder Diffraction. Adv. X-Ray Analy. 20: 27-39.

Hubbard, C.R., Swanson, H.E. and Mauer, F.A. (1975). A Silicon Powder Diffraction Standard Reference Material. J. Appl. Cryst. 8: 45-48.

Jenkins, R. and Hubbard, C.R. (1979). A Preliminary Report on the Design and Results of the Second Round Robin to Evaluate Search/Match Methods for Qualitative Powder Diffraction. Adv. X-Ray Analy. 22: 133-142.

Pyrros, N.P. and Hubbard, C.R. (1982). Rational Functions as Profiles in Powder Diffraction. J. Appl. Cryst. (Submitted for publication)

Smith, D.K. and Barrett, C.S. (1979). Special Handeling Problems in X-ray Diffraction. Adv. X-Ray Analy. 22: 1-12 .

Snyder, R.L., Hubbard, C.R. and Panagiotopoulos, N.C. (1982). A Second Generation Automated Powder Diffractometer Control System. Adv. X-Ray Analy. 25: 245-260.

Swanson, H.E., McMurdie, H.F., Morris, M.C. and Evans, E.H. (1966). Standard X-Ray Diffraction Powder Patterns. NBS Monograph 25 - Section 4: 3-4. National Bureau of Standards, Washington, D.C. 20234

PRECISION AND REPRODUCIBILITY OF LATTICE PARAMETERS FROM GUINIER POWDER PATTERNS: FOLLOW-UP AND ASSESSMENT

A. Brown

Studsvik Energiteknik AB
Nyköping, Sweden S-611 82

C.M. Foris

E.I. du Pont de Nemours & Co.
Central Research & Development Dept.
Wilmington, Delaware 19898

ABSTRACT

Lattice parameters, obtained by least-squares refinement of data from calibrated Guinier powder patterns, are compared in an effort to establish procedures for obtaining reproducible 2Θ values with Guinier-type focusing camera techniques. The calibration procedure is discussed and a method of calculation to reduce loss of precision is proposed. Effects of particle size and crushing are also discussed as well as some sources of error such as use of only selected calibrant lines, film background, and mutual interference of powder lines.

1. INTRODUCTION

It has been recognized [1] that there is a need to begin to establish procedures for obtaining reproducible 2Θ values from Guinier patterns under routine laboratory conditions. For this purpose, cell parameter data were compared for six cubic compounds exposed in three different Guinier-type focusing cameras [2]. Powdered silicon, SRM 640 (National Bureau of Standards) [3], was included in all the x-ray specimens as a calibrant. Techniques of specimen preparation, methods of reading the film, application of the calibration correction (conversion of film readings to 2Θ values), and data reduction systems were previously reported [2].

For all but four of the initial 6 x 3 measurements, a single reading of each film gave the lattice parameter with a standard deviation, σ, better than 0.005%. Moreover the difference, Δa, between any two measurements for a given compound in relation to the respective standard deviations was such that the ratio $K = \Delta a/(\sigma_1^2+\sigma_2^2)^{1/2}$ was generally less than 1.645. These differences lack statistical significance.

Three notable exceptions to this high level of agreement were the parameters for $Pb(NO_3)_2$, Mo_3Sb_7 and As_2O_3. A follow-up of the initial comparison [2], to trace possible sources of error that might reduce the ultimate reliability of Guinier data, is reported below. The supplementary work involved:

- remeasurement of selected films.
- preparation of new specimens of the earlier compounds under different conditions.
- use of a more objective numerical procedure for applying the calibration correction.
- measurement of Guinier patterns of some additions compounds.

2. EXPERIMENTAL

2.1 Photography. New Guinier patterns of the original six specimens were recorded with the 114.6 mm IPT camera at Studsvik. Specimens of materials included in the follow-up studies were NaCl and KCl, for reasons given in 2.4. Exposure times in vacuum were typically 10-15 min. with strictly monochromatic radiation. To attenuate the low-energetic fluorescence from potassium, the KCl specimen was exposed in air.

Single coated Ceaverken (CEA) 15 film was used throughout. The clear, near-colorless base of this film and the low background levels obtained in the absence of fluorescence make it particularly suitable for instrumental measurement.

2.2 Film Measurement. The films from the 114.6 mm IPT camera were measured visually at Studsvik to ± 0.01 mm and then at the Du Pont laboratory to ± 0.005 mm, with a split image comparator. Details of these methods were given earlier [2].

2.3 Calibration Procedure. A new approach has been adopted to the use of calibrant line measurements. The true value of 2Θ is expressed as the sum of the physical quantities that determine the position of a diffraction line on the film. Since one of these quantities is the error in locating zero-2Θ, the reference point chosen for the measurement becomes irrelevant.

With IPT cameras, the trace of the focal line is recorded at the beginning of an exposure as the origin for all subsequent measurements. Since, with Huber cameras, the fiducial line is not recorded, a suitable calibrant line is used as a reference point to estimate the position of zero-2Θ. With the expression now in use, any error in this estimate is corrected on the basis of the collective calibrant measurements without loss of precision.

The expression is formulated as

$$2\Theta_{true} = a_o + a_1 S + a_2 \Phi$$

where $a_o = -\Delta S \cdot a_1$ and ΔS is the error in estimating the zero-2Θ limit in a film measurement S. Similarly, $a_1 = 90/\pi R_E$ where R_E is the effective film radius after shrinkage; $a_2 = -\varepsilon \cdot a_1$ and ε is the displacement of the specimen from the focusing circle. For asymmetric Seeman-Bohlin focusing geometry relevant to the Guinier film cassette

$$\Phi = \sin 2\Theta / \cos\zeta \cdot \cos(2\Theta - \zeta)$$

where ζ is the angle between the plane of the specimen and the axis of the monochromator [4,5]. Here, 2Θ can be approximated by $a_1 S$ with negligible loss of precision. Thus, for 2R = 114.6 mm and $\zeta = 45°$, Φ simplifies to 1 - tan(45 - S/2).

With the use of a hand-held calculator [6], multiple linear regression can be applied to compute the coefficients a_o, a_1 and a_2 from calibrant line measurements. The coefficient of determination, R^2, indicates goodness of fit. Experience shows that for the IPT and Huber cameras the data used in [2] give $R^2 > 0.9_7$. No significant improvement in precision can therefore be expected by including more than the three terms already specified.

2.4 Specimen Preparation. The Huber and IPT cameras also differ as to the method chosen to present the specimen to the x-ray beam. The specimen is oscillated in the Huber camera through the path of the beam along the tangent to the film circle. In the IPT cameras, the specimen is rotated about the point where the beam penetrates the film circle. This difference in presentation imposes an upper limit on the permissible crystallite size which is lower for the Huber camera (∿15 μm) than for the IPT cameras (∿30 μm). The well-crystallized samples of $Pb(NO_3)_2$ and As_2O_3 used in the original work [2] gave satisfactory patterns in the IPT cameras after passage through 36-μm mesh. For the Huber camera, however, it proved necessary to crush the specimens slightly to obtain diffraction lines smooth enough for scanning in an automated microdensitometer.

To check the influence of light crushing on the measurements, specimens for the IPT camera were first prepared by dusting the sample through sieving silk onto the specimen holder. After exposure in the camera, the x-ray specimens were pressed firmly against a steel surface with the end of a metal rod. Inspection under the microscope at 100 X magnification showed that the originally well-formed, translucent crystals of $Pb(NO_3)_2$ and As_2O_3 were shattered by this treatment to sub-micron fragments. The diffraction patterns of the pressed specimens showed noticeable line broadening.

$Pb(NO_3)_2$ and As_2O_3 have a Moh's hardness of roughly 2. A comparison study to test the influence of line broadening on pattern measurement was made with well-crystallized samples of NaCl and KCl, which have a similar hardness. As with the $Pb(NO_3)_2$ and As_2O_3 specimens, the crystals of these materials also shattered on pressing to give comparably broadened patterns.

2.5 Additional Comparison Materials. The reproducibility of Guinier measurements of non-cubic patterns has been examined with the following materials.

α-Fe_2O_3: A sintered plate of nuclear grade material, crushed and sieved to <36 μm. The heavy fluorescence from iron, produced by CuK beam irradiation, was attenuated by placing a 30-μm thick screen of aluminum in the cassette window. The exposure was increased to 2 h.

SnO_2: Fisher Scientific with a particle size <20 μm, heat treated to eliminate crystallite size broadening.

α-Quartz: Two samples have been circulated as follows, a) a reject monochromator crystal, coarsely ground and sieved to <36 μm, b) Arkansas stone with a particle size <20 μm.

These materials were also photographed at the Du Pont laboratory and the patterns processed as described in [2]. All of the above materials have, additionally, been measured in two different powder diffractometers under a range of conditions which are described elsewhere in these proceedings [7].

3. RESULTS AND DISCUSSION

3.1 Calibration Procedure and Errors. The expression given in 2.3 has been used to recalibrate the data collected in the earlier study [2]. The results exhibit little change and the discrepancies already mentioned remain. There is negligible difference between parameters obtained with this expression and those obtained with a second-order polynomial to describe the calibration curve according to

$$2\Theta_{true} = A_o + A_1S + A_2S^2$$

Appreciable differences between the coefficients a_i and A_i for the same measurement are consistent with the difference in the way these coefficients are defined. Evidently, a fiducial zero 2Θ is unnecessary as a reference for film measurements and can be estimated, with no loss in precision, from a measurement of one line of the internal standard.

The Studsvik pattern of Mo_3Sb_7 was remeasured to give a total of six BFB* results to trace the discrepancy in the lattice parameter. Considerable variation in the value given by these measurements was independent of the method used for applying the calibration data. Back calculation of the 2Θ values for silicon showed that, in each case, discrepancies of the order of ± 0.01° were present for at least two of the calibrant lines. Deleting these lines did not improve the result. Close inspection of the film revealed that the calibrant pattern was somewhat grainy, attributable to the coarseness of the silicon powder. This effect evidently influenced measurement in the split image comparator more than visual estimation or profile fitting [8] used for the BBB* and BEB* measurements. *(See note, Table 1.)

Back calculation of the calibrant 2Θ's for the remaining patterns rarely gave discrepancies greater than 0.005°, and this was limited to no more than one reflection. This method of checking the validity of the calibration curve is clearly preferable to inspecting the value of R^2 after calculating the coefficients a_i or A_i. Thus, obtaining a new pattern is preferable to using only selected calibrant lines in view of the uncertainties involved in this approach.

3.2 Specimen Treatment. For As_2O_3 and $Pb(NO_3)_2$ the Huber patterns gave a lower parameter than the two IPT patterns, independently of method of measurement or application of the calibration curve. For As_2O_3, for example, the difference Δa is of the order of -0.0017 Å with a standard deviation, σ, of 0.0001 (K ≫ 4), comparing results for Huber and IPT (114.6 mm) data. This was traced to the need to crush specimens of these materials slightly to minimize graininess of the powder lines in the Huber camera. A follow-up study of this effect was made using the 114.6 mm IPT camera and specimens sieved and crushed on the sample holder. Sieved samples were also studied in a diffractometer with different degrees of packing in the sample holder. While the sieved specimen exhibited a slightly smaller parameter than the unsieved specimen, crushed samples afforded a decrease of 0.0024 Å or 0.02% of the lattice parameter. Results for pressed specimens of As_2O_3 and $Pb(NO_3)_2$ are in good agreement with those obtained from the Huber patterns.

Specimens of NaCl and KCl treated in the same way did not exhibit the same decrease in lattice parameter, although the extent of line broadening was very similar. Thus, the BFB results for NaCl are 5.6421(1) Å for a well crystallized specimen and 5.6419(3) Å for the specimen after crushing.

TABLE 1

CELL PARAMETERS, Å: COMPARISON OF RESULTS OBTAINED FROM VISUAL (BBB) AND INSTRUMENTAL (BFB) MEASUREMENTS OF GUINIER FILMS (CEA REFLEX 15) WITH RESULTS OBTAINED FROM DIFFRACTOMETER MEASUREMENTS

	CeO_2	ThO_2	PtP_2	$Pb(NO_3)_2$	Mo_3Sb_7	As_2O_3
BBB (to 100° (2θ))	5.412 08 (22)	5.596 83 (11)	5.696 28 (10)	7.856 85 (9)	9.568 81 (15)	11.076 54 (18)
BFB (to 100° (2θ))	2 14 (11)	7 28 (9)	6 03 (5)	6 86 (9)	8 88 (9)	5 97 (19)
Diffractometer to 158° (2θ)	1 94 (4)	7 28 (6)	--	7 90 (9)	9 06 (4)	5 95 (10)

NOTE: BFB refers to a film obtained at Studsvik (B), this film measured at Du Pont (F), and data reduction at Studsvik (B). BBB has similar meaning.

Cell parameters are a_o for the cubic lattic followed by the standard deviation, σ, in parentheses.

For KCl, the corresponding results are 6.29367(6) Å and 6.29367(14) Å; the higher values of σ for the crushed specimens can be correlated with increased line width. These parameters are based on a value of λ = 1.54051 for $CuK\alpha_1$.

Film measurements are apparently not subject to a systematic error that leads to a perceptible decrease in lattice parameter with increasing line width. However, on pressing or crushing, the lattice parameters of both As_2O_3 and $Pb(NO_3)_2$ appear to exhibit a real contraction in comparison with the value for the well-crystallized materials.

3.3 Photography with CEA 15 Film. Results for the new patterns of the cubic materials used in the earlier study [2] are listed in Table 1. Results from visual and instrumental film measurements are compared with the results obtained using diffractometer data, as described elsewhere [7]. Calibrations using both a second-order polynomial and the expression given in 2.3 show negligible differences in lattice parameter ($\sim 2.10^{-5}$ Å). Agreement of the values in Table 1 with the corresponding results in Table 5 of the earlier study [2] is good and the BFB value for Mo_3Sb_7 now conforms with the other measurements for this material. Most noticeable is a general lowering of the values for σ in comparison with those in the corresponding BBB and BFB results of the earlier study. This can be attributed to the improved peak to background (P/B) ratio afforded by the clear base of CEA 15 film. A clear film base permits peak location with better precision than was previously possible.

3.4 Mutual Interference of Adjacent Powder Lines. The lattice parameters of CeO_2 and silicon are so close that lines for the calibrant and CeO_2 up to 69°(2Θ) are separated by only 0.09, 0.17, 0.2 and 0.27°, respectively. For the BFB reading in Table 1, the CeO_2 reflections in this angular range exhibit a systematic shift towards the calibrant lines; all $\Delta 2\Theta$ (obs-calc) values are negative up to 50°, for example. Instrumental reading yields a cell parameter which is high in this instance. The effect is attributed to the overlap of adjacent profiles which are broadened by convolution with the scanning aperture.

For CeO_2 alone, the diffractometer measurement gives a noticeably lower parameter than the film measurement BFB. By comparison, the two techniques are in complete agreement for ThO_2, which gives a similar pattern. The lines of ThO_2 are, however, well separated from those of silicon. The parameter shift is marginal for CeO_2 mixed with silicon, where the full-width at half-maximum intensity (FWHM) for the lines is $\leq$ 0.06°(2Θ). The observed level of interference can therefore indicate the limit permissible for precision work with well-crystallized specimens. An alternative calibrant is needed, when specimen lines with this proximity are broad, or when specimens, such as UO_2, give small separation in 2Θ.

TABLE 2: CELL PARAMETERS FOR α-QUARTZ

CRUSHED MONOCHROMATOR CRYSTAL (<36 μm) PLUS SRM 640

	F-M-C*	a(σ) Å	c(σ) Å
[1]	B B B	4.91 29 (2)	4.40 50 (2)
	B F B	32 (3)	47 (4)
[2]	E E B	31 (1)	46 (2)
	E F B	31 (1)	47 (1)
[3]	F F B	35 (3)	45 (4)

TYPE OF CAMERA USED TO OBTAIN FILM.
[1] 114.6 mm IPT, [2] 114.6 mm HUBER, [3] 80 mm IPT
*Film-Measurement-Calculation indicator; M=instrumental for E,F

3.5 Non-Cubic Compounds. Table 2 lists lattice parameters for a coarsely crushed quartz crystal. The original measurements, determined as part of the earlier work [2] but not published, have been calibrated using the expression described in 2.3. All detectable reflections (34 to 100°) are included and agreement is particularly close for the instrumental measurements.

Parameters of three materials (particle size ≤ 20 μm), which were graphed as part of a new series are given in Table 3. For Arkansas Stone quartz and SnO_2, the results agree at the 0.003% level of precision. The somewhat higher σ values for the F films can be attributed to the cutoff at 80°(2Θ) imposed by the camera used.

The higher standard deviations for α-Fe_2O_3 demonstrate the adverse influence of background level (increased by fluorescence in this case) on precise film measurements, since they cannot be attributed to the character of the pattern. This result emphasizes the importance of a high P/B ratio as a source of high quality data.

TABLE 3

CELL PARAMETERS, Å
COMPARISON OF RESULTS FOR NON-CUBIC COMPOUNDS

	Arkansas Stone'		α-Fe_2O_3*		SnO_2+	
	a_o	c_o	a_o	c_o	a_o	c_o
B B B	4.913 04 (10)	5.404 63 (17)	5.034 91 (20)	13.748 25 (91)	4.737 31 (15)	3.186 65 (16)
B F B	3 04 (5)	60 (8)	4 53 (20)	47 21 (72)	50 (10)	32 (10)
F F F	2 80 (15)	22 (22)	4 40 (42)	51 06 (136)	36 (8)	42 (8)
F F B	3 15 (18)	76 (29)	3 99 (44)	52 75 (1 42)	13 (19)	41 (20)

See note Table 1 for origins of data, etc.

Space groups for compounds indicated are '(α-quartz) $P3_{1,3}21$ (152);

*$R\bar{3}c$ (167); and + $P4_2/m\ n\ m$ (136).

REFERENCES

1. JCPDS-International Centre for Diffraction Data, 1601 Park Lane, Swarthmore, PA 19081, private communication, proceedings of technical subcommittee meetings.

2. A. Brown, J. W. Edmonds and C. M. Foris, "Reproducibility and Precision of Guinier Patterns Using Powdered Silicon Calibrant", Adv. X-ray Anal. 24 111 (Edited by Deane K. Smith, Plenum Press 1981).

3. C. R. Hubbard, H. E. Swanson and F. A. Mauer, "A Silicon Powder Diffraction Standard Reference Material", J. Appl. Crystallography 8 45 (1975).

4. M. Möller, "On the Calibration and Accuracy of the Guinier Camera for the Determination of Interplanar Spacings", Atomenergi Report, AE-67 (1962).

5. A. Brown, "Optimal Calibration Curves for Guinier-type Focusing Cameras", Adv. X-ray Anal. 21 289 (Edited by C. S. Barrett, D. E. Leyden, J. B. Newkirk and C. O. Ruud, Plenum Press, 1978).

6. Texas Instruments, "TI Programmable 58/59, Applied Statistics" 3-8 and 5-14 (Texas Instruments, Inc. 1977).

7. A. Brown, "Precision Lattice Parameter Measurements with Guinier Camera and Counter Diffractometer: Comparison and Reconciliation of Results", Adv. X-ray Anal. 26.

8. J. W. Edmonds, "Precision Guinier X-ray Powder Diffraction Data", National Bureau of Standards Special Publication 567. Proceedings of Symposium on Accuracy in Powder Diffraction, Gaithersburg, MD., 1979 (Edited by S. Block and C. R. Hubbard, NBS, 1980).

POWDER-PATTERN: A SYSTEM OF PROGRAMS FOR PROCESSING AND INTERPRETING POWDER DIFFRACTION DATA*

Nikos P. Pyrros

JCPDS--International Centre for Diffraction Data
National Bureau of Standards
Washington, DC 20234

and

Camden R. Hubbard

Center for Materials Science
National Bureau of Standards
Washington, DC 20234

INTRODUCTION

The production of standard x-ray diffraction patterns at NBS imposes special requirements in the data processing of powder patterns. The patterns should be complete and have an overall accuracy of better than 0.01 degree two theta. To ensure completeness all the observable peaks should be indexed. To make certain that the sample is a pure phase, weak peaks have to be identified as well.

The indexing of all the peaks implies that the cell constants must be known and there should be a good agreement between all the calculated and observed peak positions. In practice this is achieved by a least-squares refinement of the unit cell parameters. This serves as a test of the assumed unit cell and also as an interpretation of the observed peaks. Finally, an attempt is made to identify the space group. This step also requires the identification of weak peaks. The agreement of a known space group with the observed reflections further confirms the purity of the sample.

Finally, the intensities are measured with special care, (Morris et al., 1981). Various sample preparation techniques are used to reduce preferred orientation. At least three samples are prepared exclusively for intensity measurements. All the samples are measured and used to derive the observed intensities and an estimate of the reproducibility of the intensities.

All these requirements impose specific demands in the automated data collection and data processing. To meet these demands, we have developed the data collection system AUTO (Snyder, Hubbard, Panagiotopoulos, 1982) and the data processing system POWDER-PATTERN. The latter system was designed to allow maximum interaction with the experienced diffractionist, but also to be intelligible to the inexperienced user as well. Various entries default to reasonable values to allow processing with minimum input. The default values also permit the use of some program modules in batch mode.

ORGANIZATION OF THE SYSTEM

POWDER-PATTERN is an interactive data processing system consisting of a number of independent modules (programs). The

Table 1. Contents of the file PKS.

1. Title.
   Cell parameters and their standard deviations.
   Crystal system and space group.
   Compound name and chemical formula.
2. Crystal Data.
3. Data collection parameters.
4. External and internal calibration parameters.
5. Processing parameters and processing 'history' flags.
6. Reflection Data.
   2θ Observed peak position and standard deviation.
   $2\theta_c$ Corrected peak position and standard deviation.
   Peak intensity and standard deviations.
   Integrated intensity and standard deviation.
   Full width at half maximum (FHHM).
   HKL
   Profile parameters.
   Identifiers: α_2 peaks, internal standard peaks, overlapping unresolved peaks.
   Flags: Profile refinement. Unit cell least squares.

structuring of the system in modular units offers advantages in updating and modifying the units. The amount of work required for the incorporation in the system of new programs and the modification of existing ones, is appreciably less than what a single unified program would require. The modules have been designed to allow repeated execution of any module.

The link in the execution of the independent modules is the peaks file (PKS). This file serves as depository of information generated by the modules and also serves as input to the various modules. The contents of the file are given in Table 1. An important aspect of the file is that processing parameters and flags are stored in it. These flags and parameters serve as a guide to the execution of the modules and also as a depository of information from previous processing. This stored processing 'history' provides the input for minimizing reprocessing in cyclic execution of the programs.

The usefulness of the file PKS is enhanced by the editing program EDIT.PEAKS which allows the manipulation of the PKS file. The program can be used to create flags and parameters that guide the profile refinement program, the unit cell least-squares refinement and other programs. It can also be used to print the contents of the file PKS, to enter new data or to delete or modify old data in the file. The program has a tree structure. At the first node after entry a number of general options are offered to the user. The user chooses one of these options and branches to the subchoices given at a second series of nodes.

THE MODULES

All the modules that currently form the system together with their input and output files are shown in Fig. 1. The files are opened, assigned unit numbers, and closed by the programs. A file naming convention helps to keep track of all the files generated during the processing of a pattern. A file name generated for the data collection is used throughout the processing. This name plus an extension is assigned by the programs to the various files generated in data processing.

THE PROGRAM PATTERN

The first program (PATTERN) reads the raw data, identifies the peaks and determines their peak height intensity and the width at half maximum. This program creates the file PKS and and stores in it all the information generated by the program. It also generates the file PLOT that contains the raw data and the background calculated by the program. The PLOT file is used for plotting and further processing by the other programs. The

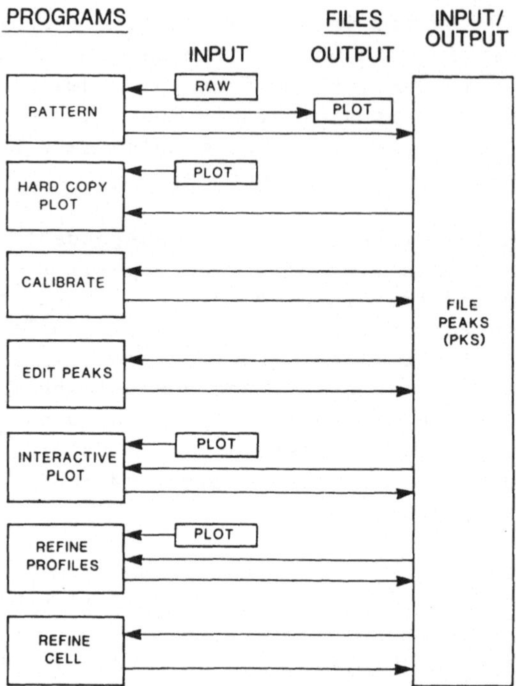

Figure 1. The modules of the POWDER-PATTERN system and their input and output files.

algorithms of Savitzky and Golay (1964) are used to smooth the data and calculate the second derivative. The peaks are initially located from the minima of the second derivative, and then their optimum positions are found with a Newton-Gregory 3 point interpolation. The data can be processed with or without α_2 elimination. For α_2 elimination, the method of Ladell, Zagofsky and Pearlman (1975) is used.

BACKGROUND DETERMINATION

The technique of locating peaks from the second derivative minima also provides a way of identifying the background. The raw data of the pattern can be considered to be composed of the signal from the reflections, a continuous but slowly varying background and noise. These three different components of the powder spectrum have different influences on the calculated second derivative. The reflections cause a sharp minimum in the second derivative, while the continuous background virtually disappears and the noise adds a noise component. For regions in the spectrum that do not contain peaks what essentially remains in the second derivative is noise, as can be seen in Fig. 2.

To locate the peaks the program PATTERN first smooths the data and calculates the second derivative. Then from the second derivative it calculates the noise amplitude and determines a noise level. The noise level is used as a cutoff to separate the signals in the second derivative due to noise from signals of true peaks. Once the peak regions and their widths have been determined the pattern is separated into regions that are occupied by peaks and regions that contain only background. The program fits a 2nd degree polynomial to intensities in the background regions (Fig. 3) as a function of 2θ. Observed points that are more than 2σ above the calculated background are dropped and the background polynomial is recalculated. This helps the fitted background to converge to a better estimate of the background. The calculated background is subtracted from the raw data to give the peak heights.

AUXILIARY PROGRAMS

The program CALIBRATE corrects the two thetas using external calibration parameters, (Snyder, Hubbard, Panagiotopoulos, 1982). It identifies and flags the standard peaks and calculates and

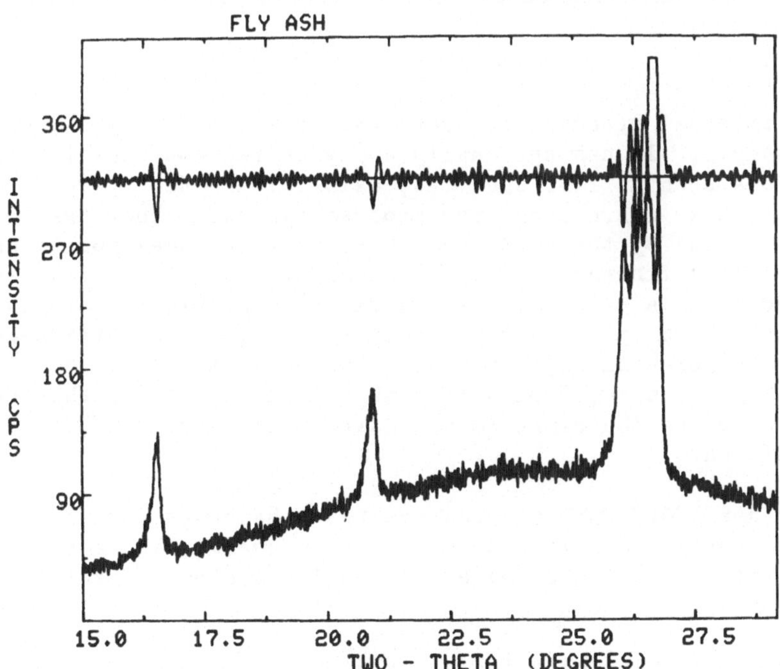

Figure 2. Noise in the 2nd derivative spectrum.
Upper part: 2nd derivative on an arbitrary scale.
Lower part: raw data.

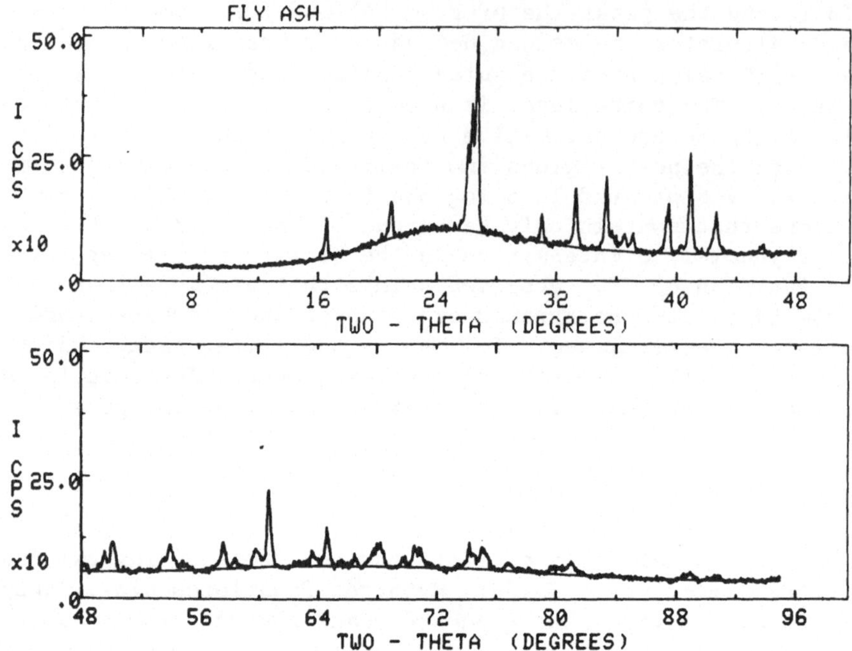

Figure 3. Raw data and calculated background for the pattern of Fig. 2 .

applies internal standard corrections, if a standard was included in the sample. It can optionally apply corrections for inherent aberrations such as flat specimen, axial divergence, transparency, etc. For these corrections the program can use either the 2θ positions found by the second derivative or the peak positions from profile refinement. Several secondary functions are also performed by this program. It searches for possible contaminant radiation. If α_2 peaks are present, the program identifies and flags these peaks as well as peaks that seem to be due to overlapping α_2 and α_1 from different reflections. For this case it can calculate the expected α_1 position from the trailing α_2 peak position.

The HARD.COPY.PLOT program creates a hardcopy plot, with a scale of 1 degree per inch and labels the peaks. The pattern is plotted twice with two different intensity scales to facilitate inspection for the weak peaks.

The interactive plotting program INTERACTIVE.PLOT displays the pattern on a graphics terminal. The user can either inspect the whole pattern or smaller regions of the pattern. The program plots the raw data, the calculated background, the theoretical profiles and the difference between observed and calculated profiles. With the help of crosshair cursors the user can locate peaks and find their peak heights. For this purpose the program can either use the nearest minimum of the second derivative or the position of the cursor. The program updates the file PKS with new information.

The file PKS is also used as an input to the NBS version of the Geological Survey's unit cell least squares refinement program (Evans, Appleman and Handwerker, 1963). Currently the EDIT.PEAKS program creates a file that is used as an input to Visser's Indexing program (Visser, 1969).

PROFILE REFINEMENT

The peak positions and their intensities are refined by the program REFINE.PROFILES. The raw data observed profiles are fitted to an empirical profile of the form:

$$f(x) = \frac{I_m}{1 + A_1x^2 + A_2x^4} \qquad (1)$$

with $x = 2\theta - 2\theta_o$ with $2\theta_o$ the position of the peak maximum and I_m the maximum intensity. The program REFINE.PROFILES uses an asymmetric form of Eq. 1 with different parameters A_1 and A_2 used to describe the profile for the low and high two theta sides of the peak. The $K\alpha_2$ profile is assumed to be identical in shape to that of the $K\alpha_1$. A simplex algorithm is used to minimize

$$R = [\Sigma\ (I_o - I_c)^2 / \Sigma\ I_o^2]^{\frac{1}{2}}$$

by varying the seven parameters I_m, r, $2\theta_o$, A_1^ℓ, A_2^ℓ, A_1^h, A_2^h where r is the intensity ratio of the α_2 to the α_1 characteristic lines.

The program divides the pattern into peak and background regions such that peaks belonging to different regions do not overlap and such that the end points of the region are far enough from the peaks to ensure that the end points belong to background. Peaks in one region can be refined with identical profile shape parameters (all the A's) or with individual profile parameters. The program can refine all the data or only selected regions. The file PKS is updated with the final parameters. The refinement can be repeated. This recycling ability coupled with the use of the EDIT.PEAKS program offers added flexibility in the refinement.

Profiles of the form in Eq. 1 can give a very good agreement with high quality data of observed profiles with R in the range of 0.02 to 0.04 (Pyrros and Hubbard, 1983). The profile refinement significantly improves the accuracy with which the peaks are located. The improvement is more significant for bands of overlapping peaks. For these bands the location of a peak position determined using only the second derivative can give poor results. In Fig. 4, we give the results of profile fitting to a band of peaks of ethylenediamine hydrochloride. The background is that determined by the program PATTERN. The first peak in the region is the 311 reflection of the silicon internal standard. This peak overlaps severely with a peak from the sample. The other ten peaks can be resolved into their α_1 and α_2 components. The refinement was stopped when R=0.055 was equal to R_L, the limiting R factor for this region. We define

$$R_L = [\Sigma(1.5\ \sigma_i)^2/\Sigma\ I_i^2]^{\frac{1}{2}}$$

where I_i is the intensity and σ_i the standard deviation at the point i of the profile. The limiting R factor indicates approximately the best value that we can expect to get in the profile fit. Profile refinement should improve significantly the accuracy of the d spacings for materials with many overlapping peaks such as those with low symmetry or large cells.

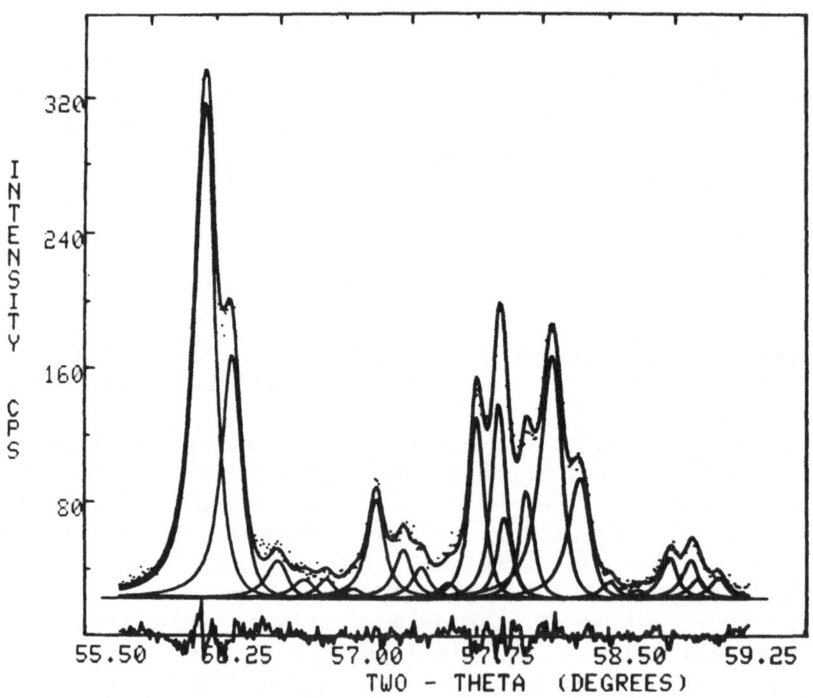

Figure 4. Profile fitted data of $C_2H_8N_2.2HCl$. Raw data (...), individual theoretical profiles and the total theoretical profile. (——) At the bottom the difference function I_o-I_c.

CONCLUSIONS

The POWDER-PATTERN system forms a flexible system for powder diffraction data processing. A modular structure of independent units makes it adaptable to extension by the introduction of new units and modifications. The units allow user intervention and hence add flexibility to the data processing. The use of flexible rational profiles with a relatively small number of parameters gives accurate peak positions and helps in the interpretation of complicated bands with overlapping profiles.

ACKNOWLEDGMENTS

A number of routines in the POWDER-PATTERN system come from the ADR system of Mallory and Snyder (1979). The development of our programs has also profited from the work of Goehner (1979), Sonneveld and Visser (1975) and from suggestions by the members of the JCPDS Associateship at NBS.

DOCUMENTATION

All the units have been written in FORTRAN 77 and take advantage of the features offered in this new version. The system has been developed for a large computer, but it may be modified to execute on a minicomputer. The only disadvantage of the minicomputer is the restriction imposed by the size of the memory which restricts the maximum size of the bands of peaks that can be refined. For some patterns we have observed overlapping peaks forming bands of peaks with widths up to 30 degrees. Documentation of the programs, users guide, and source code are available from the authors. Please supply a tape (9 track preferably) for the source code.

REFERENCES

Evans, H. T., Appleman, D. E., and Handwerker, D. S. (1963). Report #PB216188, U.S. Dept. of Commerce, National Technical Information Center, 5285 Port Royal Rd., Springfield, VA 22151.

Goehner, R.P. (1979). Specplot - An Interactive Data Reduction and Display Program for Spectral Data, Adv. in X-Ray Analysis, 23, 305-311.

Ladell, J., Zagofsky, A. and Pearlman, S. (1975). $CuK\alpha_2$ Elimination Algorithm, J. Appl. Cryst. 8, 499-506.

Mallory, C.L. and Snyder, R.L. (1979). The Alfred University X-Ray Powder Diffraction System, Technical Paper No. 144, New York State College of Ceramics, Alfred University, Alfred, N.Y. 14802.

Morris, M.C., McMurdie, H.F., Evans, E.H., Paretzkin, B., Parker, H. S., Panagiotopoulos, N.C. and Hubbard, C.R. (1981). Standard X-Ray Diffraction Powder Patterns, NBS Monograph 25, Sect. 18. National Bureau of Standards, Washington D.C. 20234

Pyrros, N.P. and Hubbard, C.R. (1983). Rational Functions as Profile Models in Powder Diffraction. Submitted for publication.

Savitzky, A. and Golay, M.J. (1964). Smoothing and Differentiation of Data by simplified Least Squares Procedures, Anal. Chem. 36, 1627-1639.

Sonneveld, E.J. and Visser, J.W. (1975). Automatic Collection of Powder Data from Photographs, J. Appl. Cryst. 8, 1-7.

Snyder, R.L., Hubbard, C.R., and Panagiotopoulos, N.C. (1982). A Second Generation Automated Powder Diffraction Control System, Adv. in X-Ray Analysis 25, 245-260.

Visser, J. W. (1969). A Fully Automatic Program for Finding the Unit Cell from Powder Data, J. Appl. Cryst. 2, 89-95.

AN EVALUATION OF SOME PROFILE MODELS AND THE OPTIMIZATION PROCEDURES USED IN PROFILE FITTING

Scott A. Howard and Robert L. Snyder

N.Y.S. College of Ceramics at Alfred University

Alfred, N.Y. 14802

ABSTRACT

This paper examines some of the concerns regarding the development of an algorithm for the refinement of X-ray diffraction profiles. The object of the algorithm is to provide a time efficient method of refinement through the choice of a suitable profile function and optimization technique.

Seven profile models were tested using a least-squares error criterion for refinement. Profile parameters were refined using non-linear Gauss-Newton, Marquardt[1] and Simplex[2] algorithms. The profiles were refined on a pattern digitally collected from an NBS 640A silicon sample.

The results of this study indicate the repetitive function evaluations are not necessarily the time consuming step in the profile fitting process. As the number of parameters needed to evaluate the profile and the number of points in the profile increases, the time required to perform the mathematics in the Gauss-Newton and Marquardt algorithms increases. Although the Simplex was most memory and time efficient, our Gauss-Newton optimization algorithm provided a more consistent set of refined values which were not as dependent on the initial estimates of the parameters.

The most favorable results were obtained by using the split Pearson VII profile with the alpha 2 reflection fixed in position and intensity with respect to the alpha 1 reflection. This method generated the lowest residual error and was found to avoid some problems resulting from the alpha 1, alpha 2 line overlap.

INTRODUCTION

Profile fitting is a familiar technique in the arsenal of the X-ray powder diffractionist and is becoming more popular as the number of automated diffraction systems increases. It is a valuable aid in quantitative analysis by giving reliable values for the integrated peak intensity, in detection of trace phases by accounting for major phases with the ability to subtract identified phases thus producing reliable residual patterns and, most recently, affording a method for the evaluation of structural parameters using whole pattern refinement.[3]

However, refinement of profiles is subject to instrumental and sample effects.[4] These effects are typically convoluted, thus leading to difficulty in the optimization of profile parameters and increased residual error after refinement. X-ray profile asymmetry strongly influences refinement. To test the effects of the Lorentz-Polarization factor and the Reitveld asymmetry correction[3], the intermediate Lorentzian was chosen. The Gauss-Newton algorithm, optimized as described below, was used to refine these profiles.

The final refinement test involved fixing the alpha 2 intensity and position, based on the alpha 1 line, during the optimization procedure. However, all other parameters in the profiles were allowed to vary independently. The split Pearson VII profile was selected for this test using the Gauss-Newton algorithm for refinement. Criteria for evaluating the refinement procedures included the memory requirement, the time involved in refinement, and the resulting residual error.

THE DIFFRACTION PATTERN

The diffraction pattern was collected on a Si NBS SRM 640A sample. A specimen was side-drifted into a sample holder and mounted onto a Norelco diffractometer. The diffractometer was configured with a 1 degree divergence slit, diffracted beam anti-scatter slit, 0.003" receiving slit and a graphite monochrometer. A step scan was taken on this automated system in the range of 20-150 degrees two-theta with a step width of 0.015 degrees. The count time at each point was 7 seconds.

THE PROFILE MODELS

Four basic profile functions with three variations constitute the set of functions evaluated. These include the Lorentzian, Pearson VII, Gaussian and Voigt[5] functions shown in Table 1.

Table 1. Profile Functions

#	Name	Form: I(x) =	Conditions:
1	Lorentzian	$I_0/(1+kx^2)$	$k = 1 / (FWHM/2)^2$
2	Mod. Lorentzian	$I_0/(1+kx^2)$	$k = 0.4142/(FWHM/2)^2$
3	Int. Lorentzian	$I_0/(1+kx^2)^{1.5}$	$k = 0.5874/(FWHM/2)^2$
4	Pearson VII	$I_0/(1+kx^2)$	$k = (2^{(1/m)}-1)/(FWHM/2)^2$
5	Split Pearson VII	$I_0/(1+kx^2)^m$ $I_0/(1+k'x^2)^{m'}$ where	for x positive for x < or = 0 $k = (2^{(1/m)}-1)/(FWHM/2)^2$ $k' = (2^{(1/m')}-1)/(FWHM/2)^2$
6	Gaussian	$I_0 exp(-kx^2)$	$k = 0.6931/(FWHM/2)^2$
7	Voigt	Lorentzian*Gaussian	Convolution product

Where:		
	$x = 2\theta_i - 2\theta_k$	Distance from the Bragg Angle($2\theta_k$)
	FWHM	Full width at half maximum intensity
	m, m'	Shape factors

The three forms of the Lorentzian (or Cauchy) function included the normal Lorentzian, the intermediate and the modified Lorentzian. The Pearson VII was used as a single profile[6] and as a split profile.[7] The split Pearson VII fits both the low and high angle sides of the profile with two half profiles having a common peak intensity and position. The Gaussian profile was included for completeness. The last function tested in fitting was the Voigt function.

Peak position, intensity, full width at half maximum, and shape factors were refined for all functions. Table 1 gives the functional forms of the profiles with the optimization parameters.

REFINEMENT TECHNIQUES

The refinement algorithms were implemented on a Honeywell 560 computer system. This is a 32 bit machine running a time sharing system and should be considered a super "mini" by today's standards. The coding of the algorithms was written in single precision and structured for the eventual implementation on a 16 bit mini computer.

The Gauss-Newton and Marquardt algorithms required large amounts of storage area for the intermediate derivatives. Since memory required for derivative storage was based on the number of parameters in the profile function, the number of profiles being evaluated, and the number of points comprising each profile, the derivatives were temporarily written to secondary storage. Past experience indicates this technique is more memory and time efficient. The Simplex algorithm, however, did not require large amounts of storage for intermediate results and so operated completely within the computer's memory.

Convergence of the non-linear Simplex algorithm was based on the change in the value for the error of the vector Ph, the point of highest error. When, after a successful reflection, contraction or expansion, the error failed to be reduced by the value 10^{-6}, then convergence was assumed. The nature of this optimization technique tends to make the final parameters somewhat dependent on the initial parameters. To minimize the effect due to initial parameter choice, a second application of the simplex algorithm was used with the refined parameters from the first as input. This reduced the problem of erratic final parameter values, but at the cost of creating a much less time efficient optimization process.

The Gauss-Newton and Marquardt algorithms both used numerical differentiation. The Marquardt algorithm was identical to the Gauss-Newton algorithm save for the intermediate step of adding a number to the diagonal elements of the normal equation matrix followed by renormalization of the matrix. The weighting of the diagonal elements, as indicated by Marquardt, tends to change the character of the solution vector from a value based on a Gauss-Newton solution, to one obtained by a gradient type solution. The numerical differentiation was optimized for profile fitting by using a 0.0001 fractional shift in parameter value. The error used in the refinement process was the residual error factor:

$$R = \frac{(\, Y_{calc} - Y_{obs}\,)^2}{(\, Y_{obs})^2}$$

PROCEDURE

Examination of the pattern of silicon revealed eleven lines ranging from approximately 28.4 to 137.5 degrees two-theta. Each line consisted of an alpha 1 and alpha 2 pair with no overlapping lines. The approximate center of the peak was found by visual examination and the region using 2.5 degrees around each side was used for refinement. The peak positions and intensities were recorded and all lines assigned a FWHM of 0.1 degrees. The shape factors were all assigned a value of 1.0. The alpha 2 components were then assigned positions and intensities that would be expected of Cu radiation. The other initial factors for the profiles were duplicated from the alpha 1 line.

The refinement proceeded by applying the profile functions one at a time to the data region selected. When all profile functions had been evaluated, the second optimization technique was applied and all profile models refined. This process was again repeated for the third optimization technique.

DISCUSSION OF RESULTS

The results of refining the line profiles is illustrated in figure 1. The solid lines indicate the residual error in the profile after refinement. The dotted lines indicate the refinement time in seconds.

The errors after refinement for the Lorentzian line shapes were not considerably different after the optimization techniques were applied, nor was there a large dispersion of times. This tended to be the trend with the Lorentzian line shapes as well as the Gaussian and Voigt profiles.

Problems were encountered with the Gaussian and Voigt profile fits. The Gaussian was a poor fit and hence the error was large and the refinement time became very long. The Voigt function did not perform as well as was expected due to the numerical method used in generating the profile. Generation of the complex error function arising from the convolution process tended to be time consuming and inaccurate.

As the number of parameters in the profile function increased, the dispersion of refinement times and residual errors tended to increase also. Figure 2 shows the results of refining the split Pearson VII function with a fixed alpha 2 intensity and position. This was the trend with the Pearson VII and the split Pearson VII.

The profile with the lowest refined error was consistently the split Pearson VII with the fixed alpha 2 reflection. The Lorentzian line shape was the best of the Lorentzians. While some authors[4-7] report the intermediate or modified Lorentzian could be considered the better diffraction profile, the basic Lorentzian must be considered the better profile for our instrumental configuration.

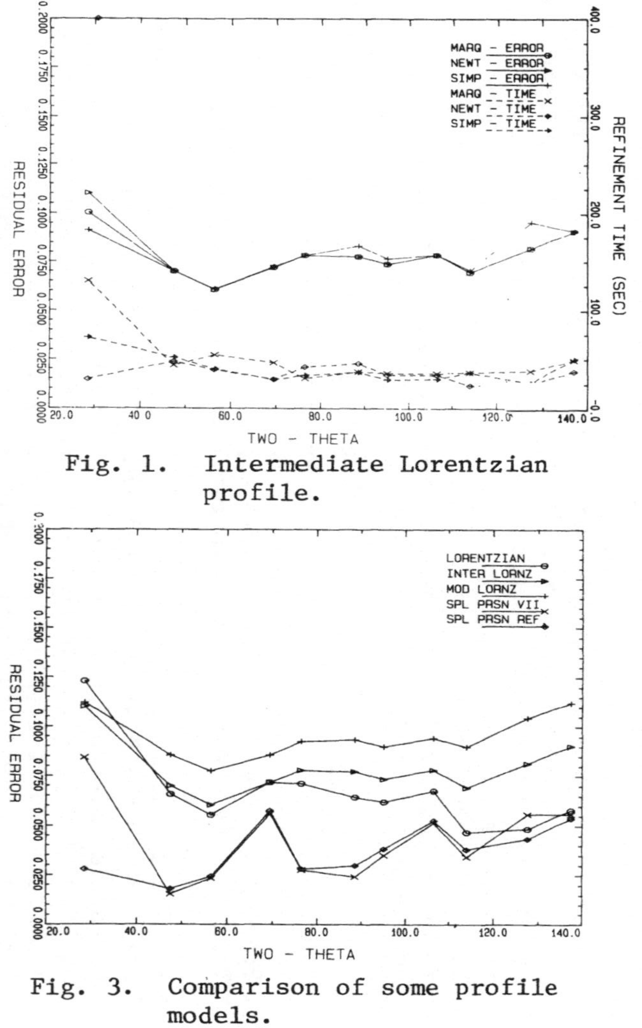

Fig. 1. Intermediate Lorentzian profile.

Fig. 3. Comparison of some profile models.

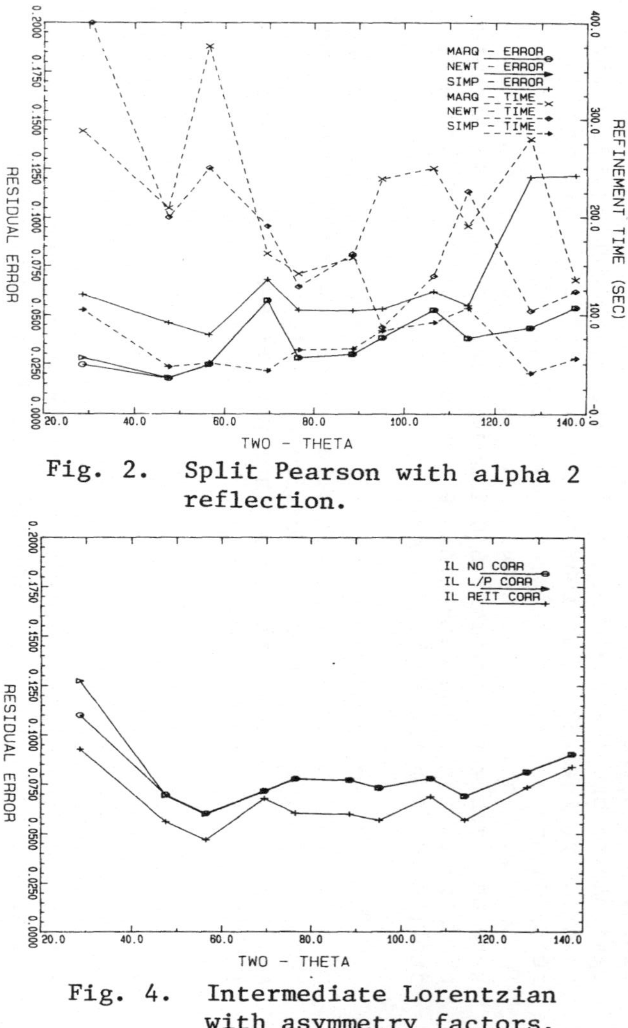

Fig. 2. Split Pearson with alpha 2 reflection.

Fig. 4. Intermediate Lorentzian with asymmetry factors.

Figure 3 illustrates the errors involved in using the various profile functions. The best fitting was the split Pearson VII with fixed alpha 2 reflection. The fixing of the alpha 1, alpha 2 position and intensities avoided a problem which manifested itself most strongly when refining profiles below approximately 70 degrees two-theta. The alpha 1 line would account for the bulk of the line area while the alpha 2 would become an appendage to account for the "bump" on the side of the profile. This was found to occur regardless of the profile or refinement technique. This is the reason for the dispersion for the residual errors at the low angles of two-theta shown in figures 1 and 2.

Correction for the Lorentz-Polarization factor was not effective in lowering the residual error resulting from refinement. The error tracked almost perfectly with that of the pure intermediate Lorentzian, except for the deviation at lowest angle. The domination of the alpha 1 line over the alpha 2 line, as noted earlier, yielded somewhat erratic results when the alpha 2 was not fixed in position. The application of the Reitveld asymmetry correction was on a profile to profile basis, rather than refining the asymmetry factor for the entire pattern. By applying the symmetry correction to the individual profile, a reduction of the residual error after refinement was achieved. Application on a pattern basis tended to force the asymmetry factor to a value such that, after application to a profile, the profile would essentially be unaltered. The end result was no change in residual error after correction.

With regards to the optimization strategies, the Gauss-Newton method was deemed the most satisfactory. Both the Gauss-Newton and the Marquardt methods would yield a consistent set of parameters regardless of starting parameters. It was found that if the fractional shift in parameters for numerical differentiation was 0.0001 then these algorithms could be made less sensitive to entrapment by local minima. The Marquardt algorithm took a longer time to converge in a system where the correlations between parameters were high. Refinement times for the Simplex algorithm were consistently the lowest, but usually gave a slightly higher residual error. The Simplex was found to provide the lowest residual error after refinement of the Voigt profiles. This was probably due to the extremely high correlation between parameters in the complex error function. These correlations violated the assumptions that the parameters were, at best, only minorly correlated for the solution of the linearized equations evaluated in the Gauss-Newton and Marquardt algorithms. The Simplex appeared to perform a search for optimium parameters despite these gross correlations.

REFERENCES

1. D. W. Marquardt, An algorithm for least-squares estimation of nonlinear parameters, J. Soc. Indust. Appl. Math., 11:431 (1963).
2. J. A. Nelder and R. Mead, A simplex method for function minimization, Comput. J., 308 (1965).
3. H. M. Rietveld, A profile refinement method for nuclear and magnetic structures, J. Appl. Cryst., 2:65 (1969).
4. T. C. Huang and W. Parrish, Accurate and rapid reduction of experimental X-ray data, App. Phys. Letters, 27:123 (1975).
5. Th. H. de Keijser, J. I. Langford, E. J. Mittemeijer, and A. B. P. Vogels, Use of the Voigt function in a single-line method for the analysis of X-ray diffraction line broadening, J. Appl. Cryst., 15:308 (1982).
6. M. M. Hall, Jr. and W. W. Henslee, The approximation of symmetric X-ray peaks by Pearson type VII distributions, J. Appl. Cryst., 10:66 (1977).
7. A. Brown and J. W. Edmonds, The fitting of powder diffraction profiles to an analytical expression and the influence of line broadening factors, Adv. in X-ray Anal., 23:361 (1970).

COMPUTER-AIDED QUALITATIVE X-RAY POWDER DIFFRACTION PHASE ANALYSIS

Raymond P. Goehner and Mary F. Garbauskas

General Electric Corporate Research and Development
P.O. Box 8
Schenectady, NY 12301

ABSTRACT

This paper describes a set of interactive computer programs written in FORTRAN IV and implemented on a PDP 11/34 equipped with RLO-2 ten Mbyte discs and an RSX-11M operating system to assist in x-ray diffraction phase analysis. The packing of the data base will be described along with the various interactive programs that access it. The search/match package is designed more as a set of aids for the diffractionist rather than as a single totally automated search/match program, the procedure adapted by many of the x-ray diffraction (XRD) equipment manufacturers. The approach taken allows the analyst to direct the analysis procedure by utilizing different search/match programs or specifying various options within a given program.

INTRODUCTION

X-ray diffraction qualitative phase analysis is essentially pattern recognition. A multiphase powder diffraction pattern is a composite of subpatterns. Due to the errors and incompleteness of the JCPDS data base, as well as the errors involved in the experimental data, the composite pattern could have either no solution, a partial solution, or a non-unique solution. The individual d&I pairs do not give a-priori insight into the identity or properties of the phase present in the sample. An excellent discussion of the search/match problem can be found in Nichols' and Johnson's paper (1).

The original development of computer search/match programs was done by Frevel (2), Nichols (3), and Johnson-Vand (4) in the later 1960's. These algorithms, along with Hanawalt type searches (5-7), still form the basis of today's current search/match programs.

The interest in computer-aided phase identification has increased dramatically due mainly to the availability of inexpensive minicomputers with mass storage capability. Automated diffractometers are now supplied from all major manufacturers of XRD equipment. In most cases these systems can be purchased with a computer search/match program as part of the automation package.

Storage of the JCPDS Data Base

The JCPDS file consists of about 39,000 entries (sets 1-31) and is divided into two main parts--inorganic phases (27,000) and organic phases (12,000). In addition, there are many subfiles such as common phases (2,900), metals and alloys (6,000), minerals (3,500), NBS patterns (1,000), common materials (200), and common minerals (200). Every year there are about 2000 new cards issued, 300 of which are replacements of existing cards. This means the storage of the data base has to be a compromise of size, retrieval speed, flexibility, and ease of maintenance. Figure 1 shows the scheme utilized at GE for the disc storage of PDF data. This scheme has been optimized for fast retrieval of card data by PDF number. The subfiles are ASCII files of PDF numbers as supplied by the JCPDS. By using ASCII files, the analyst can build and/or edit a subfile by using a system editor. The PDF data is stored in random access binary files. There is one file for each of the 31 sets currently available. These files contain all of the information

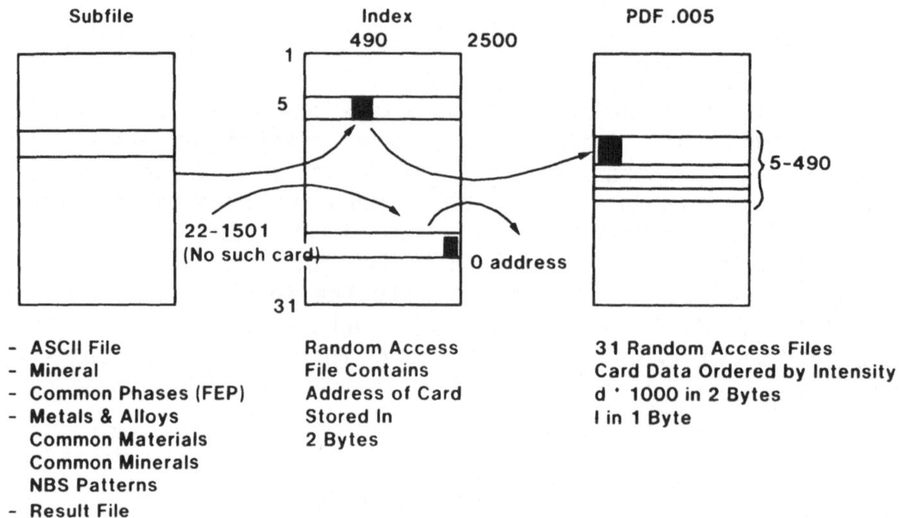

Fig. 1. Disc storage of PDF data.

found on the magnetic tape supplied by the JCPDS. The dspacing is stored as 1000*d in two bytes and the intensity is stored in one byte. The d&I data is ordered in the file by intensity with the 100 line being first. In the next packing of the PDF, the dspacings will be stored as 2θ*100 for CuKα. Using this scheme, very realistic accuracy (.01°) can be obtained over the entire measurement range in a 2 byte integer (0 to 18,000). This also has the advantage of allowing for a fixed error window. In its present form the ASCII subfile, INDEX file, and 31 PDF files occupy about 8 Mbytes on disc.

Programs

Various utility programs have been written to allow the user to access the data base. EDITPDF is used to add user cards or correct errors found in existing PDF cards. VERIFY matches a card against the observed d&I data. The operator can then choose to have the matched d&I pairs subtracted. FREAD displays a PDF card on a terminal given the file and card number. The program will write to a 80 character by 24 line CRT without scrolling. CARD is essentially the same program as FREAD except that it will sequence through any result file generated by the qualitative analysis software.

CHEM is a program which is used to build ASCII files of PDF numbers having the indicated chemical elements. The analyst has the option of requiring a single element or a group of elements to be present in each phase. This becomes useful if a certain element is known to be found in the sample in large amounts (XRF data) but no phase containing this element has been identified. The diffractionist can then restrict the search to only those phases having the missing element and any other combinations of other elements.

DSPACE does a search, within a 2θ error and above a minimum card intensity, on dspacings from an ASCII file or those typed in by the operator. The number of missed input lines allowed can be fixed or can be auto-incremented by the program. If the auto-increment option is chosen, the program will try to find phases that contain all of the entered dspacings. If none are found, the program allows one line to be missed, then two lines, etc., until at least one phase matches. This program is particularly useful if there are only a few extraneous lines left in a pattern. The analyst can then perform a DSPACE search on these few remaining lines. The DSPACE search would normally be done on a subfile generated using CHEM.

XRDQUAL is the complete search/match program. The dspace and intensity file is an ASCII file generated by the peak finding and interactive graphics routines of SPECPLOT (8,9). The first line of this file is a title line, the second is the maximum and minimum dspacings, which correspond to the scan range. The following entries are the dspacing and intensity values with the intensity normalized to 100. Since this is an ASCII file, the analyst can build the input file using an editor. This allows the operator to utilize the program on data obtained from any diffraction instrument such as manual diffractometers, Debye-Scherrer cameras, Guinier cameras, etc. Figure 2 is a simplified flow diagram of the search/match logic of XRDQUAL. The d&I data can be searched against the entire PDF file, any of the JCPDS supplied subfiles, user subfile or result file from CHEM and DSPACE. The operator can also apply chemical restrictions to this search in the same manner as is done in CHEM. If the chemistry option is chosen, no card will be considered which does not fulfill the indicated chemistry. The next search test applied is the strongest line criteria. This criteria requires that the n strongest lines within the scan range be present in the observed pattern allowing for the 2θ error. If one of the n strongest lines on the card is not within the scan range, the n+1 strongest line is checked. n is an operator option and can take on the values of 1 to 9. Cards that pass the strongest line criteria are then matched against the observed pattern and an ordering parameter is calculated. The ordering parameter has the following form:

$$OP = 100 * SD * (I)^{1/2} \left(\frac{N_c}{N_m}\right)^2 * \left(\frac{N_{ob}}{N_m}\right)^{1/2}$$

where SD is the 2θ standard deviation, N_c is the number of lines on the PDF card, N_m is the number of lines matched, and N_{ob} is the number of observed lines.

$$\text{and} \qquad I = \frac{1}{N_m} \sum_{i=1}^{N_m} \left| \frac{I_c - I_{ob}}{I_c + I_{ob}} \right|$$

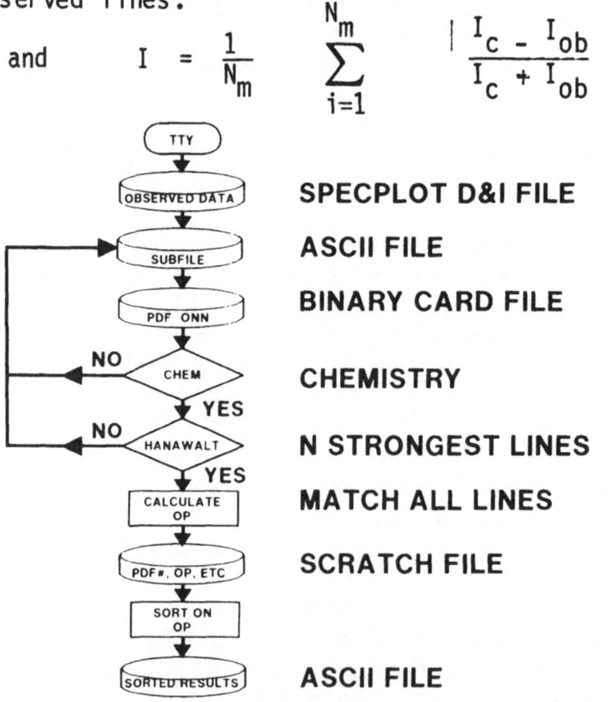

Fig. 2. Simplified flow diagram of XRDQUAL.

where I_c is the intensity on the card and I_{ob} is the observed intensity. This ordering parameter is the heart of the program since it is used to order the top 50 matches. The OP described here is biased in order to find the major phases in the sample. OP will have a value of zero for a perfect match and is usually restricted to be less than 50 for the match to have any validity at all. The match results are written to a user designated file containing the title information, search match options, PDF numbers, OP and all values used to calculate the OP. This result file could be sequenced through by the CARD utility, thus displaying the PDF data on a CRT for user verification. XRDQUAL was tried on all the JCPDS round robin data sets and was found to perform exceptionally well in that the correct phases were always at the top of the list.

The speed and accuracy of the search match program is very strongly affected by the options chosen. If only subfiles are searched, the program will take a short period of time and give good results provided that the phase to be identified is in the subfile. The program runs the fastest with a small 2θ error window, chemistry restrictions, and a large strongest line criteria (n=9). The accuracy of the analysis is not strongly affected by the 2θ error but the speed of the search is very dependent on the 2θ window used. A large strongest line criteria biases the search toward the major constituents. By applying chemistry restrictions, the user can make the analysis faster and more accurate (10).

PDQUAL is the data collection program used to collect digital data from a Rigaku diffractometer equipped with a multisample changer. If the search/match option is chosen, this task will collect the data for the first sample and then start the program FIND. This second task will then do a peak search and an automatic search/match on this first sample while data is being collected from the second sample. This batch collection and processing of diffraction data, especially for night runs, can save a considerable amount of time on routine samples. SPECPLOT (8,9), an interactive data reduction and display program, also has the capability of displaying colored markers for any PDF card on the experimental spectrum (Figure 3). This capability is ideal for verifying the results of a peak search program.

CONCLUSIONS

A set of qualitative phase analysis programs have been described which have proven to be extremely useful in an industrial service diffraction laboratory. These programs were designed to be interactive with many user-designated options, thus giving the diffractionist the flexibility needed for the particular problem at hand.

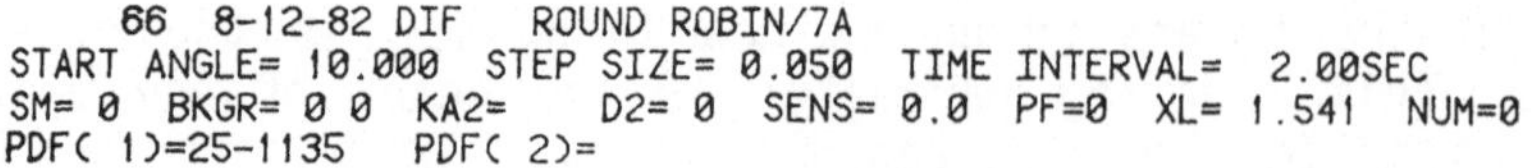

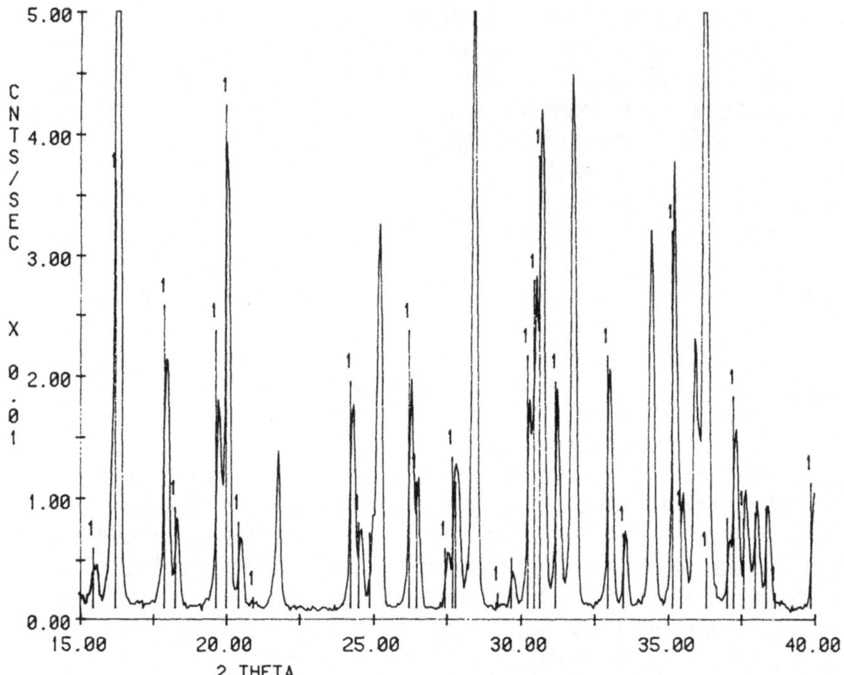

Fig. 3. SPECPLOT spectrum with JCPDS markers.

Listing of the programs described in this paper can be obtained from the authors with the exception of the data collection routines which utilizes portions of Rigaku proprietary software. The PDF file cannot be distributed due to a lease agreement with the JCPDS.

REFERENCES

1. M.C. Nichols and Q. Johnson, Adv. X-Ray Anal., Vol. 23, pp. 273-278, 1980.
2. L.K. Frevel, Anal. Chem., Vol. 37, pp. 471-482, 1965.
3. M.C. Nichols, UCRL-70078, Lawrence Livermore Laboratory, 1966.
4. G.C. Johnson and V. Vand, Ind. Eng. Chem. Vol. 59, pp. 19-31, 1967.
5. J.D. Hanawalt, H. Rinn and L.K. Frevel, Ind. Eng. Chem. Ed., Vol. 10, pp. 457-512, 1939.
6. R.L. Snyder, Adv. X-Ray Anal. Vol. 24, pp. 83-89, 1980.
7. R.A. Sparks, to be published in Adv. X-ray Anal., 1983.
8. R.P. Goehner, Adv. X-Ray Anal., Vol. 23, pp. 305-311, 1980.
9. R.P. Goehner, GE Technical Information Series, 82CRD114, 1982.
10. G.J. McCarthy and G.G. Johnson, Jr., Adv. X-Ray Anal., Vol. 22, pp. 109-119, 1979.

THE JCPDS DATA BASE - PRESENT AND FUTURE

Winnie Wong-Ng, Mark Holomany and W. Frank McClune

JCPDS-International Centre for Diffraction Data
1601 Park Lane, Swarthmore, PA 19081

Camden R. Hubbard*

Center for Materials Science
National Bureau of Standards
Washington, D.C. 20234

The Powder Diffraction File (PDF), published by the JCPDS-International Centre for Diffraction Data, is one of the most widely used scientific data bases. It currently consists of about 40,000 x-ray diffraction patterns, organized into 32 sets and 5 subfiles: metals and alloys, minerals, common phases, forensic patterns, and those from the National Bureau of Standards (NBS). New patterns are being added to the PDF at a rate of 2,000 patterns per year. The sources of these patterns are the literature, private contributions, grants-in-aid projects, and the JCPDS Associateship at the NBS.

The PDF data base is available in 3x5 inch card sets and microfiche sets, with accompanying Alphabetic, Hanawalt and Fink Search Manuals. Periodically several sets are reedited and published in bound volumes. A subset of the PDF data base is available on search/match magnetic tape. Many intermediate editorial and technical procedures are used to maintain quality, accuracy, and completeness of the data base. In 1976, the JCPDS recognized the need to convert to a computer data base for the following reasons:

1. To increase the efficiency of data entry, data editing and data review.
2. To simplify the production efforts by eliminating the need to maintain the data base in two separate forms.
3. To eliminate problems such as incompleteness of data and rounding of d-spacings on the PDF tape.
4. To have a more flexible and informative data base.
5. To enhance the assessment of quality of the data.

Systematic methods of evaluating powder patterns as well as creating the data base have been developed by the JCPDS Data Base Subcommittee. The FORTRAN program NBS*AIDS80[1], which was originally developed at NBS for crystallographic data evaluation, has been extended to include features for evaluation of powder patterns. The general evaluation scheme is summarized as follows:

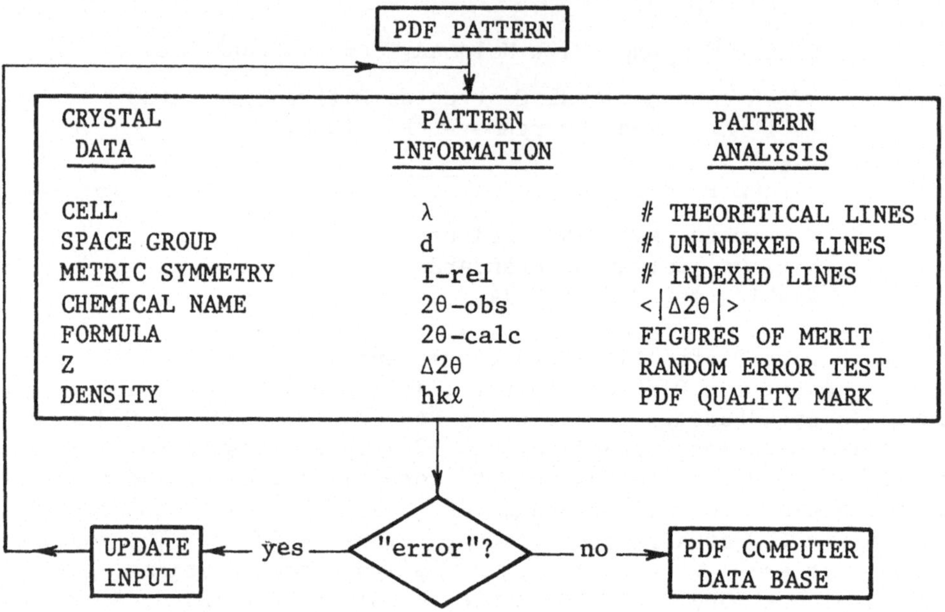

If NBS*AIDS80 finds a data or format inconsistency ("error"), then the data must be corrected and reprocessed. Any pattern with unresolvable errors will be added to a list of questionable phases that might be remeasured. The program also gives warnings of possible errors in the data, each of which are reviewed, and, if necessary, editorial changes are made.

With implementation of this program at JCPDS Headquarters, new data entering the editorial system and several sets of historical data have been processed. Sets 23 and 24 have been reevaluated and will be published in 1983 in bound book form. Reevaluation of Sets 1-32 is expected to be completed in 1983, and the new master data base will be completed in 1984. Direct product generation from this data base is planned to begin in 1985.

REFERENCES

[1]Mighell, A.D., Hubbard, C.R., and Stalick, J.K., NBS*AIDS80. A FORTRAN Program for Crystallographic Data Evaluation, Natl. Bur. Stand. (U.S.) Technical Note 1141 (1981), National Bureau of Standards, Washington, D.C. 20234.

SEARCH/MATCH IMPLICATIONS OF THE FREQUENCY DISTRIBUTION OF "d" VALUES IN THE JCPDS POWDER DATA FILE

Shozo Toyohisa, Iwao Fujiwara, Takuji Ui, and
Eiichi Asada

Toyohashi University of Technology
Tempaku-cho
Toyohashi 440, Japan

INTRODUCTION

Interplanar spacings (d values) from X-ray powder diffraction data occur over a wide range of values for the phases that have been described to date. They are not, however, uniformly distributed; rather, the d values are concentrated near certain values and thinly distributed in other regions. We have investigated the possibility of utilizing this distribution of d's to develop a search/match method for phase identification. We found that there exist single d values that can be unique or almost unique to a certain phase. Also, the number of phases that could be uniquely characterized by single d's was increased by limiting the classes of compounds that were searched. The use of such characteristic d values resulted in a simpler search/match procedure.

FREQUENCY OF D VALUES

Figure 1 shows a frequency distribution of d values from the eight most intense lines from phases in the JCPDS data base, both organic and inorganic. This is the same data used for the Fink index. Both organic and inorganic patterns were used. As expected, the large spacings occur with much less frequency than do the smaller spacings.

CHARACTERISTIC D VALUES

When the larger d spacings were examined in greater detail, we found that there were cases in which there was only one phase

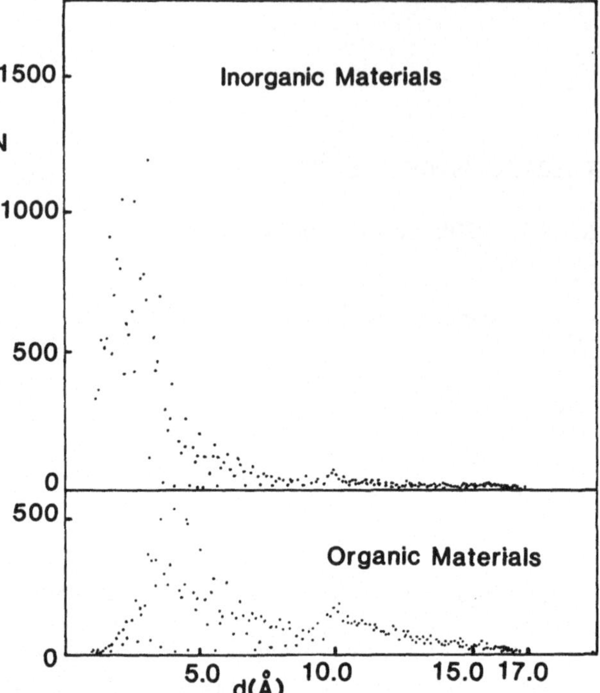

Fig. 1. Frequency (N) of d values

represented by a given d. We found 66 unique d's for the inorganic phases and 61 for the organic phases. These unique values appeared between 5 and 15 A for the inorganic phases and between 10 to 15 A for the organic phases. The d values used were limited to the second decimal place. For these cases then, a unique identification can result even when using the entire JCPDS data base.

Such characteristic d values would be expected to occur much more often when the size of the data base used is limited to a specific group of compounds. We chose two such limiting groups for the purposes of this study, one consisting of calcium phosphates, and the other made up of aluminum calcium silicates. The characteristic d values of these two groups were investigated by using all the d values present in the original JCPDS data base; the Fink index was not utilized. We found that there were 149 d spacings that were characteristic of single calcium-phosphorous-oxygen compounds. Some of these compounds had more than one characteristic d value; the final total of uniquely identifiable compounds was 111. Since the total number of phases contained in this restricted group was 168, a single d spacing was sufficient to identify about 66% of the total. It is estimated that by employing such characteristic d values, identification of members of this group can be greatly facilitated. In the silicate case, the number of unique d values was 103 with 99 compounds giving rise to them; this corresponds to about 13% of the 780 phases that comprise this group.

UTILIZATION OF CHARACTERISTIC D VALUES FOR SEARCH/MATCH METHODS

Characteristic d values should prove beneficial for phase identification. To test this hypothesis, the JCPDS data base file as distributed on magnetic tape was rearranged into a new file composed of subsets. Each subset was characterized by the d value it represented and contains the JCPDS numbers of the phases possessing this nominal d value. The search/match method is based on picking up these JCPDS numbers as the d's for the unknown pattern are input. Those JCPDS numbers that match d's with a frequency greater than some previously determined value are kept as possible matching phases. These candidate phases are arranged in the order of their 'Total Reliability .' The reliability factors employed in this study were the 'Reliability Factor' (R.F.) and the 'Total Reliability Factor'(T.R.F.). The frequency of the d value is represented by a symbol N, and the R.F. is regarded as the value represented by the reciprocal of N corresponding to the d value matched with each other, multiplied by 100. The T.R.F. is the sum of the R.F.'s.

TEST DATA

Table 1a shows the results of such a procedure for a mixture of potassium chloride and calcium chloride. We employed a Fink index data base for this example with an allowable variation in d of (+-) 0.01 A. The two phases with the highest total reliability factors were found to be the true phases. However, these total reliability factors may not always be so obvious. The reliability numbers can be quite similar. This happens when there are not so many characteristic d values in a given group. On the other hand, when the number of phases in a given group is restricted by chemical or other considerations, a higher proportion of characteristic d values exists and a unique identification is more likely to result. Table 1b shows the results for the group of calcium-phosphorous-oxygen phases. By employing the above mentioned search/match method, it can be easily recognized whether characteristic d values exist or not. A result of 100 for the R.F. indicates an exact match.

Table 1. Search/Match of (a) KCl and $CaCl_2$ Mixture and (b) $Ca_2P_2O_7 \cdot 4H_2O$

(a)

Test data d (A)	KCl 40587* R.F.	$CaCl_2$ 120056* R.F.	PrTe 210738* R.F.	Mo 40809* R.F.	U(NH)Cl 201330* R.F.
3.15					
3.14	0.0855		0.0855		
3.13					
2.65					
2.64		0.0963			
2.63					
2.50					
2.49		0.2506			
2.48					
2.40					
2.39		0.1271			
2.38					
2.23					
2.22	0.1136		0.1136	0.1136	
2.21		0.1587			
1.82					
1.81	0.1076				
1.80					
1.65					
1.64		0.1080			
1.63					
1.58					
1.57	0.1451		0.1451	0.1451	0.1451
1.56					0.1227
1.41					
1.40	0.1972		0.1972		
1.39					
1.29					
1.28	0.2016		0.2016	0.2016	
1.27					
1.24					
1.23		0.1880			0.1880
1.22					0.1513
1.06					
1.05		0.4132	0.4132		
1.04	0.6098				0.6098
0.91					
0.90		0.6897		0.6897	
0.89					
0.85					
0.84	0.6329		0.6329	0.6329	
0.83					
	2.0935**	2.0316**	1.7893**	1.7830**	1.2169**

* indicates JCPDS No. and ** Total Reliability Factor

(b)

Test data	$Ca_2P_2O_7$ $4H_2O$ 220537*	$Ca_8H_2(PO_4)_6$ $5H_2O$ 261056*	$Ca_3(PO_4)_2$ 290359*	$Ca_5(PO_4)_2SiO_4$ 210157*
6.22				
6.21	100.00			
6.20				
5.57				
5.56	33.33			
5.55				
5.14				
5.13	100.00			
5.12				
3.95				
3.94	25.00			
3.93				
3.62				
3.61	16.67			
3.60				
3.28				
3.27	5.56			5.56
3.26				
3.20		50.00		
3.19	8.33			
3.18		3.03		3.03
3.11		4.76		
3.10	4.00			
3.09				5.88
3.02				
3.01	5.88	5.88		5.88
3.00				
2.95				
2.94	4.54	4.54		4.54
2.93				
2.81		3.70		3.70
2.80	5.26			
2.79				11.11
2.76				
2.75	5.88		5.88	
2.74		8.33		
2.69				
2.68	3.57			
2.67		4.34		
2.56		3.13	3.13	
2.55	5.26		5.26	
2.54		9.09	9.09	
2.49				
2.48	5.56	5.56		
2.47		5.00		
2.45		11.11	11.11	
2.44	7.69		7.69	
2.43				
2.28		3.45	3.45	
2.27	3.45	3.45	3.45	
2.26				
	339.99**	152.66**	49.06**	39.71**

CONCLUSIONS

The method mentioned above is neither perfect nor complete. However, we wish to point out that the use of characteristic d values or those that are similar to them can be used to supplement rather than supplant existing search/match methods. This method can be an extremely reliable one provided that the target phase belongs to the group of restricted phases considered. It should be pointed out that the following two problems exist concerning the utilization of this method: (1) the JCPDS file does not always have d spacings that are accurate to the second place after the decimal; this makes them less than useful for our method. (2) Full knowledge concerning any solid solution present is indispensable.

COMPUTER SEARCH/MATCH OF STANDARDS

CONTAINING A SMALL NUMBER OF REFLECTIONS †

T. C. Huang and W. Parrish

IBM Research Laboratory, San Jose, California 95193

B. Post

Polytechnic Institute of New York, Brooklyn, NY 11201

ABSTRACT

A simplified and effective computer Search/Match method is described in which no more than 12 largest d-value reflections regardless of intensities are used for the Powder Diffraction File standards. A quantitative procedure utilizing the figure-of-merit in the IBM Search/Match method was employed to show the applicability to inorganic, mineral and organic mixtures. The analyses gave the same results as the complete File but the standard databases and analysis time are reduced by about a factor of two. Search/Match without using intensities also gave the same results; this may be useful when instrument geometry and/or preferred orientation cause large errors in the relative intensities. The error limits in measuring the d-spacings were found to vary from $\pm 0.1^\circ$ to 0.5° (2θ) depending on the degree of overlaps.

INTRODUCTION

The most commonly used standard file for Search/Match (S/M) is the Powder Diffraction File (PDF) of JCPDS (1). There are about 38,000 patterns in sets 1-31. The File is growing at an annual rate of about 2000 patterns which would double the size by the end of this century. The rapidly expanding use of computers for automation is likely to increase the growth rate even more. Much larger computer memories and longer S/M times will be needed to process the entire PDF. Methods of effectively reducing the size of the File and improving analysis speed are thus of considerable interest especially for analyses using the restricted memory and relatively slow mini- and microcomputers (2,3).

---

† Part of this work was presented at the ACA meeting, NBS, March 1982.

METHOD

The distribution of patterns in the PDF based on the number of reflections per pattern is shown in Fig. 1. The average number of reflections in a pattern is about 30. For the conventional computer S/M, all the d-spacings (d's) and intensities (I's) of the standards are used to match the unknown. The larger the number of reflections, the larger the database and the longer the analysis time.

A new approach is described here for computer S/M using a reduced standard File in which only the 12 largest d-value reflections of the patterns are included. Its size is about 43% of the regular File and thus the S/M time is greatly improved because there are fewer d's and I's to be processed.

The IBM S/M program (4) was used to test the effectiveness of this approach. The algorithm used for calculating the figure-of-merit (FOM) is based on five factors:

$$FOM = f(F1, F2, F3, F4, F5)$$

where F1 and F2 = goodness of match in d's and I's respectively, F3 = % of the sum of the standard I's matched in the experimental range, F4 and F5 = % of standard and unknown reflections matched in the range considered respectively. F2 and F3 are omitted for matching without I's. The value of the FOM lies between 0 and 100, the higher the FOM, the higher the probability of a match. This program has been used successfully for analyzing a wide variety of materials with the true components of the mixtures identified by the highest FOM's.

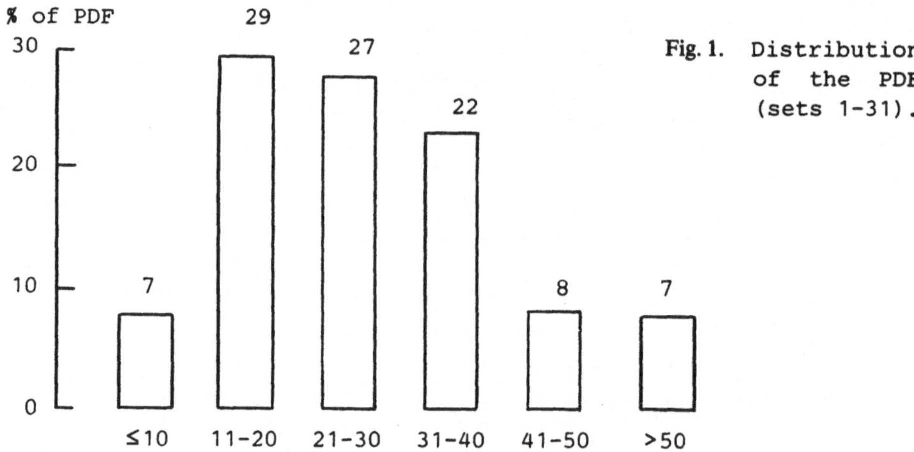

Fig. 1. Distribution of the PDF (sets 1-31).

RESULTS

The reduced standard File has been used for the analyses of a large number of mixtures prepared in our laboratories and patterns published in the literature. Three of the examples are listed below:

1. Three-Phase Inorganics Mixture

The d's and I's of this pattern were used by JCPDS to demonstrate the Johnson S/M Program (5). The pattern was stated to be "Guinier quality", and elemental information was included to assist the analysis.

S/M parameters used in this study included an error window of 0.2° (2θ), and chemical restrictions of Si, Cl, K and Br as present and elements H to Na as undetermined. This elemental information can be obtained by X-ray fluorescence, and is equivalent to those used by JCPDS. The results of S/M of the FEP (frequently encountered phases) section of the IN (inorganic) subfile are given in Table 1. The standards are listed according to their FOM's. The three correct phases: K_2SiF_6, LiCl and NaBr (printed in bold face) were ranked with the highest FOM's. The fourth phase Na(Cl,CN) may be eliminated after the subtraction of the top three phases using the interactive facility of the program (6). The last column of Table 1 gives the number of reflections in the unknown pattern (NM) which match those in the standard pattern (NS). It took 0.8 second for an IBM 3033 processor to complete the S/M of the FEP section (2332 standard patterns), this was 43% faster than using all the reflections of the inorganic standards. The results were identical to those obtained with the full standard patterns (4).

2. Four Minerals Mixture

The data were synthesized by JCPDS for the second round robin test (7). They consist of four standard mineral patterns (calcite, aragonite, vaterite and smithsonite) with added random errors of $\sigma=0.05^\circ$. A total of 86 reflections were listed between 5 to 82.46 $^\circ 2\theta$, and elemental restrictions gave Ca and Zn present, H to Mg undetermined.

Table 1. S/M Results for 3-Phase Inorganics Mixture

#	Standard Formula / Name	PDF #	FOM	NM/NS
1	**K_2SiF_6 / Hieratite Syn**	7-217	88.6	11/12
2	**LiCl**	4-664	74.4	5/5
3	**NaBr**	5-591	64.2	6/8
4	Na(Cl,CN)	2-778	47.1	4/6

Table 2. S/M Results for Four Minerals Mixture

#	Standard Formula / Name	PDF #	FOM	NM/NS
1	$CaCO_3$ / Aragonite Syn	5-453	90.8	12/12
2	$ZnCO_3$ / Smithsonite	8-449	88.3	12/12
3	$CaCO_3$ / Calcite	24-27	87.8	12/12
4	$CaCO_3$ / Calcite Syn	5-586	86.4	12/12
5	$CoCO_3$ / Sphaerocobaltite	11-692	82.9	12/12
6	$CaCO_3$ / Vaterite	25-127	70.5	10/12
7	$CaCO_3$ / Vaterite Syn	24-30	67.3	10/12

Table 2 lists the top seven FOM patterns found by S/M of the MI (mineral) subfile without using chemical information. #3 and 4 are patterns of the same phase calcite. #5 $CoCO_3$ and #2 $ZnCO_3$ are isomorphous, and have almost the same unit cell dimensions. $CoCO_3$ can be easily eliminated using elemental restrictions given above, or after the subtraction of the $ZnCO_3$. #6 (PDF 25-127) is the replacement vaterite pattern of 13-192 used originally by JCPDS to synthesize the unknown pattern. #7 is a calculated pattern of the same phase. The computer time was 1.52 second for the S/M of 3049 patterns of the MI subfile, a 49% of reduction in time.

3. Three-Phase Organics Mixture

Organic patterns generally are more complex and usually have various degrees of preferred orientation. In tests conducted by JCPDS a number of participants reported difficulties with organics (7,8). The third example is a 3-phase organics mixture suggested by one of us (Post). It has a complex pattern with 88 reflections between 5 to 40.15 °2θ with standard deviation σ estimated to be 0.025° (Fig. 2).

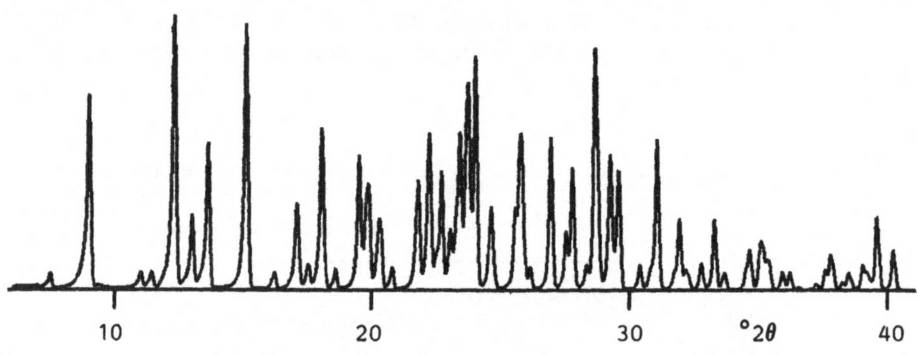

Fig. 2. XRD pattern of 3-phase organics mixture.

Table 3. S/M Results for 3-Phase Organics Mixture

#	Standard Formula / Name	PDF #	FOM	NM/NS
1	**Ampicillin Hydrate**	29-1546	92.8	12/12
2	**Acetanilide**	30-1506	90.0	12/12
3	**Triaminoguanidine Hydrochloride**	28-1520	89.9	11/11
4	Poly Trans 2 Butene Oxide	13-845	56.4	5/7

Table 3 lists the top four FOM phases found by S/M of the FEP section (1345 patterns) of the OR (organic) subfile. The three correct phases (printed in bold face) were ranked with the highest FOM's. The S/M time was 0.7 second, a 43% improvement.

DISCUSSION

Most of the difficulties in using the intensities for phase identification with the PDF arise from the instrumentation and/or specimen preparation. The various instrument geometries have inherently different sets of relative intensities, as for example, reflection and transmission specimen diffractometers, Guinier and Debye-Scherrer cameras. Differences in preferred orientation in the preparation of the unknown and the standard samples, particularly in the organics, may cause large differences in the relative intensities, especially when combined with different instrument geometries.

Our S/M method provides an option to use only the d's and omit the I's for identification. All test examples were also run without using I's and only the 12 largest d's. All correct phases were identified with the highest FOM's. The standard files of 12 d's were then only about 22% of the regular PDF.

To determine the practical operating ranges for this method, A random number generator was used to add various errors in d's and I's of unknown patterns synthesized from JCPDS patterns. The workable error limits for d's were found to depend on the complexity of the pattern as shown in Table 4. The synthesized patterns with generated errors ranged from simple patterns with only a few overlaps to the complex patterns similar to that shown in Fig. 2. The error limits allowed are somewhat smaller if the I's are not used, and they are also decreased as the degree of overlaps increase.

Table 4. Approximate Error Limits (2σ) in $^{\circ}2\theta$

Pattern	S/M with I's	S/M without I's
Simple	0.50	0.35
Intermediate	0.25	0.20
Complex	0.15	0.10

The minimum number of reflections per standard needed for successful phase identification depends on the complexity of the unknown pattern. It was found that S/M of standard patterns with only the 12 largest d-value reflections is sufficient for various types of patterns routinely used in phase identification.

CONCLUSIONS

Tests have been successful using reduced standard Files containing only the 12 largest d's in the IBM S/M method. The analyses are about twice as fast and the databases are smaller. The same results were obtained without using the I's, and this may be useful for those cases when the intensities of the standards are doubtful due to preferred orientation or different instrument geometries. The workable error limits (2σ) vary from 0.1 to 0.5° (2θ) depending on the degree of overlaps.

REFERENCES

(1) JCPDS - International Centre for Diffraction Data, 1601 Park Lane, Swarthmore, Pennsylvania 19081.

(2) M. C. Nichols, Minicomputer Search/Match: The Pros and Cons, Norelco Reporter 27:23 (1980).

(3) T. M. Hare, J. C. Russ, and M. J. Lanzo, X-Ray Diffraction Phase Analysis Using Microcomputers, Adv. X-Ray Anal. 25:237 (1982).

(4) T. C. Huang and W. Parrish, A New Computer Algorithm for Qualitative X-Ray Powder Diffraction Analysis, Adv. X-Ray Anal. 25:213 (1982).

(5) G. G. Johnson, Jr., "User Guide: Data Base and Search Program," JCPDS, Swarthmore, PA (1975).

(6) W. Parrish, G. L. Ayers and T. C. Huang, A Versatile Minicomputer X-Ray Search/Match System, Adv. X-Ray Anal. 25:221 (1982).

(7) R. Jenkins and C. R. Hubbard, A preliminary Report on the Design and Results of the Second Round Robin to Evaluate Search/Match Methods for Qualitative Powder Diffractometry, Adv. X-Ray Anal. 22:133 (1979).

(8) R. Jenkins, A Round Robin Test to Evaluate Computer Search/Match Methods for Qualitative Powder Diffractometry, Adv. X-Ray Anal. 20:125 (1977).

COMPARISON OF THE HANAWALT AND JOHNSON-VAND COMPUTER SEARCH/MATCH STRATEGIES

Satyam C. Cherukuri and Robert L. Snyder

N.Y.S. College of Ceramics, Alfred University
Alfred, N.Y. 14802

Donald W. Beard

Siemens Corporation
1 Computer Drive, Cherry Hill, N.J. 08034

INTRODUCTION

Over the past fifteen years two basic computer search/match strategies have evolved. The exhaustive search approach of Johnson and Vand (1) uses a sequential file structure whereas Nichols (2) developed a strategy which uses an inverted file, examining only those patterns containing lines of interest. Frevel (3) was the first to attempt to relate the quality of the reference patterns to the search strategy using a very restricted data base. These "first generation" search/match algorithms were forced to use very wide d and I windows due to the poor quality of the unknown and reference patterns.

Snyder (4) wrote the first "second generation" search/match procedure which takes advantage of high quality of data in the JCPDS data base when it is present. Recently, a minicomputer optimized version of the Johnson-Vand strategy has been incorporated into this search system enabling a chance to compare these two strategies under similar conditions.

GUIDELINES OF SEARCH/MATCH

Figure 1 shows a schematic representation of the Siemens Search Match System (5). A detailed description of the data base and program logic is given in reference 4. The hierarchical Hanawalt search strategy uses these data bases, called micro, mini and maxi files, sequentially. The micro file consists of the 300 most common phases based on the work of Frevel (3). This may also contain user defined, abberrant and proprietary patterns of interest to a specific laboratory. To avoid any bias the micro file was unaltered in this study. The mini file consists of the common phases defined by the JCPDS where as the maxi file consists of 31 sets of the JCPDS data base. All data bases are binary packed for maximum storage efficiency.

The figure of merit (FOM) as shown in Fig. 1 determines whether a particular pattern is a match or not based on Fmin which is the minimum figure of merit specified by the user. The form of the FOM used by the Hanawalt and Johnson-Vand search/match strategies differs considerably, reflecting the nature of these two strategies. The FOM for the Hanawalt search is given by (4)

$$FOM = d_R \times I_R^2 \times d_U$$

where d_R = percent of reference lines with $I \geq I_m$; where I_m is the intensity of the lowest matched line in the unknown for the range considered.

I_R = percent of the reference intensity matched

d_U = percent of unknown lines matched.

It should be noted here that the FOM in the Hanawalt strategy does not consider the closeness of the d agreement. I_R is given added weight and assumes that the intensities are not distorted too drastically. The factor d_U encourages the subtraction of the major phase before the minor phases. The FOM for the Johnson-Vand strategy is given by (1)

$$FOM = A\left[1 - \frac{\sum^{N}|\Delta D|}{IW \times N}\right]\left[1 - \frac{\sum^{N}|\Delta I| - K}{\sum^{N} I}\right]$$

where A = percent of the line match in d range considered, above background.

ΔD = differences between d_{unk} and d_{ref} in appropriate integer d-code units.

N = number of lines under consideration.

ΔI = difference between I_{unk} and I_{ref}

K = scale factor.

IW = d error window in d-code units.

PRACTICAL CONSIDERATIONS

Several patterns of varying degrees of complexity were used in this study. The first set of data presented in Table 1 consists of various combinations of β-spodumene, corundum, mullite and wollastonite with α-quartz as an impurity. The powder mixtures were not spray dried to ensure a typical pattern. The data were collected on an automated Siemens D-500 diffractometer and reduced to d-I files by Alfred University data reduction programs (6). The peaks that were not reported by the peak search program and observed false peaks were corrected by an interactive color graphics program implemented on a Tektronix-4027A terminal. It should be pointed out that each of these phases is complex by itself with about 100 peaks in a 2θ range of 10 to 70 degrees. Only the top twelve matches in the Johnson-Vand search were considered for comparison.

Information is also drawn from several round robin samples (7), JCPDS test data sets and numerous routine analyses at Alfred University and Siemens Corporation.

RESULTS

The notation followed in the tables is as follows:

x = hit
- = miss
L = low FOM. In the case of Johnson-Vand this stands for the presence of incorrect phases above the phase of interest.
I = impurity
* = incorrect phase

Table 1 shows the results obtained for the mixtures of β-spodumene, corundum, mullite and wollastonite. For sample No. 2 in this table, it should be noted that though sillimanite is a possible impurity in the mixture, its pattern is quite close to mullite's which interfered with the identification of mullite. Table 2 shows the results for three round robin samples with creditable performance by both the strategies. Particularly sample No. 3B shows the influence of elemental information on the Johnson-Vand search. The round robin samples (7) were synthesized from PDF patterns existing at the time of that study. They were not intended as reference patterns for testing search algorithms. Since their creation a number of the PDF patterns used have been updated, introducing errors into the round robin data. We therefore recommend that these data no longer be used for the testing of search programs. Table 3 shows the results for two arbitrarily chosen samples from 78 JCPDS test data sets. The nature and quality of these sets is unknown.

CONCLUSIONS

Based on numerous samples run through the Johnson-Vand and Hanawalt search algorithms, the following conclusions are drawn:

Table 1

Phases present	Without Chemistry J-V	Without Chemistry HAN	With Chemistry J-V	With Chemistry HAN
Sample No. 1				
β-Spodumene	x(L)	x	x	x
Corundum	x	x	x	x
Sample No. 2				
β-Spodumene	x(L)	x(L)	x	x(L)
Corundum	x	x(L)	x	x(L)
Mullite	-	-	-	-
Sillimanite (I)	-	-	x	-
α-Quartz (I)		x(L)	-	x(L)
Sample No. 3				
β-Spodumene	-	x(L)	x	x(L)
Corundum	-	x(L)	x	x(L)
Mullite	-	-	-	-
Wollastonite	-	x(L)	x	x(L)
Sillimanite (I)	-	-	x	-
α-Quartz (I)	-	-	-	-

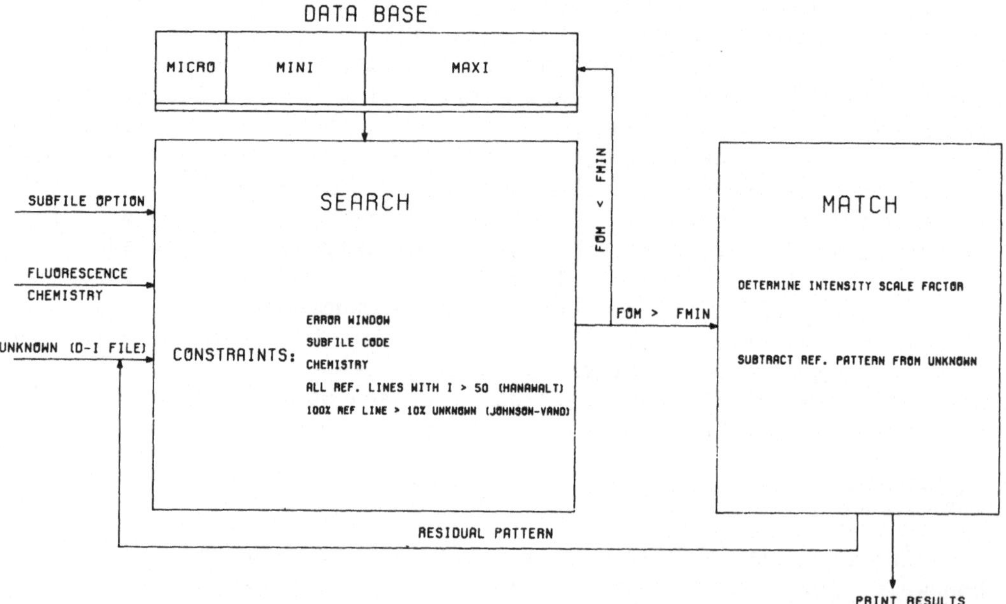

Fig. 1. Schematic representation of Siemens Search/Match System.

Table 2

Phases present	Without Chemistry J-V	Without Chemistry HAN	With Chemistry J-V	With Chemistry HAN
RR No. 1B				
Fe_2S	x	x	x	x
FeS_2	-	x	x	x
CuS	x	x	x	x
$CuFe_2O_4$	x	x	x	x
Cu_2O			*	
RR No. 2B				
$PbBr_2$	x	x	x	x
$PbCl_2$	-	x	-	x
$Pb_2(SO_4)O$	x	x	x	x
$Pb_3O_2Cl_2$	-	x	x	x
$Cu_{0.95}V_2O_5$		*		
PbO		*	*	*
PbO_2			*	
$PbSO_3$			*	
PbS_2O_3			*	
RR No. 3B				
$BaSO_4$	x(L)	x	x	x
$BaCO_3$	x(L)	x	x	x
$SrSO_4$	-	x	x	x
$CaSO_4$	-	x	x	x
BaS_2O_7			*	
$BaSO_3$			*	

Table 3

Phases present	Without Chemistry J-V	Without Chemistry HAN	With Chemistry J-V	With Chemistry HAN
Sample No. 1				
$CaTiO_3$	-	x	x	x
TiO_2	x	x	x	x
$Ca_4Ti_3O_{10}$			*	
Sample No. 2				
$Ca_3Si_2O_7$	-	x	x	x
Fe_2SiO_4	x	x	x	x
SiO_2 (low)	-	x	x(L)	x
$MgSiO_3$	-	x(L)	-	x(L)

a) There is negligible change in the performance of the Hanawalt search with or without chemistry except for a great reduction in the search time.
b) There is a drastic improvement in the performance of the Johnson-Vand search when supplied with chemistry.
c) With chemistry Johnson-Vand often performs better than Hanawalt.
d) For poor quality patterns Johnson-Vand works better than Hanawalt.
e) A hit by the Hanawalt strategy is very seldom wrong. This is obvious from the strict constraints employed in this search/match strategy.
f) In the case of the Johnson-Vand results, user evaluation is essential.

DISCUSSION

There will be no significant difference in the results even if we increase the number of patterns considered in the Johnson-Vand search/match output. As the quality of the JCPDS data base gets better there will be an improvement in the performance of the Hanawalt strategy which may not be the case for the Johnson-Vand strategy.

REFERENCES

1. Johnson, G. G., "The Johnson-Vand Search/Match Algorithm," Norelco Reporter, 26-3, 15-18 (1979)
2. Nichols, M. C., "A Fortran II program for the Identification of X-ray Powder Diffraction Patterns," UCRL-70078, Lawrence Livermore Laboratory, Oct. 1966.
3. Frevel, L.K., "Computational Aids for Identifying Crystalline Phases by Powder Diffraction," Anal. Chem., 37, 471-482 (1965)
4. Snyder, R. L., "A Hanawalt Type Phase Identification Procedure for a Minicomputer," Adv. X-ray Anal., 24, 83 (1981)
5. The Siemens Powder Diffraction Search Match System, proprietary information, Siemens Corporation, 1 Computer Drive, Cherry Hill, N.J. 08034.
6. Mallory, C. L. and Snyder, R. L., "The Alfred University X-ray Powder Diffraction Automation System," N.Y. S. College of Ceramics Technical Paper 144 (1979).
7. Jenkins, R. and Hubbard, C. R., "A Preliminary Report on the Design and Results of the Second Round Robin to Evaluate Search/Match Methods for Quantitative Powder Diffractometry," Adv. X-ray Anal. 22:133-142 (1979).

A COMPARISON OF METHODS FOR REDUCING PREFERRED ORIENTATION≠

L. D. Calvert, A. F. Sirianni, and G. J. Gainsford*

Chemistry Division
National Research Council of Canada
Ottawa, K1A0R9, CANADA

C. R. Hubbard

Center for Materials Science
National Bureau of Standards
Washington, D.C. 20234

INTRODUCTION

Preferred orientations in powder diffraction specimens can cause large errors in measured intensities. An extreme case is shown in Figure 1. Smith and Barrett (1979) reviewed the various methods which have been proposed for reducing this effect. Subsequently, two methods which are used commercially for aggregating finely divided solids have been proposed for preparing powder diffraction specimens (Smith, Snyder, and Brownell, 1979; Calvert and Sirianni, 1980). In one of these, spray drying, a finely divided solid is suspended in a liquid together with small quantities of a deflocculent and a binder. This mixture is pulled by venturi action through a nozzle into a heated chamber. The spherically shaped aggregates dry before falling to a collection surface. The apparatus is fairly large (3 X 3 X 4 ft. at NBS), and operating parameters must be carefully chosen. In the other process, liquid phase spherical agglomeration (LPSA), a finely divided solid is

---

*Permanent address: D.S.I.R., Petone, New Zealand.

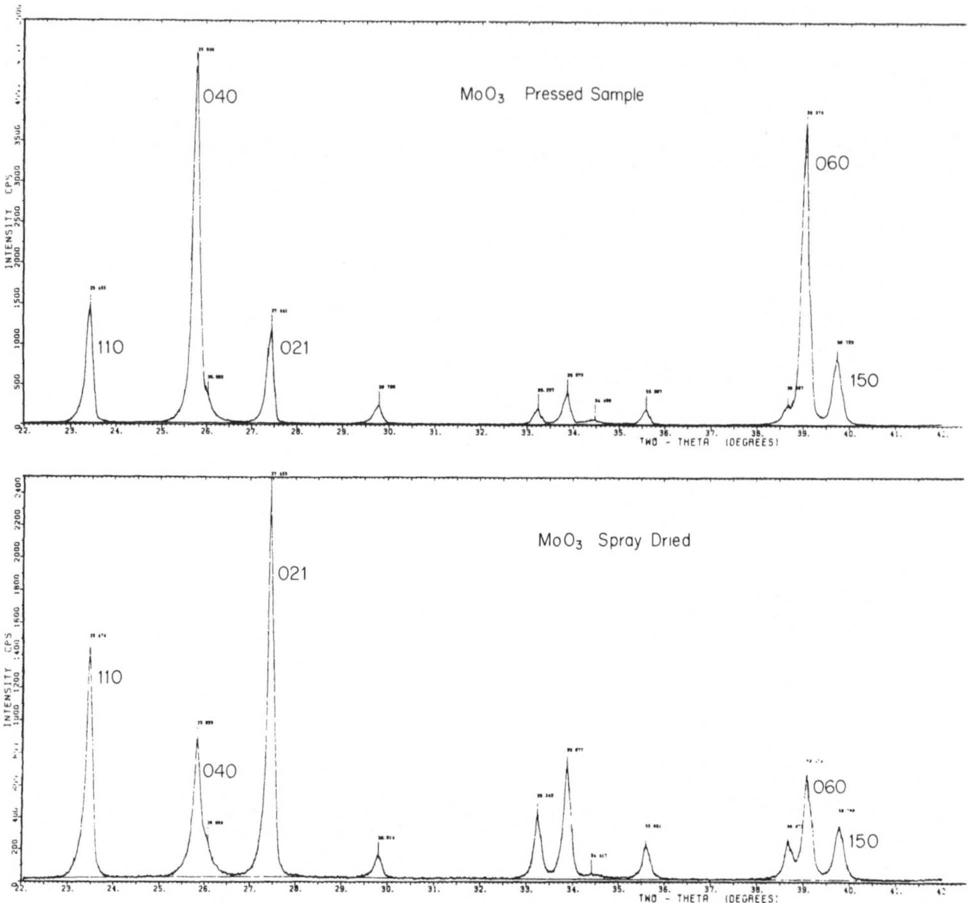

Figure 1. Powder patterns of MoO_3 produced on a reflection para-focussing diffractometer (NBS).

suspended in a liquid and a small amount of an immiscible liquid, which preferentially wets the solid, is added. Shaking the mixture produced spherical agglomerates (Figure 2a). Care is necessary in the choice of the amount of added immiscible liquid. Both methods were tested under normal conditions in our laboratories to assess their relative characteristics in comparison with side drifting methods which can routinely achieve reproducibility of intensities in the one to five percent range (NBS Monograph 25, 1981). For the side drifted samples, the ground powder was mixed with ≈ 50 percent by weight silica gel, which serves as a diluent for reducing orientation.

Ideally, materials to be used in testing the reduction of preferred orientation should have the following characteristics.

Their crystallites should have a variety of clearly defined shapes (plates, needles, blades, octahedra, etc.). The material should be stable, safe to handle, and have well known crystal structures. They should have well resolved powder patterns with several lines showing the effects of preferred orientation. The ideal material should be available as well crystallized fine particles (0.1-5 μm) to avoid damage due to grinding.

EXPERIMENTAL

For this limited test, the materials chosen were MoO_3 (blade-like needles; Figure 2b) and fluorophlogopite (plates). At NBS, powder patterns were prepared using a diffractometer with para-focussing reflection geometry and monochromatized CuKα radiation. Peak height intensities were measured. At NRC, patterns were recorded on a diffractometer with an incident beam graphite focussing monochromator and MoKα radiation (Debye-Scherrer geometry). Specimens were mounted in or on cylindrical capillaries and integrated intensities were obtained by a profile fitting algorithm.

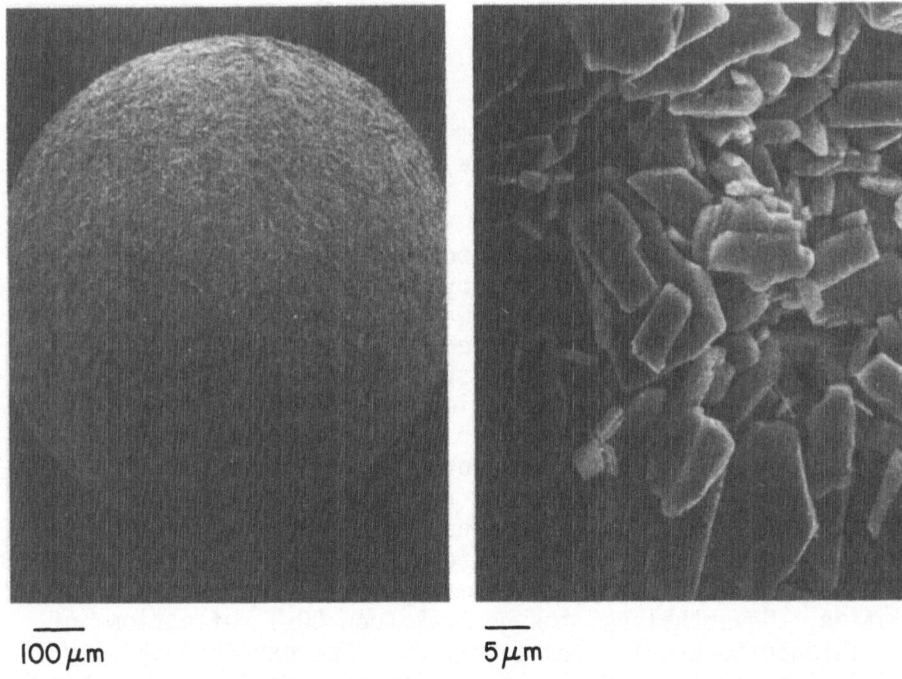

Figure 2a. Spherical agglomerate of MoO_3 produced by liquid phase spherical agglomeration.

Figure 2b. The starting material used to produce the sphere of 2a.

For the LPSA of MoO_3, ≈ 1 gm of MoO_3 was suspended in 25 ml of a 1:1 mixture of hexane and octane containing 0.02 ml of MeOH. This suspension was agglomerated by shaking with approximately 1 ml of H_2O in a Spex blender. For the spray drying of MoO_3, 10 gm of powder was suspended in 10 ml water with 0.5 ml Darvon C added as a deflocculent. A binder, 0.2 gm of polyvinyl alcohol, was also added. This mixture was blended in a micronizing mill for one minute just prior to spray drying.

RESULTS

The relative intensity data are given in Tables 1-4. The calculated relative intensities represent random orientation and are based on the MoO_3 crystal structure (Kihlorg, 1963) and Fluorophlogopite structure (McCauley et al., 1973). For MoO_3, the spheres produced by the LPSA process were 200-400 μm in diameter and could not be mounted in the normal way in reflection holders, as the necessary pressure, although small, crushed the spheres. Moreover, it was expected that their size would preclude precise measurements. For the NRC measurements, both sample preparation methods gave results in good agreement with the calculated relative intensities. For spray dried MoO_3 and fluorophlogopite, the spheres were ≈ 50 μm in diameter. These spheres were easily mounted in a cavity holder.

CONCLUSIONS

The relative intensities from side drifted MoO_3 plus silica gel are in good agreement with the calculated values, except for the 040 and 060 reflection (Table 1). For this mounting method, no special equipment or expertise is required. The simplicity of the side drifting technique has much to recommend it for routine use. Spherical agglomerates, prepared by liquid phase spherical agglomeration or spray drying, were shown to be free of preferred orientation, even though preferred orientation produced up to ten-fold changes in intensity in oriented specimens. For MoO_3, the comparison between calculated and experimental relative intensities for both reflection and Debye-Scherrer geometries is quite good. For fluorophlogopite, the agreement for Debye-Scherrer geometry and $MoK\alpha$ radiation is better than for reflection geometry and $CuK\alpha$ radiation. As expected, Debye-Scherrer geometry with a more penetrating radiation is more robust to the effects of preferred orientation than is the reflection geometry with a less penetrating radiation. Nevertheless, the general (non-00ℓ) reflections of fluorophlogopite are in good agreement. The excess intensity of the 003 probably indicates that perfect spherical agglomerates were not formed due to a too large crystallite size.

Table 1

MoO_3 (Reflection, CuKα)

hkl	110	040	021*	111	060	200	002
		Relative Intensities (Peak Height)					
Pressed	141	406	100	27	224	61	122
Side Drifted	79	52	100	27	27	10	16
Spray Dried	72	39	100	26	19	9	15
Calculated	76	38	100	26	21	8	15
PDF 5-508	82	61	100	35	31	13	21

*Scale reflection.

Table 2

MoO_3 (Debye-Scherrer, MoKα)

hkl	111*	060	200	002
	Relative Intensities (Integrated)			
Rolled	28	60	15	20
As Is	28	41	13	23
LPSA	28	50	13	20
Calculated	28	23	9	19

Rolled = capillary rolled in a thin layer of MoO_3 spread on a slide.

As Is = starting material loaded into an 0.2 mm capillary.

*Scale reflection.

Table 3

Fluorophlogopite (Reflection, CuKα)

hkl	003	130 + 20$\bar{1}$	13$\bar{1}$ + 200
	Relative Intensities (Peak Height)		
Pressed	100*	< 1	< 1
Spray Dried	100*	31	86
Calculated	56	31	86*

*Scale reflection.

Table 4

Fluorophlogopite (Debye-Scherrer, MoKα)

hkl	003	130 + 20$\bar{1}$	13$\bar{1}$ + 200
	Relative Intensities (Integrated)		
Rolled	281	41	100*
As Is	28	38	100
Lightly Ground	84	39	100
LPSA	64	28	100
Spray Dried	54	31	100
Calculated	52	35	100

*Scale reflection.

We believe that for best results, both spherical agglomeration methods require that the starting material be five μm or less in order to produce aggregates of a useful size ($\leq$ 50 μ). For LPSA, and with small amounts of material, preparation of aggregates of this size is difficult to achieve without extensive trial preparations. Aggregates produced by LPSA do not have a binder and are structurally weak.

REFERENCES

Calvert, L. D. and Sirianni, A. F., 1980, J. Appl. Cryst., 13:462.

Kihlborg, L., 1963, Ark. Kemi., 21:357-63.

McCauley, J. W., Newnham, R. E., and Gibbs, G. V., 1973, Amer. Miner., 58:249-54.

NBS Monograph 25, Section 18, 1981, Morris, M. C., McMurdie, H. F., Evans, E. H., Paretzkin, B., Parker, H. S., Panagiotopoulos, N. C., and Hubbard, C. R., p. 3-4, National Bureau of Standards, Washington, D.C. 20234.

Smith, D. K. and Barrett, C. S., 1979, Adv. X-Ray. Analy., 22:1.

Smith, S. T., Snyder, R. L., and Brownell, W. E., 1979, Adv. X-Ray Analy., 22:77.

THE DRAMATIC EFFECT OF CRYSTALLITE SIZE ON X-RAY INTENSITIES

James P. Cline and Robert L. Snyder

N.Y.S. College of Ceramics
Alfred University
Alfred, N.Y.

INTRODUCTION

Preferred orientation has long been considered the primary source of systematic error involved in quantitative analysis by X-ray powder diffraction. Techniques of spherical agglomeration have been shown to eliminate preferred orientation[1] provided that the agglomerate size is made sufficiently larger than the particle size. These techniques invariably employ the surface energy minimization of a liquid phase dispersed within a second fluid to create the spherical form desired. Spray drying has been the only method to date which has been successfully used to prepare spherical agglomerates suitable for X-ray diffraction. This study was undertaken to investigate possible deleterious effects of spray drying as a diffraction sample preparation technique.

The spray drying procedure involves the suspension of the powder of interest in a non-dissolving liquid. A deflocculant is added to regulate the surface charge (zeta potential) on the particles thus assuring that a mutual repulsive force exists between them. This repulsion insures that the collisions of particles in suspension will not result in adherence of particles into flocs, it also allows for the highest possible solids content at a given slip viscosity. A binder consisting of long chain organic molecules is also added to impart strength to the agglomerates after drying. These components are then mixed or milled jointly to yield a homogeneous suspension. The slip is then atomized into a heated chamber where the suspending medium contained in the droplets is rapidly driven off. The resulting agglomerates retain their spherical form and settle to the bottom of the dryer for collection.

An aspect of this work involved an analysis of the dynamics of microstructure formation of the agglomerates during the drying process[2]. The evaporation of the suspending medium initiates a flow pattern in the droplet that carries the more mobile species to the surface. This results in an agglomerate with an interior consisting of coarser particles surrounded by an exterior layer of fines. If this finer phase were to absorb X-rays appreciably it may shield the interior and thus distort diffracted intensities. Water filtration technology has shown the foremost factor on particle migration to be the ratio of the dimensions of the interstices of the larger particles to the diameters of the smaller particles. To test this an experimental design was devised that included varying the particle size of the starting materials. Intra-agglomerate inhomogeneities so induced were found to have an immeasurable effect on X-ray intensities. The particle (crystallite) size on the other hand, did have a dramatic effect on X-ray data. This effect is the focus of this paper; other spray drying considerations will be addressed elsewhere[3].

EXPERIMENTAL PROCEDURE

Parameters concerning the powder preparation and spray drying operations were varied to affect the state of homogeneity of the agglomerates. In contrast to a procedure used to produce homogeneous agglomerates, a worst case situation was devised to maximize particle migration. These procedures were used in conjunction with materials selected to test the effects of absorption contrast between the two materials used in each series. Silicon and alumina, with a linear absorption coefficient of $141.2 cm^{-1}$ and $126.1 cm^{-1}$ respectively, were used in the first three series. Silicon was used with titanium carbide, $\mu = 814.1 cm^{-1}$, for the last two series.

The particle size of all starting powders was measured using a light scattering sedigraph. Figure 1 indicates the particle size distribution of the material in the Si-Al_2O_3 "homogeneous" and "no grind" series. The silicon used in this study was ground from zone refined single crystal boules and wet sieved through a 20-micron screen. The size distribution measured from the resulting powder is indicated as "Si Prepared" and is the exact distribution of the Si contained in the "no grind" agglomerates. Preparation of the "homogeneous" slips called for a five minute mill time, in order to simulate the comminution of this procedure, pure Si was milled in the same manner prior to measurements. The as received alumina, one micron Linde C, was de-agglomerated by milling prior to all slip preparations and measurements. Figure 2 indicates the size distribution of the Si and TiC contained in the Si-TiC "homogeneous" series. Figure 3 is that of the material used in the "worst case" preparations.

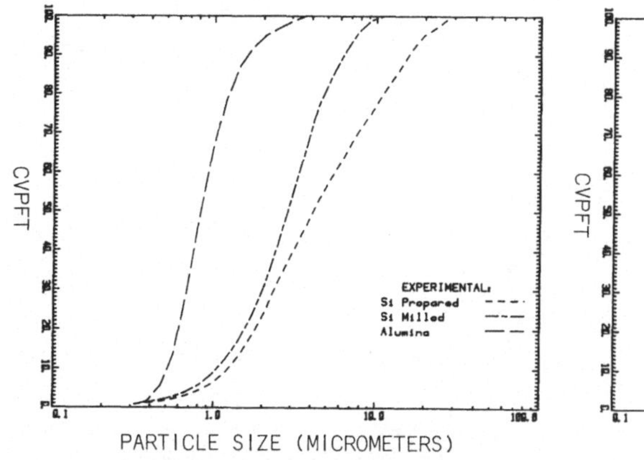

Fig. 1. Particle size of the Si and Al_2O_3 contained in the Si-Al_2O_3 "homogeneous" and "no grind" series.

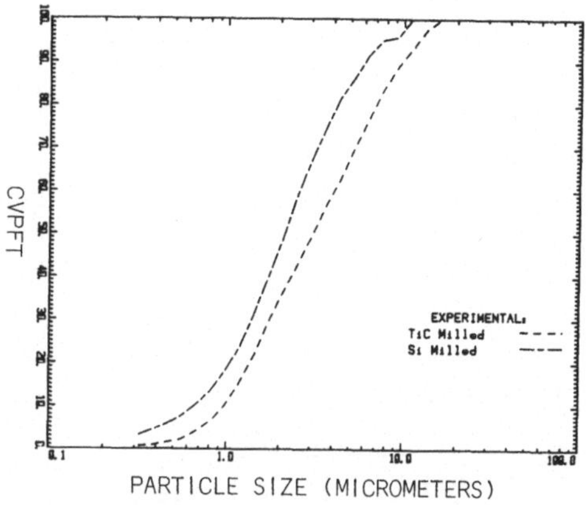

Fig. 2. Particle size of the Si and TiC contained in the Si-TiC "homogeneous" series.

All X-ray data was collected at the National Bureau of Standards using the Auto control algorithm[4]. Integrated intensities were measured with a precision of up to 1% with the time for each peak limited to an optimized 15 minute scan as calculated by Auto's subroutines. Six Al_2O_3 lines were measured in this manner while three lines were measured from the Si and TiC phases. Data was processed via NBS*QUANT82[5] to yield a Reference Intensity Ratio, RIR, (cf, I/I corundum) for all combinations of reflections. The RIR is a ratio of the diffraction intensities of the standard to the unknown scaled with respect to composition and relative intensity. For analyte α and standard s, the RIR is given by

$$RIR_{\alpha,s} = \frac{I_{i\alpha}}{I_{js}} \; \frac{I_{js}}{I_{i\alpha}^{rel}} \; \frac{X_s}{X_\alpha}$$

where i and j denote diffraction lines for phases α and s, respectively, X being the weight fraction, with I_{js}^{rel} and $I_{i\alpha}^{rel}$ indicating the relative intensities. The RIR is thus a constant corresponding to the slope of the internal standard curve. Random counting errors were tracked throughout data reduction yielding a sigma value associated with each RIR determination.

RESULTS AND DISCUSSION

Figure 4 is a plot of the relative intensities of the 3 Si lines and 6 Al_2O_3 lines used for the RIR determination in the

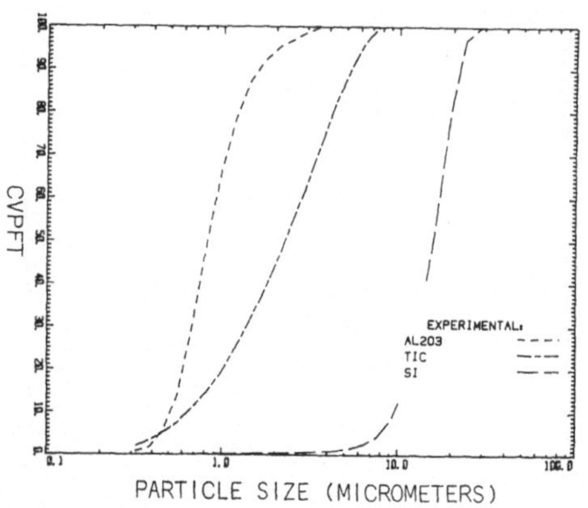

Fig. 3. Particle size of Si, TiC, and Al_2O_3 used in the Si-Al_2O_3 and Si-TiC "worst case" series.

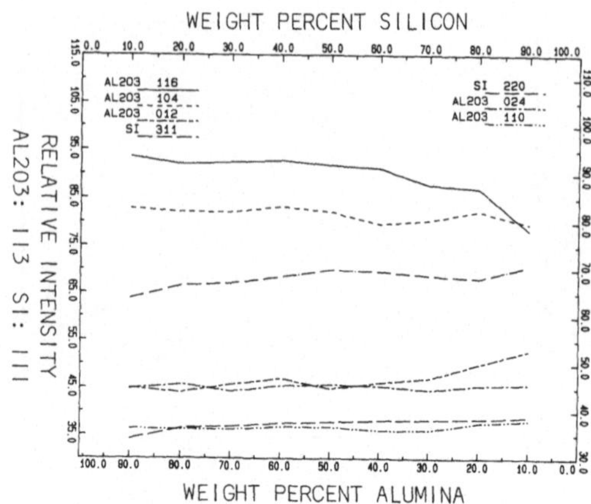

Fig. 4. Relative intensity data from the Si-Al_2O_3 "homogeneous" series.

Si-Al_2O_3 "homogeneous" series. Random counting errors ranged from 2 to 10%. The lack of variation of relative intensity with each sample is indicative of the consistent elimination of preferred orientation by the spray drying method. Of the two Al_2O_3 lines that do deviate, one was caused by a peak overlap that induced errors in the background determination. The second was caused by a diffuse peak of an anomalous Si hydration phase beneath the Al_2O_3 peak. Both these lines were eliminated from the RIR calculation with a very small effect on its value.

Shown in Figure 5 is a plot of RIR (Al_2O_3/Si) vs. composition for the three Si-Al_2O_3 series. A difference in the relative positions of the lines is noted; RIR values vary from .21 to .32, a 34% change. If a shielding effect had been operative, reflections of the Al_2O_3 would have been enhanced and the "worst case" values would have been greater, not less than the "homogeneous" values. As the 1 micron particle size of the alumina is considered to be constant for each series, the effects shown are believed to be caused by the variation in the particle size of the silicon. Comparison of the particle size and RIR data indicates that with the increase in the size of the silicon particles a corresponding increase in the reflected intensity occurs. As the particle size of the silicon is well above that of the alumina in all preparations the true RIR value, one determined in the absence of particle size effects, would lie well above any determined in this study.

A particle size related enhancement of reflected intensities is also noted in the Si-TiC data shown in Figure 6. Inspection of Figure 2 reveals that the size distribution of the TiC is somewhat greater than that of the Si in the "homogeneous" series. This

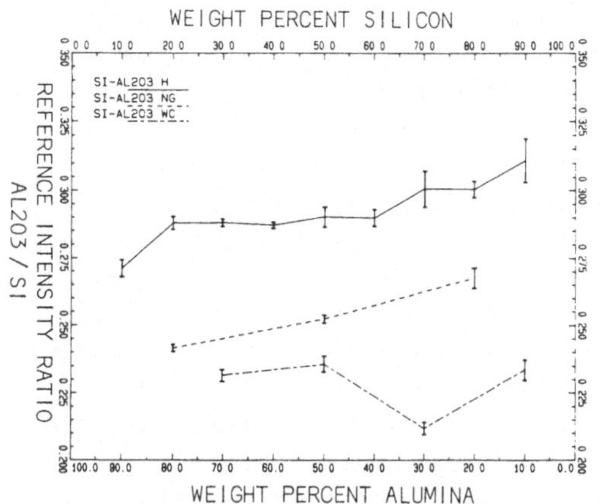

Fig. 5. Reference intensity ratio data from the three Si-Al_2O_3 series.

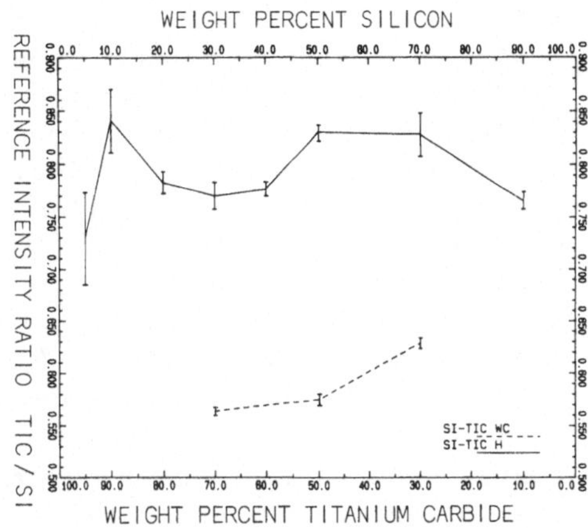

Fig. 6. Reference intensity ratio data from the two Si-TiC series.

condition is reversed in the "worst case" series with the Si being much coarser than the TiC. Thus the true RIR value would be between these two data sets, consideration of the particle size data would indicate it to be just below the "homogeneous" values.

The positive slope noted in 3 of the 5 RIR vs. composition data sets is indicative of a systematic effect related to the change in each mixture's average absorption coefficient. The RIR values of the "homogeneous" and "no grind" series in the system of Si-Al_2O_3 are both seen to approach the true RIR as the absorption coefficient of the mixture approaches that of the larger particles. The nearly indicernible curvature possessed by these data sets and the nearly parallel nature of them indicate that the true RIR will never be reached by either particle size condition regardless of composition. Thus the effects noted are directly related to the particle size and absorption coefficients of the phases present and are influenced by the absorption coefficient of the mixture. The trend of the "worst case" Si-TiC series also approaches the true RIR value in the same manner as did the two Si-Al_2O_3 series. While the variation in the data for the "worst case" Si-Al_2O_3 and Si-TiC "homogeneous" series cannot be explained at this time, it is noteworthy that the two points containing the highest fraction of Si for both series indicate an approach to the true RIR value.

G. W. Brindley considered the effect of particle size on the intensity of X-ray reflections in his 1945 analysis[6]. He considered a multiphase system of coarse polycrystalline particles. When a crystal within a particle is in the reflecting position, the incident and reflected rays will be unduly influenced by the absorption coefficient of the particle as opposed to the mean absorption coefficient of the bulk mixture. This absorption is termed microabsorption as opposed to the absorption of the mixture, macroabsorption. A correction factor termed the "Particle Absorption Factor" is derived to compensate for this lack of absorption homogeneity. Brindley's theory predicts that if the reflecting phase has a small absorption coefficient relative to the mixture that microabsorption will enhance its reflected intensities. This is seen to be the case in the Si-TiC system where the "worse case" RIR values were dropped by an increase in the coarse grained Si diffraction intensities. But consideration of the Si-Al_2O_3 data indicates that Brindley's model fails for this situation. A direct relationship between diffracted intensity and particle size is noted despite the fact the coarser grain species exhibits a higher absorptivity. The data predicted by the Brindley model is the inverse of that observed with respect to both line position and slope. Furthermore, the silicon and alumina used in this study are single crystal particles, thus the phenomenon of microabsorption in the sense that Brindley and other workers have defined it is inapplicable to this data. It is believed that an increase in crystallite size results in an increase in diffracted intensity regardless, but not independent of, the relative absorptivity of the phases present.

CONCLUSION

The results of this study have shown that X-ray powder diffraction intensities are strongly influenced by the crystallite sizes of the phases present. This effect will be enhanced by an increase in the absorption contrast between the reflecting and non-reflecting phases. Present theories do not address the data presented; no inference is made as to their applicability to the systems they are modeled after. But it is believed the effects these theories do address will be compounded by the ones isolated in this work. Quantitative analysis by X-ray powder diffraction can be in error by a factor of 2 or more as a result of crystallite size effects. This can be true even if the powders used are sub 325 mesh in accordance with accepted procedures. In order to minimize these errors, the X-ray diffractionist should match the crystallite size and quantity of each phase in the standard to those in the unknown when determining the calibration curve or RIR. Particular attention should be paid to particle size in systems that contain phases of highly differing absorption coefficients.

REFERENCES

1. Smith, S. T., Snyder, R. L. and Brownell, W. E., "Minimization of Preferred Orientation in Powders by Spray Drying," Adv. X-ray Anal., 22:77 (1979).
2. Submitted to J. Am. Cer. Soc.
3. Submitted to J. Appl. Cryst.
4. Snyder, R. L., Hubbard, C. R. and Panagiotopoulos, N. C., "AUTO: A Real Time Diffractometer Control System," NBS Rept. NBSIR 81-2229, National Bureau of Standards Washington, D. C. 20234 (1981).
5. Snyder, R. L. and Hubbard, C. R., NBS*QUANT82, NBS Special Publication, in press (1982).
6. Brindley, G. W., "The Effect of Grain on Particle Size on X-ray Reflections from Mixed Powders or Alloys," Phil. Mag. (7) 36, 347 (1945).

X-RAY DIFFRACTION INTENSITY OF OXIDE SOLID SOLUTIONS: APPLICATION TO QUALITATIVE AND QUANTITATIVE PHASE ANALYSIS

Ronald C. Gehringer, Gregory J. McCarthy and R.G. Garvey, Department of Chemistry, North Dakota State University, Fargo, ND 58105 and Deane K. Smith, Department of Geosciences, The Pennsylvania State University, University Park, PA 16802

INTRODUCTION

Solid solutions are pervasive in minerals and in industrial inorganic materials. The analyst is often called upon to provide qualitative and quantitative X-ray phase analysis for specimens containing solid solutions when all that is available are Powder Diffraction File (PDF) data or commercial standards for the end members. In an earlier paper (1) we presented several examples of substantial errors in accuracy of quantitative analysis that can arise when the crystallinity and composition of the analyte standard do not match those of the analyte in the sample of interest. We recommended that to obtain more accurate quantitative analyses, one should determine the analyte composition (e.g., from XRF on grains seen in a SEM or from comparison of cell parameters with those of the end members) and synthesize an analyte standard with this composition and with a crystallinity approximating that of the analyte (e.g., as determined from peak breadth or α_1/α_2 splitting).

In this paper, we address the possibility of avoiding the synthesis step by using instead the composition determined for the analyte and intensity data for the end members, or other available solid solution members, to approximate the correct intensity data of the analyte. This possibility is especially relevant to mineral groups such as olivines, garnets, pyroxenes, amphiboles and feldspars where syntheses are difficult, microprobe analyses of solid solution composition are common and sufficient solid solution compositional members can be obtained, even if none matches that of the analyte under study.

Four oxide solid solution systems were examined:

MgO-NiO. There is complete miscibility in this rocksalt structure solid solution.

NiO-ZnO. ZnO has the hexagonal wurtzite structure. It is miscible in rocksalt structure NiO to about $(Ni_{0.7}Zn_{0.3})O$.

$SrTiO_3$-$BaTiO_3$. $SrTiO_3$ has the ideal cubic perovskite structure and $BaTiO_3$ has a tetragonal distortion of this structure at room temperature. X-ray intensity data for compositions from pure $SrTiO_3$ to $(Sr_{0.2}Ba_{0.8})TiO_3$ with the tetragonal structure were obtained.

Cr_2O_3-Fe_2O_3. There is complete miscibility in this corundum structure solid solution series.

The results described here include the compositional variation of peak intensities, as used in the search/match and subtraction steps of qualitative analysis, and integrated intensities ratioed to an internal standard, as used in quantitative analysis.

EXPERIMENTAL PROCEDURES

The four solid solutions were synthesized in our laboratories from reagent grade starting materials using routine ceramic procedures (mix and grind oxides, pelletize, fire at high temperature and, if required, regrind, pelletize and refire to increase crystallinity and complete the reaction). End members as well as the intermediate compositions were processed identically so they would not exhibit different crystallinity due to different thermal histories. The starting materials and processing conditions for each solid solution were:

MgO-NiO. $MgO + NiO$ fired at 1350°C for 72 hours.

NiO-ZnO. $NiO + ZnO$ fired at 1200°C for 18 hours.

$SrTiO_3$-$BaTiO_3$. $SrTiO_3$ synthesized from $SrCO_3 + TiO_2$; $SrTiO_3 + BaTiO_3$ fired at 1200°C for 38 hours, then reground and fired at 1330°C for 12 hours.

Cr_2O_3-Fe_2O_3. $Cr_2O_3 + Fe_2O_3$ fired at 1200°C for 18 hrs. then reground and fired at 1200°C for an additional 22 hours.

High sensitivity X-ray scans were taken to determine that the solid solution products were phase pure. The internal standards α-Al_2O_3, CaF_2, ZnO and CeO_2 were reagent grade chemicals. The α-Al_2O_3 source had to be fired at 1200°C for 72 hours to remove minor amounts of other Al_2O_3 phases. A fourth internal standard, $(Sr_{0.5}Ba_{0.5})TiO_3$ was synthesized from $SrTiO_3$ and $BaTiO_3$ at 1350°C.

All solid solution products were ground to a fine powder first in an agate mortar and pestle, and then in a McCrone "Micronizing Mill" using either corundum or agate grinding elements and an ethanol medium. For the intensity ratio measurements, the solid solution was first ground, then mixed with the internal standard and put through a similar grinding and homogenization treatment. It was assumed that these procedures would yield well mixed powders having grain sizes below 5 µm as required for reproducible intensity ratios and for minimizing microabsorption effects. As noted below, the results supported this assumption.

Specimen preparation was by the National Bureau of Standards "side drifting" method (2) using Al sample holders. Multiple loadings with additional mortar and pestle grinding of the powders between loadings were made in all cases to check for the effects of preferred orientation or sample inhomogeneity. The reproducibilities of intensity ratios were in the very satisfactory ±1.0-3.0% range. This result indicates that preferred orientation was minimized by the specimen preparation.

Peak relative intensity data were read from 1°2θ/min. strip chart recordings after three or more separate specimen preparations of each phase. Integrated intensity data were collected using the scaler-timer of a Philips diffractometer equipped with a long fine focus Cu tube, theta-compensating slit and graphite diffracted beam monochromator. The intensities collected from individual reflections varied from 5,500 to 82,000 counts but were typically 20,000 counts. These count levels kept the counting statistics errors below 1% in most cases. A minimum of three repetitions of each intensity measurement was made.

RESULTS AND DISCUSSION

Figures 1-7 summarize compositional effects on relative intensities and intensity ratios. The raw data used to prepare these figures as well as additional results not presented here are found in reference 3, copies of which may be obtained from either of the first two authors.

Qualitative Analysis -- Relative Intensities

Relative peak intensities (I^{rel}) for three of the solid solution series are shown in Figs. 1-3. (I^{rel} data were not

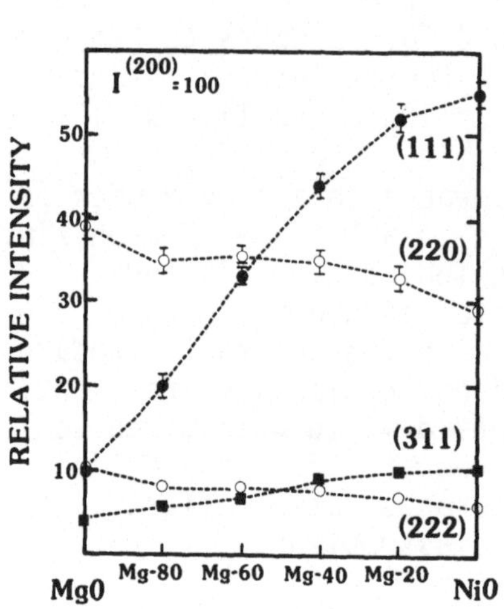

Fig. 1. Relative peak intensities in the MgO-NiO solid solution series.

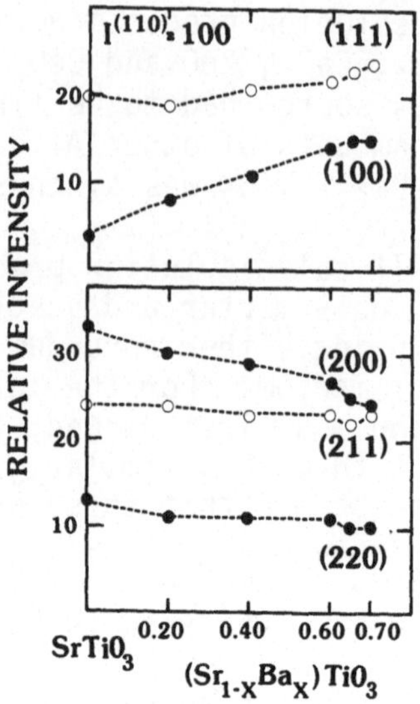

Fig. 2. Relative peak intensities in a portion of the $SrTiO_3$-$BaTiO_3$ solid solution series.

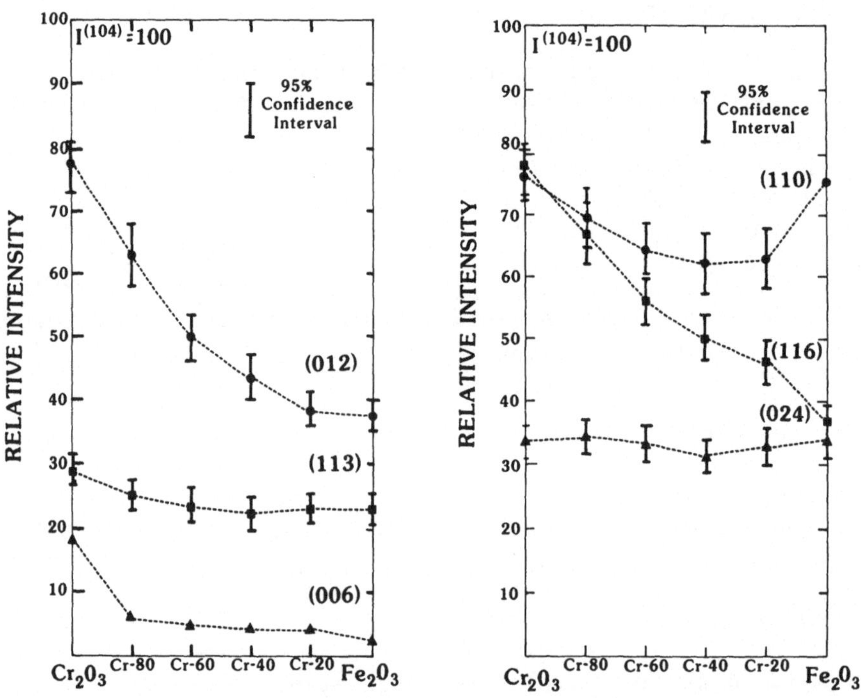

Fig. 3. Relative peak intensities in the Cr_2O_3-Fe_2O_3 solid solution series.

obtained for the limited cubic range of the NiO-ZnO series). All I^{rel} data have been converted to fixed slit values. The error bars or data points themselves represent the 95% confidence interval obtained from the pooled data.

In the MgO-NiO series, the I^{rel} of the (111) exhibits the greatest dependence on composition. Consideration of the structure factor expression for the rocksalt structure shows that reflections having odd Miller indices should be much more dependent on the scattering power of the metal atoms than those with even indices. Relative to the (200), the (111) intensity increases from 10% in MgO to 55% in NiO. In the other two systems, there are also relative intensity variations with compositions that can amount to 100% or more across the solid solution series.

Two points relevant to qualitative analysis can be inferred from these results. The first is the necessity for setting larger "windows" for an I^{rel} match in search/match procedures when solid solutions are expected. The second concerns the possibility that a linear interpolation between end member I^{rel} data could be used to approximate the I^{rel} values of an intermediate of known composition. The I^{rel}-composition plots in Figs. 1-3 are generally within 10-20% of a linear interpolation between end members. An analyst, upon deriving a tentative identification of one end member, but recognizing from chemical analyses and shifts in d-spacings that solid solution with another phase is present, could estimate the actual composition and use this linear interpolation of I^{rel} in his manual or computer subtraction process and thereby reduce or eliminate residual intensity.

Quantitative Analysis -- Intensity Ratios

Internal standard methods of quantitative X-ray phase analysis use ratios of the intensity of the analyte to the intensity of an internal standard added in a fixed proportion. Figures 4-7 show how the integrated intensity ratios for several reflections in three of the solid solutions vary as a function of composition. The mass ratio of solid solution phase to internal standard is 1:1 in all cases. Note that these intensity ratios have not been converted to fixed slit values.

The increase of the MgO-NiO intensity ratio with increasing NiO concentration is approximately linear, with the (111) exhibiting the greatest increase (Fig. 4). Fluorite rather than α-Al_2O_3 was chosen as the internal standard to avoid overlap with the (200) reflection of the solid solution. In the rocksalt structure portion of the NiO-ZnO system, the (111) intensity showed an 18% linear increase and the (200) a 7% linear increase in intensity relative to the (220) of the fluorite internal standard as the composition changed from NiO to $(Ni_{0.7}Zn_{0.3})O$. In both solid solutions, the

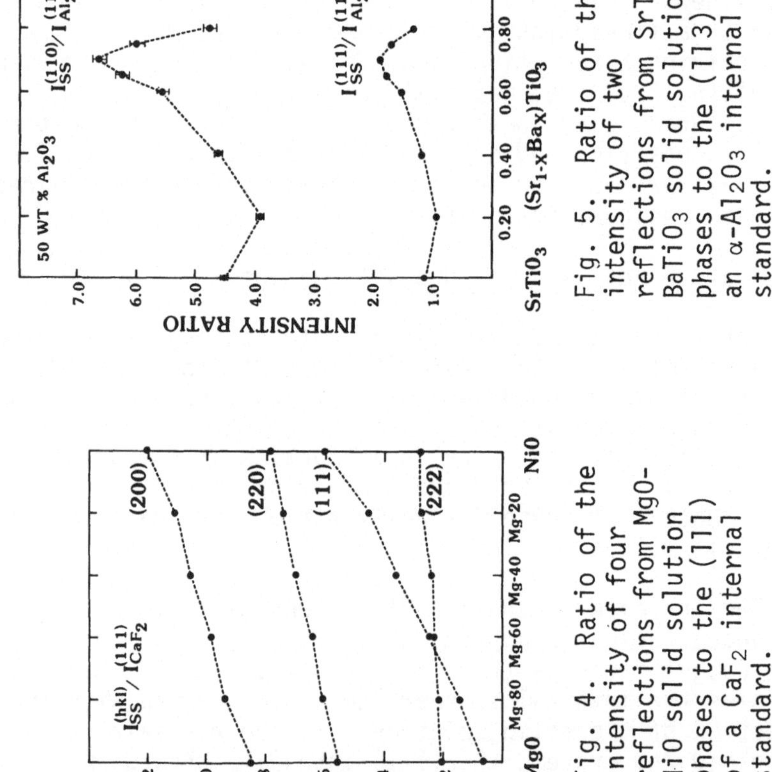

Fig. 4. Ratio of the intensity of four reflections from MgO-NiO solid solution phases to the (111) of a CaF_2 internal standard.

Fig. 5. Ratio of two intensity of two reflections from $SrTiO_3$-$BaTiO_3$ solid solution phases to the (113) of an α-Al_2O_3 internal standard.

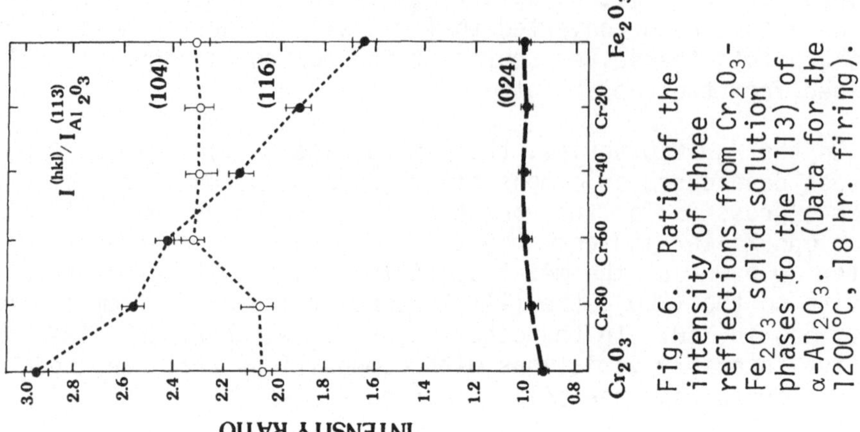

Fig. 6. Ratio of the intensity of three reflections from Cr_2O_3-Fe_2O_3 solid solution phases to the (113) of α-Al_2O_3. (Data for the 1200°C, 18 hr. firing).

greater compositional dependence of the (111) intensity compared to the (200) intensity is predicted by the structure factor expression. The magnitude of the increase is much smaller in the NiO-ZnO system because of the much smaller difference in the scattering power of Ni and Zn as compared to the difference of Mg and Ni.

We initially thought that the addition of ZnO to NiO would result in clustering or other structural disruption due to the tendency of Zn to favor 4-fold coordination and that this disruption could affect the shape of intensity ratio-composition curves. Instead, the dependence is essentially linear. Such is not the case with another solid solution in which a structure change is approached as composition is changed. As Ba is substituted for Sr in the A-site in $ATiO_3$, the cubic perovskite structure becomes increasing unstable at room temperature until, at about 70 mole % substitution, the structure undergoes a tetragonal distortion. Figure 5 shows the dependence of the intensity of two reflections ratioed to the (113) of α-Al_2O_3. The first effect of substituting Ba for Sr is a decrease in intensity ratio followed by an increase until the composition at which the structure becomes tetragonal is reached. The intensity ratio then decreases sharply with increased Ba for Sr substitution. In the tetragonal region, the (110) splits into a (110) and (101) and it is the total integrated intensity of both reflections that is plotted in Figure 5.

Figure 6 is a plot of the intensity ratios of three reflections in the Cr_2O_3-Fe_2O_3 system. There is a marked contrast in the behavior of the (116) compared to the (104) and (024). Note, however, that in each instance, the deviation from a linear interpolation from end member values does not exceed 10% at any composition.

Microabsorption, which can have a profound effect on intensity ratios (4,5), would be expected to be more of a factor in a mixture where the linear absorption coefficient (μ) of analyte and internal standard differ greatly (4). In the Cr_2O_3-Fe_2O_3 system, μ varies from 950 to 1153 cm^{-1} while the μ of α-Al_2O_3 is only 125 cm^{-1}. This large a difference could give microabsorption problems, especially if there were a large proportion of particles above about 2μm in size (4). To explore this possibility, the (116) intensity ratio data were obtained with three additional standards: ZnO (μ = 287 cm^{-1}), $(Sr_{0.5}Ba_{0.5})TiO_3$ (μ = 1033 cm^{-1}, near the mid-range of the solid solution) and CeO_2 (μ = 2082 cm^{-1}). As Figure 7(a) shows, the intensity ratio curves drop as more absorbing internal standards are used, but the percentage decrease across the series remains about the same. If microabsorption were an important factor, the behavior with the much less absorbing standard (α-Al_2O_3) would not trend the same as that from a much more absorbing standard (CeO_2) (4).

In our earlier paper (1), we reported that the intensities obtained from solid solutions could be strongly dependent on firing time. We heated the Cr_2O_3-Fe_2O_3 phases for an additional 22 hr. at 1200°C to see if crystallinity and intensity ratio would change. There was no significant difference in crystallinity as measured by peak breadth and α_1/α_2 splitting in the same phase fired for 18 or 40 hrs. However, with products from both firings the crystallinity did decrease with the $(Cr_{0.2}Fe_{0.8})_2O_3$ and Fe_2O_3 phases compared to the other four compositions. This may be due to slight nonstoichiometry of oxygen in the Fe-rich phases. The intensity ratios measured with the $ATiO_3$ and CeO_2 internal standards are shown in Figure 7 (b and c). While there are differences that fall outside of each other's 95% confidence interval (the size of the data points), for the most part the extra firing had little effect on intensity ratio.

There are several points relevant to quantitative analysis that can be inferred from these results. First, use of an end member intensity ratio when the actual analyte composition is well into the solid solution series could lead to major errors, depending on the reflection chosen. For example, using pure Cr_2O_3 as the standard when the actual composition is $(Cr_{0.5}Fe_{0.5})_2O_3$ would give a 30% error with the (116), a 10% error with the (104) and an 8% error with the (024). Including data for pure Fe_2O_3 and using a linear interpolation between the Cr_2O_3 and Fe_2O_3 values would reduce these errors to 1%, 7% and 4% respectively. If intermediate compositions are available (e.g. with minerals), fitting a curve to observed intensity ratios and interpolating or extrapolating to the analyte composition could yield quite accurate intensity ratio values, especially in unusual systems such as $SrTiO_3$-$BaTiO_3$.

Finally, the results presented here support the recommendation of Hubbard, Robbins and Snyder (6) to incorporate multiple reflections of the analyte and the internal standard in the analysis. Not only will use of multiple reflections minimize particle statistics and preferred orientation errors, but it can also reduce errors that could arise in solid solutions when exact composition is not known and the intensity of the reflection chosen is a strong function of the solid solution composition. In some cases, e.g. the (116) and (104) of Cr_2O_3-Fe_2O_3, the increases and decreases in intensity with composition may partially cancel each other.

SUMMARY AND RECOMMENDATIONS

It has been shown that relative intensities used in qualitative analysis and integrated intensity ratios used in quantitative analysis can exhibit a wide range of compositional dependence across a solid solution series depending on the particular (hkl) being considered. These dependences reduce the "goodness of fit" in

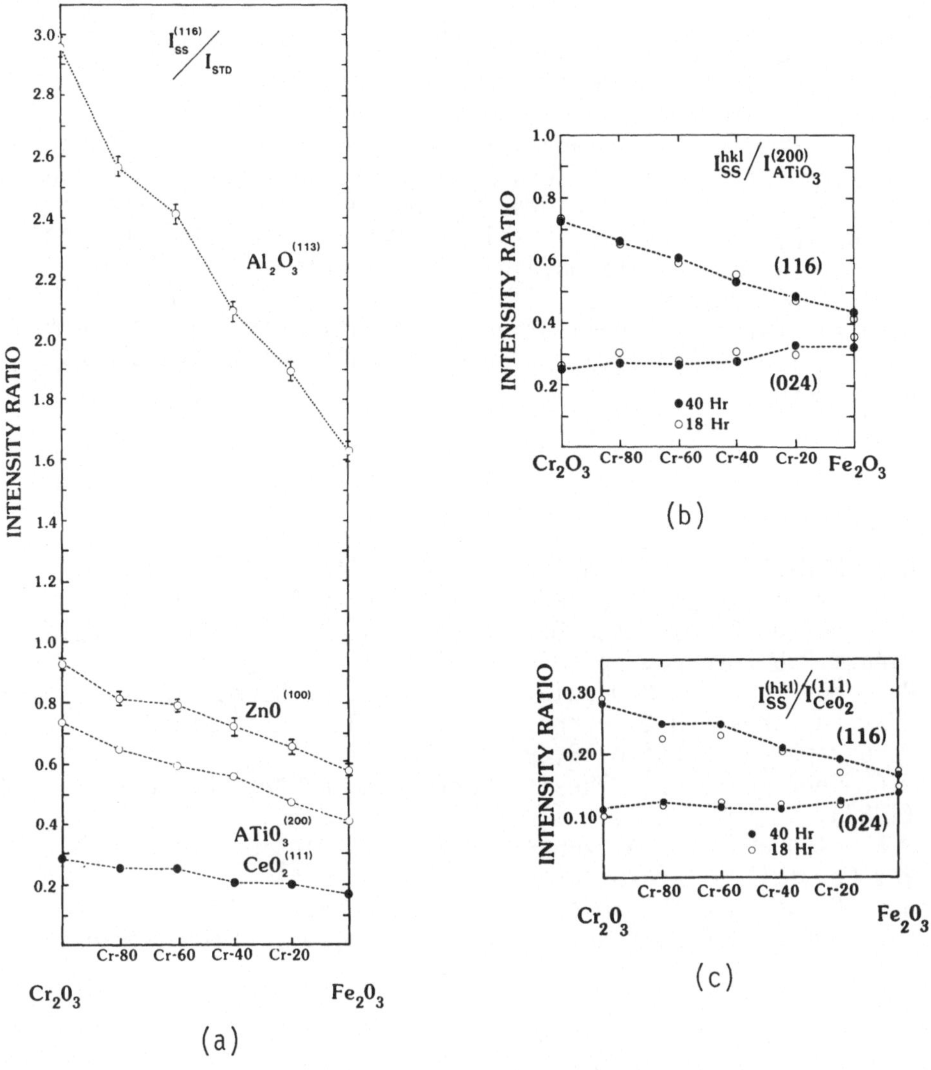

Fig. 7. Intensity ratios from Cr_2O_3-Fe_2O_3 solid solution phases. (a) Ratio of the (116) to reflections from four internal standards that cover a wide range of linear absorption coefficients. (b and c) Effect of firing time at 1200°C on selected intensity ratios using two of the internal standards. ($ATiO_3$ = $(Sr_{0.5}Ba_{0.5})TiO_3$).

search match results and can produce uncertainties of 25% or more in quantitative analysis if data for an end member composition are used for a solid solution member. In the four oxide systems studied, compositional dependence of I^{rel} or intensity ratio did not deviate by more than about 10% from a linear interpolation between end member values. We suggest that for qualitative analysis where approximate composition is known and end member standards are available, that a linear interpolation (i.e. a weighted average) of end member I^{rel} can be used in match and subtraction steps of the search/match process in lieu of synthesis of the actual solution phase and measurement of its I^{rel} data Similarly for quantitative analysis, interpolation or extrapolation of intensity ratio data from end members or other solid solution compositions would result in greatly reduced errors compared to use of the ratio from a single solid solution composition or end member. Averaging results obtained from multiple reflections would further reduce errors. Those developing algorithms for computer automation of qualitative and quantitative analysis may wish to consider solid solution subroutines based on the concepts discussed here.

REFERENCES

1. G.J. McCarthy, R.C. Gehringer, D.K. Smith, V.M. Injaian, D.E. Pfoertsch and R.L. Kabel, Adv. X-Ray Anal., 24, D.K. Smith, C. Barrett, D.E. Leyden and P.K. Predecki, Eds., Plenum Publ. Corp., pp. 253-264 (1981).
2. M.C. Morris, H.F. McMurdie, E.H. Evans, B. Baretzkin, H.S. Parker and N.C. Pyrros, National Bureau of Standards Monograph 25, Section 18, p. 3 (1981).
3. R.C. Gehringer, M.S. Thesis, North Dakota State University (1983).
4. H.P. Klug and L.E. Alexander, X-Ray Diffraction Procedures, 2nd Ed., Wiley Interscience, New York (1974).
5. J.P. Cline and R.L. Snyder, (this volume).
6. C.R. Hubbard, C.R. Robbins, and R.L. Snyder, (this volume).

QUANTITATIVE ANALYSIS OF PLATELIKE PIGMENTS BY X-RAY DIFFRACTION

P. Kamarchik and J. Ratliff

PPG Industries, Inc.
P.O. Box 9 Rosanna Drive
Allison Park, PA 15146

INTRODUCTION

X-ray diffraction has been shown to provide an accurate qualitative and quantitative determination of the pigment composition of paint films.[1-6] This analysis, however, is frequently complicated by the presence of pigments with plate-like crystallites which show a marked tendency to orient in preferred directions in drying paint films. The variability in the degree of orientation causes line intensity variations not attributable to pigment concentration.

A technique based on a summation of the intensities of many lines has been used to correct for the effects of preferred orientation in the analysis of retained austenite in certain steels.[7,8] This method is shown to be applicable to the quantitative analysis of mica in paint films.

THEORY

The method commonly employed to eliminate the effect of the absorption coefficient of a mixture is to consider only ratios of intensities of pairs of components.[4-6] In this way the intensity ratio is related to the weight ratio through the proportionality constant K_{iL}^{jR}. Diffraction lines are indicated by i and j, and components by L and R. The constants are determined by measuring intensity ratios for mixtures of known composition. If only one diffraction line contributes to the intensities, preferred orientation will affect these ratios and therefore the composition calculated.

Although this technique is adequate for samples of randomly oriented materials it can lead to severely erroneous results on highly oriented specimens. As an example, consider the results presented in Table I comparing mixtures of rutile (TiO_2) and zinc oxide and mixtures of mica and TiO_2. The average absolute error for zinc oxide, a non-orienting phase, is 0.6% while the average absolute error for mica is 6.3% when using the (002) mica line and is 4.2% when using the (006, 024) mica line.

Table I
Comparison of Analysis Results of Orienting and Non-Orienting Materials

	Method A*		Method B*	
% Mica Theory	% Mica	Abs. Error	% Mica	Abs. Error
12.4	17.5	5.1	12.5	0.1
34.2	38.9	4.7	37.5	3.3
70.8	66.0	4.8	66.4	4.4

% ZnO Theory	% ZnO Calculated	Abs. Error
25.1	25.7	0.6
74.2	73.9	0.3

* Method A: (002) mica line; Method B: (066, 024) mica line at 0.334 nm.

The ratio method described above requires intensities which are unaffected by orientation changes. Such an intensity can be calculated if one chooses a particular reflection for each phase as a reference and then calculates what the observed intensity of this line would be if all crystallites were reoriented so as to contribute only to this reflection.

Consider a phase exhibiting J different reflections and let n_j be the number of crystallites of this phase with the set of planes giving rise to reflection j being oriented parallel to the specimen surface i.e., in correct reflection orientation. For a given orientation distribution, the total number of crystallities, N, within the x-ray depth is proportional to the sum of all reflecting orientations. For a specimen with random orientation, all orientations with respect to a reference plane (surface) are equally likely, therefore all n_j's are equal. The intensity observed for a particular reflection of a particular phase, I_j is the sum of the reflection intensities from all crystallities, n_j, contributing to that reflection. The intensity of a single crystallite, i_j, is a function of crystal and instrument characteristics that can be held constant so that all crystallities yield the same i_j. The total reflection intensity can then be given as:

$$I_j = n_j i_j \qquad (1)$$

for diffraction line j. For a randomly oriented specimen I_j can be written as I_j^R and the subscript on n can be dropped since they are defined as equal for a random distribution. If all reflecting crystallities in an observed pattern for some phase are assumed to be reoriented to contribute intensity to a reference line I_s, the total intensity will be given by;

$$I_s = I_s^o + n_1^o i_s + n_2^o i_s + \ldots n_j^o i_s \qquad (2)$$

where the superscript ^{o} refers to a specimen of any orientation distribution. Rearranging and substituting from equation 1 for the case of a non-random orientation equation 2 becomes;

$$I_s = I_s^o + \sum_{j=1}^{J}{}' \frac{i_s}{i_j} I_j^o \qquad (3)$$

where the sigma prime means that the summation is carried out over all reflections except the reference reflection.

Since n's are equal for a random distribution

$$I_s^R / I_J^R = i_s / i_j \qquad (4)$$

defining $C_j = I_s^R / I_J^R$ and substituting into the equation above gives;

$$I_s = I_s^o + \sum_{j=1}^{J}{}' C_j I_j^o \qquad (5)$$

The intensity I_s will be independent of orientation distribution in a specimen but will depend on the total number of crystallites of a particular phase in the x-ray beam. The I_j^o's are those obtained from the observed pattern. And the C_j's are obtained from a pattern in which $n_1 = n_2 \ldots n_j$, that is, a randomly oriented specimen. A means of obtaining a randomly oriented specimen for this work is discussed in the experimental section.

The total indensity I_s is then used for quantitation as if it were the intensity of a single line for mica.

EXPERIMENTAL

All diffractometery was done on a Siemens D500 operated in a step scan mode using steps of 0.02° 2θ and a step duration of 1.2 sec. The divergence and scatter slits are 1.0° and the receiving slit is 0.15°. A curved graphite monochrometer as well as pulse height analysis are used. Integrated intensities minus background are provided by the IDENT program of the Siemens software. Copper K_α radiation was used in the analyses with the tube operated at 40 KV and 30 mA.

A. Randomly Oriented Mica

Mica was dispersed at high shear at the 10% level in a high viscosity epoxy resin. The curing catalyst was added and shearing stopped when the viscosity became too high to get effective mixing. The mixture was cast in a block and was essentially solid within 10 minutes. This block was cut parallel, perpendicular, and obliquely to the vertical direction established during curing. The exposed surfaces were examined by light microscopy for the number of crystallites in a given area with basal planes roughly parallel to the exposed surface. These counts indicated that the orientation was random. Diffraction patterns were obtained on these faces and relative intensities observed at each surface for each line. These average relative intensities for the three surfaces are given in Table 2 where they were compared with values reported by Yoder and Eugster.[10] These values were then used to evaluate the summed intensity of all mica peaks in subsequent analyses by use of equation (5).

B. Sample Preparation

With the exception of the samples prepared as blocks of solid epoxy, all samples were prepared as paint films. These paint films were prepared by grinding appropriate amounts of the pigment in a "Wiggle-bug"[11] with an alkyd paint vehicle. The largest pigment particles were reduced to less than 25 um. This "paint" is then prepared as a 0.0015 in. film on mylar.

RESULTS AND DISCUSSION

In order to obtain the proportionality constants K_{iL}^{jR} for mica with rutile taken as the reference material, a series of nine samples were made up and measured ranging from roughly 10% mica/ 90% rutile to 90% mica/10% rutile. The total mica intensity was calculated using equation (5) and using the (006, 024) line as that of total reorientation. The proportionality between intensity ratio and weight ratio was found by regression analysis to be 0.327. This constant was then used to calculate the composition of each of the standards from the intensity data of the twelve lines listed in Table 2. Actual and calculated values are compared in Figure 1. With the exception of a rogue data point for sample 2 the agreement is quite good over this extensive range. The average error using this method is 2.5% while it is 5% if the single (006, 024) line is used and 6% if the (002) line is used. In all subsequent analyses the proportionality constant calculated, 0.327, is used in the calculation of composition.

Nine additional samples have been prepared to test the validity of this method. Actual versus calculated compositions are shown in Figures 2 and 3. These compositions were chosen in order to

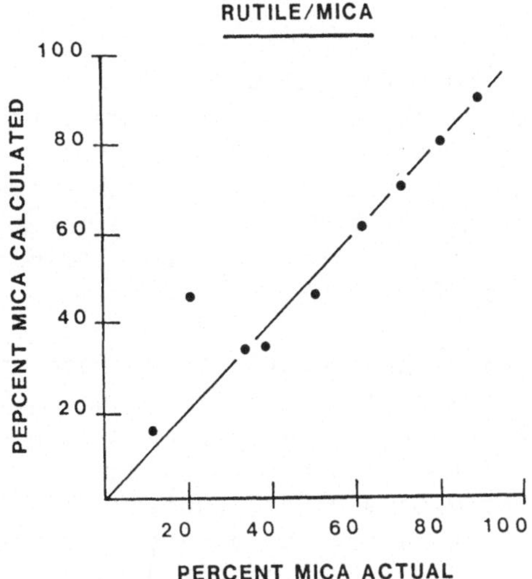

Fig. 1. Comparison of actual and calculated compositions.

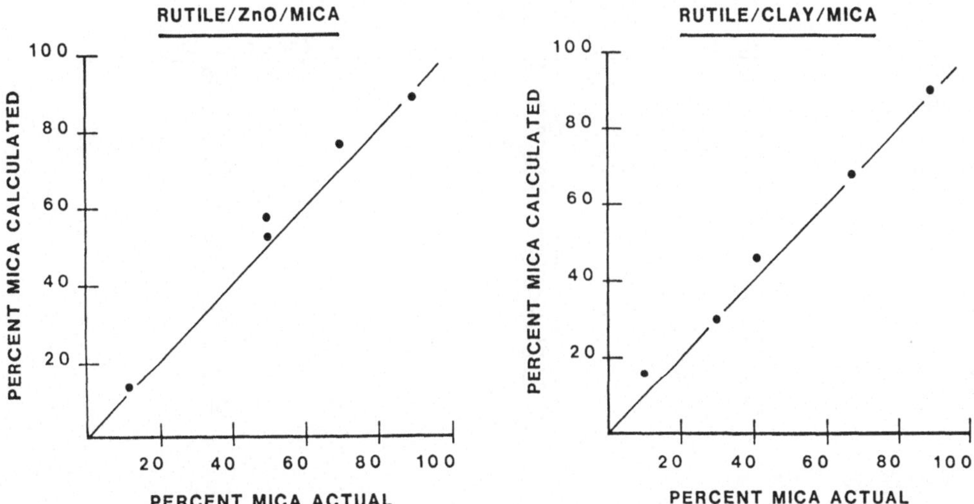

Fig. 2. Comparison of actual and calculated compositions.

Fig. 3. Comparison of actual and calculated compositions.

place the mica particles into two very different environments. These are matrices consisting mostly of either platey or spherical pigment particles. The degrees of orientation are quite different among these samples.

CONCLUSION

This summation method which assumes a reorientation of all crystallites to a particular orientation is an effective means of avoiding erroneous quantitation results which may arise due to preferred orientation. In typical samples the average error is reduced by a factor of two or three while the gross errors found in certain cases or extreme orientation are eliminated so that no more error than usual is found.

This method differs significantly from those approaches in which quantitation is done using several lines individually and finally averaging the results. That approach in effect averages erroneous data in order to produce an improved result. An accurate analysis depends on fortuitous circumstances. The summation approach presented here attempts to arrive at an accurate result through theoretically correct manipulation of the data.

Table 2 Relative Intensities of The Diffraction Lines Of Randomly Oriented Muscovite

"d Spacing" nm	hkl	% Of Max. This Study	% Of Max. Yoder & Eugster
1.004	002	33	100
0.504	004	21	55
0.448	110,020	59	55
0.390	113	28	37
0.374	023	27	32
0.350	114	41	44
0.334	006,024	100	100
0.321	114	42	47
0.300	025	39	47
0.287	115	26	35
0.278	116,200	20	22
0.257	113	67	90

Although this technique has been applied only to mica, there is no reason to believe that it wouldn't be applicable to any phase that tends to orient in a preferred sense. The primary difficulty in the use of the method is the attainment of the relative intensities that correspond to a purely randomized sample. It may be possible to overcome this difficulty by recourse to calculated patterns.

REFERENCES

1) P. Kamarchik, Jr., Journal of Coatings Technology, 52 79 (1980)
2) P. Kamarchik, Jr., and Glenn P. Cunningham, Progress in Organic Coatings, 8, 81 (1980)
3) R. W. Scott, Journal of Paint Technology, 41, 422 (1969)
4) F. H. Chung, Journal Appl. Cryst., 7, 519 (1974)
5) F. H. Chung, Journal Appl. Cryst., 7, 526 (1974)
6) F. H. Chung, Advances in X-ray Analysis, 17, 106 (1974)
7) M. H. Dickson, Journal of Appl. Cryst., 2, 176 (1969)
8) R. D. Arnell, Journal of the Iron and Steel Inst., p. 1035, Oct. 1986
9) H. P. Klug and L. E. Alexander, "X-ray Diffraction Procedures for Polycrystalline and Amorphous Materials," John Wiley and Sons, 2nd Ed., New York, 1974
10) H. S. Yoder and H. P. Eugster, Geochimica et Cosmochimica Acta, 8, 225 (1955)
11) Wiggle Bug vibrating ball mill, Crescent Dental Mfg., Co., 7750 West 47th St., Tyons, IL 60534

PREPARATION AND CERTIFICATION OF STANDARD REFERENCE MATERIALS TO BE USED IN THE DETERMINATION OF RETAINED AUSTENITE IN STEELS

George E. Hicho and Earl E. Eaton

Fracture and Deformation Division
National Bureau of Standards
Washington, DC 20234

INTRODUCTION

In the steel hardening process, steel is heated to a temperature where a face-centered-cubic solid phase called austenite is formed. After a stabilization period, the steel is quenched into a medium which transforms the austenite into a metastable, body-centered-tetragonal solid phase called martensite. On occasion the austenite is not entirely transformed into martensite and some austenite remains. This untransformed (retained) austenite is sometimes detrimental to the finished product, and often there are requirements as to the amount of retained austenite permitted in the finished product.

X-ray diffraction procedures (XRD) are normally used to determine the amount of retained austenite and this paper describes the preparation and characterization of the Standard Reference Materials used to calibrate x-ray diffraction units.

PROCEDURE

These Standard Reference Materials (SRM's) are composed of two constituents whose structures are metallurgically different. The components are AISI 310 stainless steel (austenitic) and AISI 430 (ferritic) stainless steel powders. The 310 is a highly stable stainless steel (24.99 weight percent Cr and 20.41 weight percent Ni) requiring a substantial change in composition[1] to produce a metallurgical structure other than austenite. The 430 stainless steel contains 16.03 weight percent Cr and virtually no nickel - .09 weight percent.

The main concept of the SRM lies in the fact that the nickel content for each compact is related directly to the austenite content.

Because of the significant difference in the nickel content of the austenite component and that of the ferrite component, it was possible to use X-ray fluorescence (XRF) to obtain a very precise measurement of the total nickel near the surface of the compact. The total nickel counts were corrected[2,3] to weight percent nickel.

The powders used for each reference material, the 5% (SRM 485a) and 30% (SRM 487) austenite in ferrite, were blended in a "V" blender, pressed, sintered, repressed, and vacuum annealed. Only one surface of the compact was polished and it is that surface which is certified as to the austenite content. No surface preparation of the compact is necessary, in fact damage to the surface renders the certification void. The fabrication procedure is shown in Figure 1.

Following the metallographic polishing of the compacts, the weight percent nickel present on each compact's surface was determined using XRF. The compacts were then ranked from lowest to highest weight percent nickel and a number of compacts, encompassing the total population, were removed and used as calibration specimens. The characterization steps are shown in Figure 2.

In order to establish calibration curves, it was necessary to initially determine the surface porosity for those selected compacts. Quantitative microscopy was used to measure the area percent porosity. Having determined the porosity, the compacts were stained with Murakamis reagent, and it was now possible to measure the area percent austenite using quantitative microscopy.

Calibration curves plotting weight percent nickel, as determined by XRF analysis, versus volume (area)[4] percent austenite, were developed and are shown in Figures 3 and 4 for each of the respective SRM's. The equations developed for the 5% and 30% SRM were

$$\text{Volume percent austenite} = 4.758 \cdot W - 1.121$$
$$\text{Volume percent austenite} = 4.737 \cdot W - 2.056$$

where W represents the weight percent nickel as determined by XRF analysis.

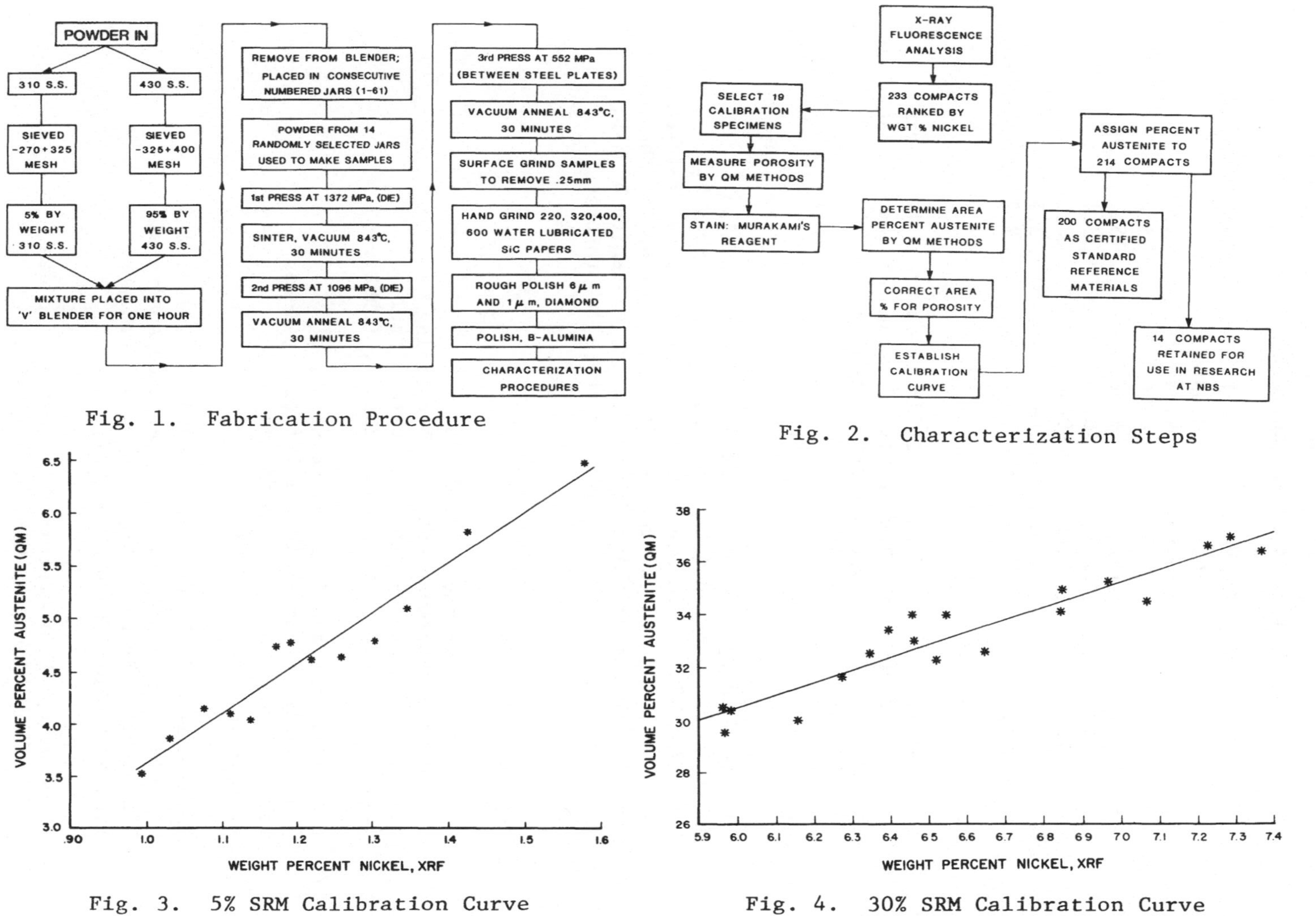

Fig. 1. Fabrication Procedure

Fig. 2. Characterization Steps

Fig. 3. 5% SRM Calibration Curve

Fig. 4. 30% SRM Calibration Curve

Table 1. Comparison of NBS SRM values and XRD results. (Volume percent austenite).

Sample	NBS	Lab 1	Lab 2
80	4.55	3.83	4.8
93	5.66	4.63	4.6
280	15.05	14.86	15.4
388	14.42	14.90	15.2
110	32.84	30.03	33.1
173	32.66	29.40	28.0
C7	2.62	-	2.5
C10	2.52	-	2.2

The certified value for the 5% SRM is accurate to within ±0.25 percent austenite. The certified value for the 30% SRM is accurate to within ±0.50 percent austenite.

RESULTS

In addition to the 5% and 30% SRM's, NBS has also certified a 15% standard (SRM 486) and is preparing a 2½% standard (SRM 488). Two specimens from each of these reference materials were sent to two independent laboratories for the purpose of determining their volume percent austenite by XRD. Table 1 shows the results obtained by these laboratories.

REFERENCES

1. C. G. Interrante and G. E. Hicho, "A Standard Reference Material Containing Nominally Fifteen Percent Austenite," Natl. Bur. Stand. (U.S.) Spec. Publ. 260-73, (1982).

2. S. D. Rasberry and K. F. J. Heinrich, "Calibration of Interelement Effects in X-ray Fluorescence Analysis," in "Analytical Chemistry", 46:81 (1974).

3. L. S. Birks, J. V. Gilfrich, and J. W. Criss, "A Fortran Program for X-ray Fluorescence Analysis," X-ray Optics Branch, NRL, Washington, D.C., Project Number DOD-0006T, (1977).

4. A. Delesse, "Pourdeterminer la Composition des Roches," in "Ann. des Mines B," (1848).

PROFILE FITTING FOR QUANTITATIVE ANALYSIS IN X-RAY POWDER DIFFRACTION

Walter N. Schreiner

Philips Laboratories
Briarcliff Manor, NY 10510

Ron Jenkins

Philips Electronic Instruments
Mahwah, NJ 07430

Quantitative phase analysis by powder diffractometry requires accurate measurement of the integrated intensities of the diffracted lines. When lines are isolated and on simple backgrounds, count integration techniques work very well. However, when one or more lines overlap the line of interest, or a complex background is present, profile fitting techniques are required in order to eliminate interferences.

Profile fitting involves choosing a mathematical model to represent the expected profile shapes. Experience has shown that the profile shapes obtained with a parafocusing powder diffractometer are not easily described and many models have been tried with varying degrees of success*[Ref 1,2]. Generally the more free parameters allowed in the model the better the fits, although, aesthetically one would like to keep the number of free parameters to a minimum.

There are, however, reasons other than aesthetics to keep the number of parameters to a minimum. In this paper we describe a model we have found to be as successful as any previously reported, but which requires fewer parameters, all of which represent physical

*Among the functions which have been tried are Gaussians, Lorentzians (Cauchy), modified and intermediate Lorentzians, Pearson VII, Voigt and other variations on these basic forms, including convolutions of these functions.

features of the diffraction profile, and therefore, behave in a predictable way. The model employs 12 intrinsic parameters to describe the instrument aberration and wavelength dependent contributions to the profile, and 3 parameters to describe the sample dependent variables of line position, height and line broadening. Since the 12 intrinsic parameters behave systematically with 2Θ, they can be fixed, given an initial estimate of an α_1 peak position. This leaves only the 3 sample related parameters to be fit to an experimental peak.

This approach to profile fitting has been developed in the past by other authors. In particular, Parrish, et. al. have used seven overlapping Lorentzian functions with 21 parameters to model the intrinsic (wavelength and instrumental) component of the profile [2]. Others have attempted the use of directly stored experimental profiles for the same purpose. In any case, coefficients at several discrete angles are stored as reference values, and the corresponding coefficients at intermediate angles are obtained by linear interpolation. Unfortunately, the interpolated values may not be self consistent because the coefficients are strongly coupled to one another, and hence, quite sensitive to counting statistics and other small effects. The strong coupling results from the use of an excessive number of parameters to describe the intrinsic profile.

In order to avoid this problem, we attempted to develop a model in which each intrinsic parameter was chosen to represent a well-defined visible characteristic of the profile. Several benefits are realized from this approach. First, such parameters will tend to be linearly independent (i.e. uncoupled) and hence a least squares program to fit them will converge faster and to a more well defined solution. Second, each parameter should exhibit a distinct and clear trend as a function of 2Θ. If, in least squares fitting the parameters at a given value of 2Θ, convergence is not complete or a secondary minimum is reached instead of the correct one, this will be evident from a plot of each parameter vs 2Θ as a point far off the trend line.

In Fig. 1 we show the parameters used in our model. Four peaks are modeled at angles Θ1, Θ2, Θ3 and Θ4. The actual positional parameters are Θ1 and the relative positions of Θ2, Θ3 and Θ4 with respect to Θ1, namely, Δ_{12}, Δ_{13}, Δ_{14}. The height of Θ1 is parameterized as H and the other three peaks are ratioed to H as R2, R3 and R4. Note that SQRT(I) is used as the ordinate. In SQRT(I) space all data points in the experimental profile have the same size (statistical) error bar and this obviates the need of weighting the least squares fit. The profile shape is a bifurcated Lorentzian in SQRT(I) space, with a HWHM of W_A below the peak of Θ1 and Θ2, and W_B above the peak. The third and fourth peaks are modeled as symmetric Lorentzians of width W_3 and W_4, respectively. In all, there are 12 free parameters in the intrinsic profile.

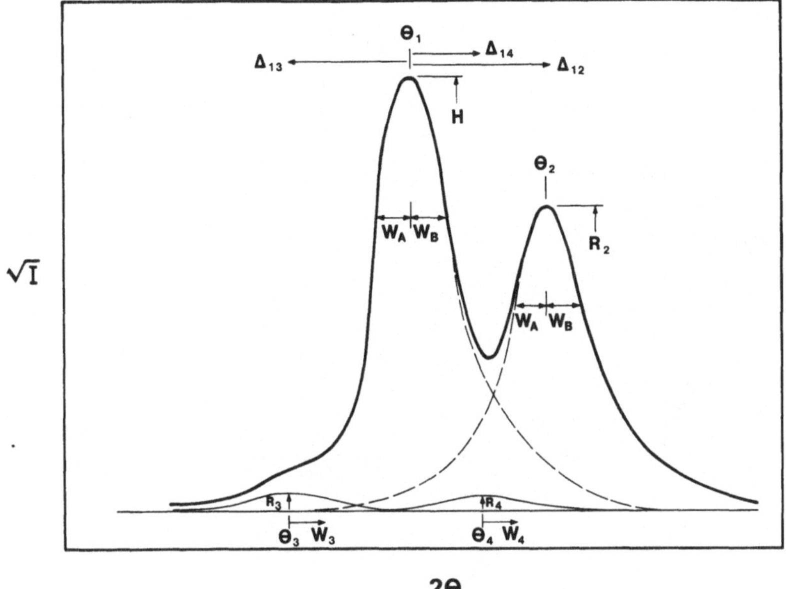

Fig. 1 Asymmetric Profile Parameters

In order to determine the 12 intrinsic parameters and their dependence on 2Θ, we fit each line of carefully prepared and measured samples of Si with our model. The Si data was measured for 10 sec at each point and a step size of 0.02° 2Θ was used. A Marquardt non-linear least squares algorithm was used to carry out the fits. A 13th free parameter was allowed in each fit to represent a constant background. R values in the range 1.5-2.0% were obtained in each case. No significant misfit was observed anywhere in the profile, including the tails. Each parameter was then plotted as a function of 2Θ. Two such plots are seen in Figs. 2 and 3.

Fig. 2 shows the variation of W_A with 2Θ. It is seen that W_A shows a general increase over the entire range, a behavior well known in powder diffraction. Typical error bars are shown on three points. The point at 47° 2Θ seems rather low and is an example of how sensitive even these parameters, which were chosen to be as independent as possible, are to small perturbations in the measured data. The straight line is the representation of the W_A parameter which we use to fix W_A, given a 2Θ value of an experimental peak, instead of attempting to interpolate between the scattered points. Use of this line effectively represents an averaging over all of the fits and, therefore, should be better than any single point alone.

In the second example seen in Fig. 3, the variation of Δ_{12} with

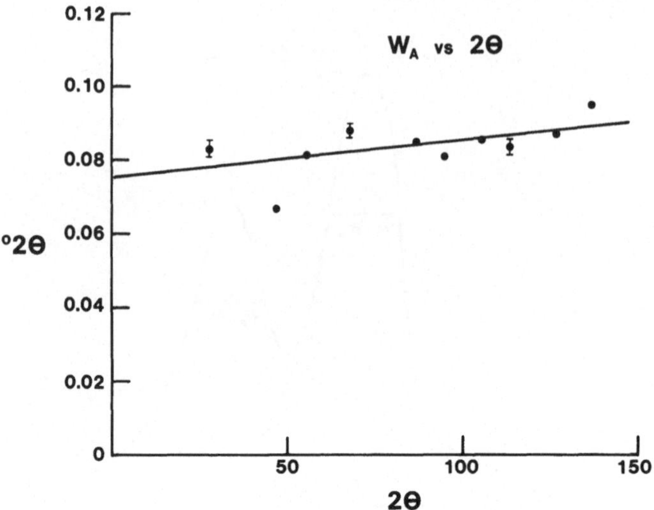

Fig. 2 Variation of W_A with 2Θ

2Θ is plotted in terms of $\Delta\lambda_{21}/\lambda_1$, the apparent fractional wavelength separation between α1 and α2. Because of the asymmetry in the diffraction profile, this value is always larger than the theoretical value of 2.485×10^{-3}. At about 103° 2Θ, where the axial divergence term goes through zero, the profile is symmetric and the apparent wavelength separation equals the true value. Once again, the behavior of $\Delta\Theta_{12}$ makes physical sense and its variation with 2Θ can be represented with confidence.

Once a representation has been obtained for each of the 12 intrinsic parameters, these can be fixed in fitting an experimental

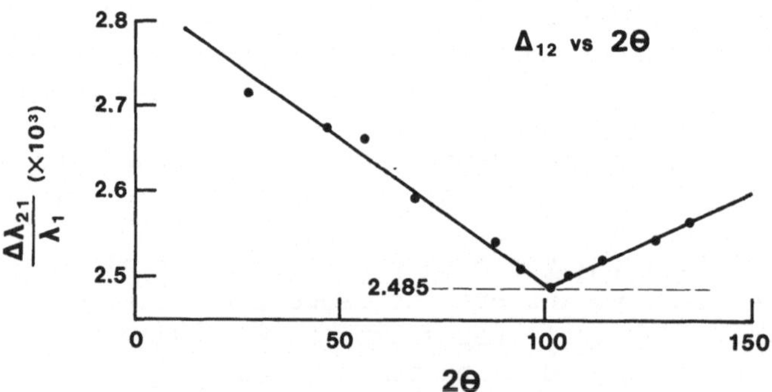

Fig 3 Variation of Δ_{12} with 2Θ

Table I: Profile Parameterization

Peak	Position	Height	FWHM
1	Θ_1	H	$(W_A+\Delta W)+(W_B+\Delta W)$
2	$\Theta_1 + \Delta_{12}$	$H \cdot R_2$	$(W_A+\Delta W)+(W_B+\Delta W)$
3	$\Theta_1 + \Delta_{13}$	$H \cdot R_3$	$2(W_C+\Delta W)$
4	$\Theta_1 + \Delta_{14}$	$H \cdot R_4$	$2(W_D+\Delta W)$

Free: Θ_1, H, ΔW
Fixed: $\Delta\Theta_{1i}$, R_i, W_i
Same linear background under all peaks

diffractogram with broader peaks and/or overlapping peaks. This leaves only three highly independent parameters per peak which require fitting. In Table I, the 12 profile parameters are shown divided into fixed and free components. The three variables are Θ1, H, and ΔW, a convolution half-width which is added to the intrinsic half-widths, W_A, W_B, W_C and W_D. Although convolution of the sample broadening factor, ΔW, with the Lorentzian intrinsic half width does not yield an exact Lorentzian of width W + ΔW because of the models' asymmetry, the approximation is quite good and causes no noticeable side effects.

Examples of the use of this model to fit experimental profiles demonstrate its applicability. In Fig. 4, the quartz quintuplet is

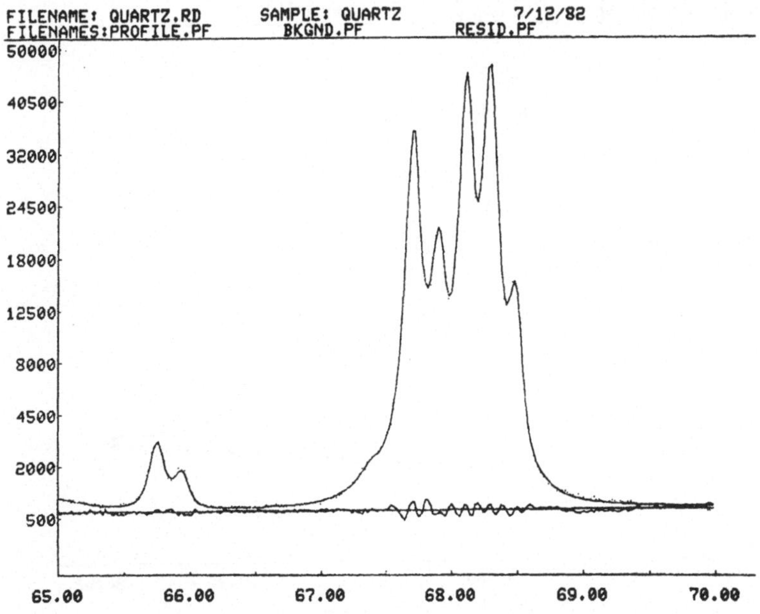

Fig. 4 Profile Model Fitted to Quartz Quintuplet

fit to three overlapping profiles and a single linear background. The data is plotted on a SQRT(I) scale which makes high and low count values equally visible. A rather large range of data from 65 to 70° 2θ, is fit in order to demonstrate how well the tails match. The residue from the fit (fit-measurement) is plotted as the jagged line centered on the background line. It is seen that all of the trends of the data are accounted for, including the tails.

In this example, 7 peaks were used; three for the quintuplet, one for the peak at 66°, another for a small peak just below 65°, and a third for the enhancement at 67.5°. The seventh peak is a broad one centered at about 68° and about 0.5° wide. We have found such broad peaks to be necessary to a varying degree in many samples. A possible explanation for these peaks are contributions from Thermal Diffuse Scattering (TDS).

In another test two broad peaks were measured in a sample of nylon (Fig. 5). The fit required four profiles and the linear background to fully match the data points. The left peak required one profile, and the other required two profiles. A fourth broad bump was found to be necessary to account for exceptionally gentle curvature between the peaks. The three main profiles were each found to have a FWHM of about 1° 2θ.

In the least example, seen in Fig. 6 two small crystalline peaks exhibiting particle size broadening are fitted on top of two large amorphous peaks. The sample was a 300 Angstrom layer of me-

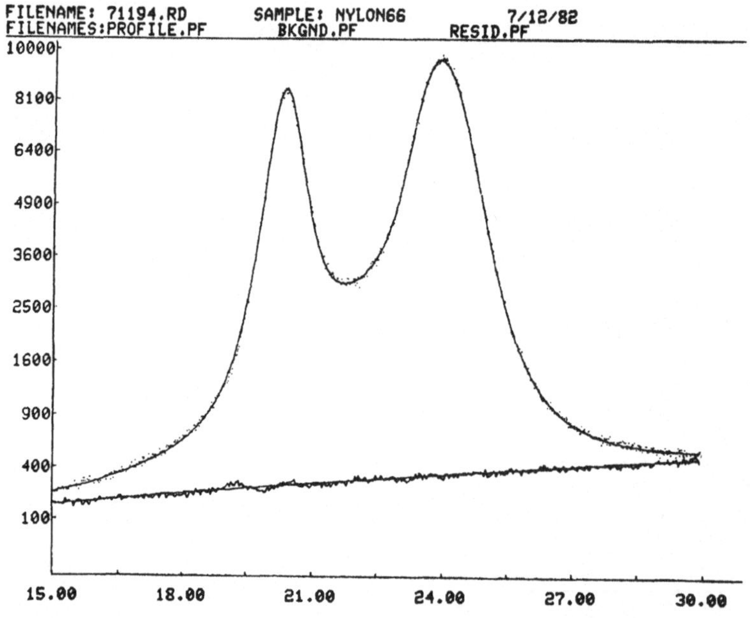

Fig. 5 Profile Model Fitted to a Nylon Sample

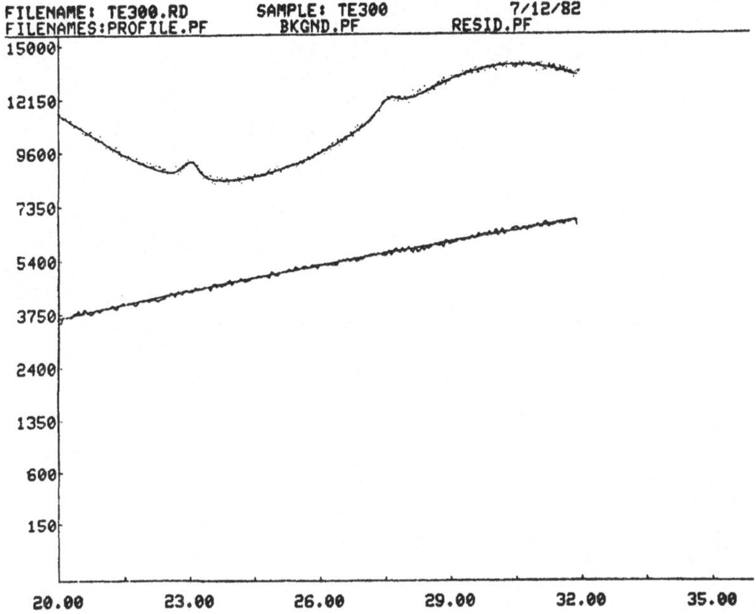

Fig. 6 Thin Layer of Tellurium on a Plastic Substrate

tallic Te vacuum deposited on a PMMA substrate for a video disk. Four peaks were used to fit the data, one for each crystalline peak and one for each amorphous peak.

Because we have found that amorphous peaks, such as these, fit our model as well as crystalline peaks, a simple linear background has been found sufficient in many cases where curved backgrounds had been used previously. This results in better fits because the background is more decoupled from the peaks.

Conclusions: A profile model has been developed using asymmetric Lorentzians to fit the instrument and wavelength related components of a powder diffraction profile in SQRT(I) space. The model parameters are chosen to represent specific characteristics of the profile in order to decouple them as much as possible and to permit representing their 2Θ dependence with simple functions. These dependencies are used to fix the (intrinsic) parameters, leaving only three sample related parameters to be fit to an experimental profile. The model has been successfully used to fit both crystalline and amorphous peaks and to effect a separation of overlapping peaks for quantitative analysis.

References

[1] M. M. Hall Jr, V. G. Veeraragharan, H. Rubin, P. G. Winchell, J. Appl. Cryst., Vol 10, 66-88 (1977)

[2] W. Parrish, T. C. Huang, G. L. Ayers, Trans. Am. Cryst. Assoc., Vol 12, 55-74 (1976).

XRD QUANTITATIVE PHASE ANALYSIS USING THE NBS*QUANT82 SYSTEM*

Camden R. Hubbard and Carl R. Robbins

Center for Materials Science
National Bureau of Standards
Washington, D.C. 20234 USA

and

Robert L. Snyder

New York State College of Ceramics
Alfred University
Alfred, New York 14802 USA

INTRODUCTION

X-ray powder diffraction (XRD) is widely used for qualitative analysis of the phases in multiphase mixtures. To extend the characterization to the level of quantitative analysis (QA) requires solution of many challenging problems such as elimination of preferred orientation (1,2,3); selection of an appropriate reference standard which closely matches the analyte phase in crystallite size, thermal history, stoichiometry etc. (4); effectively collecting the experimental intensities (5,6); and finally, performing the data reduction and analysis.

The desire to solve the last problem in a general way and to incorporate the "best" procedures developed in several laboratories into one package was one of the stimuli for developing the NBS*QUANT82 system of programs. This system is integrally coupled with the data collection system AUTO (6,7).

The AUTO system allows intensity data collection for any number of lines of the analyte and reference phases. One can repeat the data collection any number of times for one or more sample mountings. When practicable, we measure three or more lines of each phase to be analyzed and make three or more sample mountings. If an internal standard is being used the standard may be present in each mounting at any known weight fraction. Coupled with spray drying of the mixture or careful side drifting (8), the use of multiple lines and multiple mountings provides useful information on the absence or presence of preferred orientation effects on the experimental intensities. If preferred orientation is shown to be absent the replicate data yield meaningful estimates of the analyte weight fractions and their estimated standard deviations.

To use multiple lines requires a calibration constant for each line or requires the relative intensities (I^{rel}=100 for the strongest intensity line) and a single calibration constant. We have chosen the latter approach since relative intensities and a single calibration constant are readily understood. For the internal standard method of QA the single calibration constant is known as the REFERENCE INTENSITY RATIO (RIR). The $RIR_{\alpha,s}$ for analyte α and standard s is given by

$$RIR_{\alpha,s} = \frac{I_{i\alpha}}{I_{js}} \frac{I^{rel}_{js}}{I^{rel}_{i\alpha}} \frac{X_s}{X_\alpha} \qquad (1)$$

where i and j denote lines i and j for phases α and s, respectively, X is a weight fraction, and $I_{i\alpha}$ and I_{js} are the experimental intensities. These intensities should be integrated values if one wishes to share results with other laboratories or to compare with theoretical calculations (9). Ratios of peak height intensities may be used as approximations to ratios of integrated intensities in some cases. If the internal standard is α-Al_2O_3 (corundum) then $RIR_{\alpha,s}$ is also known as I over I-corundum (I/Ic). Numerous I/Ic values are available from the Powder Diffraction File.

PROGRAM DETAILS

The package of programs consists of four main programs and twelve subroutines. Many of the subroutines are used by each of the four main programs. All of the routines are written in standard FORTRAN 77 and have been compiled and tested successfully on a Univac 1182 and a Honeywell computer. The memory requirements of the programs may be readily modified through changes in the FORTRAN PARAMETER statements at the top of each routine. Currently the programs are set for 25 independent diffraction lines and 20 data files. Output should be directed to a 132 column printer. All programs are intended to be run interactively. The NBS*QUANT82 system of programs and documentation can be obtained from the first author (CRH).

PROGRAMS IN NBS*QUANT82

The four main programs are SPIKE, RUNFIL, QUANT and COBRAG. Program SPIKE performs analysis by the spiking method. The intensity ratio and internal standard methods of QA are implemented by programs QUANT and COBRAG. Program RUNFIL performs several functions including calculation of relative intensities and calibration constants from calibration data files produced by AUTO. When calibration files for several phases are input to program RUNFIL it combines the diffraction line information for the several phases together and creates an AUTO run file for data collection from the multiphase mixture. The resulting run file contains all the two-theta regions, data labels, I(rel) and calibration constants derived by RUNFIL. This avoids transcription errors by the user. All this data is transferred by AUTO to the data file containing the raw data for the analytical sample. This raw data file is then processed by program QUANT. All the user enters to program QUANT is the data file name. QUANT automatically calculates the weight fraction of each analyte phase for each resolved line. QUANT also averages over all data files, all reference lines and finally, over all lines for a given phase. Throughout the programs two error estimates are calculated. These are the propagation of counting statistical uncertainties and the root mean square(RMS) deviations from the average. The propagation of the counting statistical uncertainties provides a lower limit to the analytical precision. The RMS deviations, if a sufficient number of lines and mountings are measured, include uncertainties arising from sample inhomogeneity, variations in preferred orientation, instrument instability, particle statistics etc.

For the internal standard method if line overlap is present or if elemental or weight fraction constraints are to be employed then program QUANT writes a data file which is read by program COBRAG. This program implements the equations of Copeland and Bragg(10). For a mixture plus internal standard

$$\frac{I_{i\alpha}}{I_{ns}} = \frac{C_{i\alpha}}{C_{ns}} \frac{X_\alpha}{X_s} \tag{2}$$

where line n of internal standard s is a resolved line free of any overlap. When the intensity of each line of a pattern, I_j, is assumed to be the sum of intensities from each contributing phase k, then

$$\frac{I_j}{I_{ns}} = \sum_{k=1}^{m} \frac{C_{jk}}{C_{ns}} \frac{X_k}{X_s} \tag{3}$$

If phase k does not contribute to pattern line j then $C_{jk} = 0$. The equations 3, j = 1,2...n-1, form a set of linear equations in

X_k/X_s (k=1, ...m). When the number of phases m is less than n-1, equations 3 can be solved by least-squared methods for the desired weight fraction ratios. As shown in Appendix 1 the matrix elements are simply related to the relative intensities and reference intensity ratio by

$$\frac{C_{i\alpha}}{C_{ns}} = \frac{I^{rel}_{i\alpha}}{I^{rel}_{ns}} \cdot RIR_{\alpha,s} \quad . \tag{4}$$

Because the least-squares method solves a set of equations linear in the weight fractions of the analyte phases, results of quantitative elemental analyses can be added to the system of equations. If all phases containing element A are included in the XRD phase analysis then the percent of element A in the mixture can be related to the weight fractions X_i and percent of element A in phase i, $(\%A)_i$, by

$$(\%A)_{mixture} = (\%A)_1 X_1 + (\%A)_2 X_2 \ldots (\%A)_n X_n . \tag{5}$$

Dividing both sides of equation 5 by the known X_s yields the equation in the unknowns X_k/X_s:

$$\frac{(\%A)_{mix}}{X_s} = \sum_{k=1}^{m} (\%A)_k \frac{X_k}{X_s} \tag{6}$$

Different weights could be used for equations 3 and 6 to make the phase analysis closely fit the elemental information or the XRD intensity ratios.

Besides addition of elemental information to the least-squares analysis one can impose physical constraints. For example, each weight fraction and the sum of weight fractions must satisfy

$$0 \leq X_i \leq 1$$

and

$$0 \leq \Sigma X_i \leq 1 \tag{7}$$

Equations 7 are known as inequality constraints. The user may also require that the solutions satisfy the equality constraint

$$\Sigma X_i = N \tag{8}$$

where N is a positive number equal to or less than 1.0. At the user's option these constraints can be imposed on the least-squares solution. To implement equations 7 and 8 the LSEI package (11) for constrained least squares has been used.

EXAMPLE ANALYSIS

A synthetic fly ash sample was prepared by blending finely ground mullite, quartz, hematite and glass (50:15:10:25). This mixture exhibits few resolved single phase lines and many overlapping lines. Each of the crystalline phases were separately mixed with a known weight fraction of silicon as an internal standard and then spray dried. AUTO run files were created for collection of the calibration data for the three binary mixtures. Each resulting data set was processed by program RUNFIL. Example results for the quartz plus silicon mixture are shown in Tables 1 and 2. Note that both peak height and integrated intensity measurements were used (in this case as a test). Phase names are restricted to six alphanumeric characters. After the average relative intensity value, <I(rel)>, Table 1 lists: (1) the estimated standard deviation of <I(rel)>, E.S.D., which is based solely on the propagation of counting statistical errors; (2) the root mean square deviation; RMS DEV, of the individual I(rel) values from the average; and (3) the number of replicate observations, #. In Table 2, the reference intensity ratio for each quartz line averaged over all silicon lines is given under <RIR>. The final row of Table 2 lists the overall average RIR value using all five lines of quartz, all three lines of silicon and all three mountings. The silicon reference lines 111 and 311 were used throughout the analysis. No systematic trends in the results were present. Table 3 shows the analytical results as determined by program QUANT (no overlap lines used). These results are in good agreement with the known weight fractions in the synthetic mixture except for the hematite 104 line. Because of an unspecified overlap with the mullite 220 line the hematite analysis based on the 104 line is in error. The mullite 220 line was inadvertantly omitted when the run file for the binary mixture mullite plus silicon was generated. Use of multiple lines helped to quickly detect this error and avoid reporting an incorrect analysis. The results from program COBRAG are quite similar.

Table 1. Summary of Relative Intensities

Phase	H K L	Method	<I(rel)>	E.S.D.	RMS DEV	#
Quartz	1 0 0	Integrt	17.65	.02	.63	3
Quartz	1 0 1	Integrt	100.00	.00	.00	3
Quartz	1 0 1	Peak Ht	93.63	.06	1.09	3
Quartz	1 1 1	Integrt	3.01	.03	.06	3
Quartz	2 0 0	Integrt	4.83	.02	.06	3

Phase	H K L	Method	<I(rel)>	E.S.D.	RMS DEV	#
Silicn	1 1 1	Integrt	100.00	.00	.00	3
Silicn	3 1 1	Integrt	30.70	.06	1.32	3
Silicn	3 1 1	Peak Ht	22.53	.03	.41	3

Table 2. Summary of Reference Intensity Ratio Determination

Phase	H K L	Standard	H K L	<RIR>	E.S.D.	RMS DEV	#
Quartz	1 0 0	Silicn	All	.9278	.013	.038	9
Quartz	1 0 1	Silicn	All	.9278	.001	.018	9
Quartz	1 0 1	Silicn	All	.9285	.005	.017	9
Quartz	1 1 1	Silicn	All	.9279	.010	.026	9
Quartz	2 0 0	Silicn	All	.9282	.007	.017	9
Quartz	All	Silicn	All	.9278	.001	.017	45

Table 3. Synthetic Fly Ash Analysis

Analyte	hkl	#	X_i (exper.)	RMS DEV.	X_i (known)
Mullite	110	6	0.452	0.030	0.425
Quartz	100	6	0.146	0.015	0.128
Hematite	012	6	0.089	0.002	0.085
Hematite	104	6	0.142*	0.002	0.085
Hematite	214	6	0.080	0.002	0.085

*See explanation in text.

REFERENCES

1. Smith, S.T., Snyder, R.L. and Brownell, W.E., 1979, Minimization of Preferred Orientation in Powders by Spray Drying, Adv. X-Ray Analy., 22:77.
2. Calvert, L.D., Sirianni, A.F., 1980, A Technique for Controlling Preferred Orientation in Powder Diffraction Samples, J. Appl. Cryst. 13:462.
3. Calvert, L.D., Sirianni, A.F., Gainsford, G.J. and Hubbard, C.R., 1983, An Interlaboratory Comparison of Methods for Reducing Preferred Orientation, Adv. X-Ray Analy., 26:000.
4. McCarthy, G.J., Gehringer, R.C., Smith, D.K., Injaian, V.M., Pfoertsch, D.E. and Kabel, R.L., 1981, Internal Standards for Quantitative X-Ray Phase Analysis: Crystallinity and Solid Solution, Adv. X-Ray Analy., 24:253.
5. Szabo, P., 1978, Optimization of the Measuring Time in Diffraction Intensity Measurements, Acta Cryst, A34:551.
6. Snyder, R.L., Hubbard, C.R. and Panagiotopoulos, N.C., 1982, A Second Generation Automated Powder Diffractometer Control System, Adv. X-Ray Analy., 25:245.

7. Snyder, R.L., Hubbard, C.R. and Panagiotopoulos, N.C., 1981, AUTO: A Real Time Diffractometer Control System, NBS Internal Report 81-2229, National Bureau of Standards, Washington, D.C. 20234
8. Morris, M.C., McMurdie, H.F., Evans, E.H., Paretzkin, B., Parker, H.S., Panagiotopoulos, N.C. and Hubbard, C.R., NBS Monograph 25 - Section 18, National Bureau of Standards, Washington, D.C. 20234.
9. Hubbard, C.R., Evans, E.H., and Smith, D.K., 1976, The Reference Intensity Ratio, I/I_c, for Computer Simulated Powder Patterns, J. Appl. Cryst. 9: 169.
10. Copeland, L.E. and Bragg, R.H., 1958, Quantitative X-Ray Diffraction Analysis, Anal. Chem., 30:196.
11. Hanson, R.J. and Haskel, K.H., Sandia Laboratories Tech Reports SAND77-0552 (1978) and SAND78-1290 (1979).

APPENDIX 1. Derivation of Equation 4.

For a flat specimen thick enough to give maximum intensity, the integrated intensity of reflection i for component α in a mixture m is given by

$$I_{i\alpha} = \frac{K_{i\alpha} X_\alpha}{\rho_\alpha \mu_m^*} = C_{i\alpha} X_\alpha \tag{A1}$$

where μ_m^* is the mass absorption coefficient of the mixture (Alexander and Klug, 1948. Anal. Chem. 20; 886). Hubbard, Evans and Smith (9) showed that

$$K_{i\alpha} = K\mu_\alpha \gamma_\alpha \; I_{i\alpha}^{rel} \tag{A2}$$

where K is a constant, μ_α is the linear absorption coefficient for phase α and γ_α is a scale factor relating the relative intensity scale to the absolute intensity scale. From equations A1 and A2

$$C_{i\alpha} = \frac{K\mu_\alpha \gamma_\alpha I_{i\alpha}^{rel}}{\rho_\alpha \mu_m^*} \tag{A3}$$

A similar equation can be written for C_{ns}. Dividing $C_{i\alpha}$ by C_{ns}

yields

$$\frac{C_{i\alpha}}{C_{ns}} = \left(\frac{\mu_\alpha \gamma_\alpha / \rho_\alpha}{\mu_s \gamma_s / \rho_s}\right) \frac{I_{i\alpha}^{rel}}{I_{ns}^{rel}} \tag{A4}$$

The term in parenthesis in equation A4 was shown to be the reference intensity ratio, $RIR_{\alpha,s}$ (9). Substituting $RIR_{\alpha,s}$ into equation A4 gives equation 4.

SCRIP - FORTRAN IV SOFTWARE FOR QUANTITATIVE XRD

Edward R. Wong
John Yeko
Philip Engler
Richard A. Gerron

Sohio Research Center
4440 Warrensville Center Road
Cleveland, Ohio 44128

INTRODUCTION

X-ray diffraction is ideal for quantitative crystalline phase determination in solid mixtures. Each phase of the mixture will produce its characteristic lines independent of other phases that may or may not be present when diffracting an X-ray beam and will be "proportional" to the amount of the phase present.

Since Klug and Alexander's work in the 40's quantitative X-ray diffraction techniques have abounded. Quantitative X-ray diffraction (QXRD) techniques may be broken down into five main categories[1]:

1. External standards (or direct method)
2. Internal standard (or indirect method)
3. Spiking or doping method
4. Dilution method
5. "No" standards method

Although QXRD internal standards techniques are well published,[2,3,4] several problems do exist[5]:

1. Reduction of analyte concentration
2. Inhomogeneities of mixing of dopant or diluent

3. Nonlinearity of diffraction ratios with respect to concentration
4. Required time for multi-standard calibrations (i.e. sample preparation and data acquisition)
5. Selection of additions that do not interfere with the analyte

With these thoughts in mind, efforts to develop and implement software utilizing an external standard technique were undertaken; SCRIP (Sohio Core Rock Identification Program), a FORTRAN IV program resulted.

THEORY

SCRIP, a program requiring approximately 12K words of memory, utilizes an iterative algorithm to directly solve a system of linear equations. Although SCRIP was written on a PDP 11/23 running V 3.2 of RSX11M, SCRIP is compatable with systems supporting FORTRAN IV. Also, although developed for core rock analysis, SCRIP may equally be applied to analysis of any powder mixture.

The original algorithm[6] encompasses the basic equations by Klug and Alexander[4] formulating X-ray peak intensities. From Klug and Alexander we know that:

$$\frac{I_{ij}}{(I_{ij})_o} = \frac{\chi_j \mu_j}{\overline{\mu}} \qquad (1)$$

where I_{ij} and $(I_{ij})_o$ represent the intensities of the ith line of the jth component of the mixture and pure phase respectively; χ_j is the weight fraction of the jth component in the mixture; and μ_j and $\overline{\mu}$ represent the mass absorption coefficients of the jth component and the sample respectively. The μ's were calculated using a simple weighted average technique[7] for each phase comprising the mixtures:

$$\mu_j = \sum_{i=1}^{N} \chi_i \mu_i \qquad (2)$$

where χ_i is the atomic or elemental weight fraction of each element in the jth component and μ_i is the elemental mass absorption coefficient. With this basic formulation, one, with rather minimum input, could calculate an intensity for each of the components. However, as noted in many recent publications,[8,9] consideration of grain size and surface contributions to the measured intensity must be given. In fact, with some thought, it is quite apparent that crystallites on the surface of a sample do indeed deserve differential treatment from those in the bulk of the mixture. The diffracted beam intensity from grains directly on the surface of the sample is not subjected to reabsorption from other grains within the sample as are grains in the bulk. One may write an intensity equation[6] for the total observed peak area percent intensity, to include these considerations:

$$\frac{I_j^T}{(I_j^T)_o} = \frac{I_j^S}{(I_j^S)_o}(SP) + \frac{I_j^B}{(I_j^B)_o}(1-SP) \quad (3)$$

where S, and B represent the intensities from the surface and bulk grains respectively, T represents the total peak area percent intensity and SP signifies the calculated percentage of the diffracted beam intensity due to the surface grains. Hence $I_j^S(SP)/(I_j^S)_o$ represents that portion of the measured intensity due to grains on the surface and $I_j^B(1-SP)/(I_j^B)_o$ represents that portion of the measured intensity from the bulk of the sample.

Incorporating these equations, SCRIP iteratively calculates an intensity based upon weight fraction, χ_i, to within a specified error window. SCRIP performs the following functions:

1. Assigns an initial "guess" for the weight fractions of each component.
2. Checks for any amorphous component depending on what flags have been set by the user. If any amorphous component is found, values for its μ and ρ are calculated.
3. Standardizes the input intensities to 100% to correct for self-absorption if the flags dictate doing so.
4. Calculates surface and bulk intensities using the current weight fraction values.
5. Calculates a total I for the ith line of the jth component.
6. Applies the user specified error window comparing I_{calc} to I_{obs} and either generates a new χ_i or displays the results of its iterations.

SOFTWARE

Parameters for SCRIP are entered on line with the program. Essential parameters for SCRIP are:

Item 1: Number of phases present
Item 2: The error window
Item 3: Median particle size of the mixture in centimeters
Item 4: Mass absorption coefficient, density, analyzing angle and name of each phase

Item 5: Ascii run name
Item 6: Flags for handling amorphous components
Item 7: Peak area percent intensities

At this point, it is necessary to discuss two crucial parameters used by SCRIP, namely particle size and peak area percent intensity, in greater detail.

As the treatment of X-ray intensities from grains on the surface and in the bulk of the sample differ, accurate sizing of the grains is essential. Scrip assumes a model of layers of uniformly sized grains. SCRIP equates the path length of the beam through each layer to the cosecant of the analyzing angle times the particle size. To illustrate the importance of particle sizing, a 43.6% quartz - 56.4% muscovite mixture whose median particle size was 3 µm was analyzed inputting 2,3,4,5,6,7 and 10 µm as the particle size. The data in Fig. 1 show the results. This range of particle sizes resulted in calculated weight fractions from 41.0% to 44.1% for quartz and 55.9% to 59.0% for muscovite. It is important to note that the analysis using the correct particle size yielded the calculated weight fractions closest to theoretical values. Although not terribly large, these errors must be minimized as they are not the sole source of error for the software. Average particle sizing of grains may be obtained by either Coulter counter analysis or image analysis.

Peak area percentage intensities are calculated from diffraction data using the rather obvious equation below:

$$\frac{I_{ij}}{(I_{ij})_o} \times (100\%) = \%\ \text{intensity} \qquad (4)$$

where I_{ij} represents the ith line peak area of the jth component in the mixture and $(I_{ij})_o$ represents that same i^{th} line of the pure j^{th} component. It is crucial to note here that peak areas and not peak heights are utilized. Using peak areas compensates for differences in line breadth between the sample and external standards. Common causes of these differences are discrepancies in degree of crystallinity and crystallite size for minerals coming from different formations.

SOFTWARE EVALUATION

SCRIP was evaluated by quantitatively analyzing five different weight fraction mixtures of quartz, dickite, muscovite, kaolinite, calcite and/or calcium montmorillonite. Each component was crushed to pass through a 325 mesh screen and weighed before physical mixing in a SPEX Mixer/Mill. Each mixture was then rear loaded into an

DATA TABLE 1

MIXTURE	ACTUAL WT.%	CALC. WT.%	ABSOLUTE ERROR
QUARTZ	51	50	-1
DICKITE	49	50	1
QUARTZ	44	43	-1
MUSCOVITE	56	57	1
QUARTZ	72	73	1
KAOLINITE	28	27	-1
QUARTZ	44	44	0
DICKITE	35	34	-1
MUSCOVITE	21	22	1
QUARTZ	25	25	0
KAOLINITE	25	19	-6
ILLITE	30	32	2
CALCITE	5	9	4
MONTMORILLONITE	15	15	0

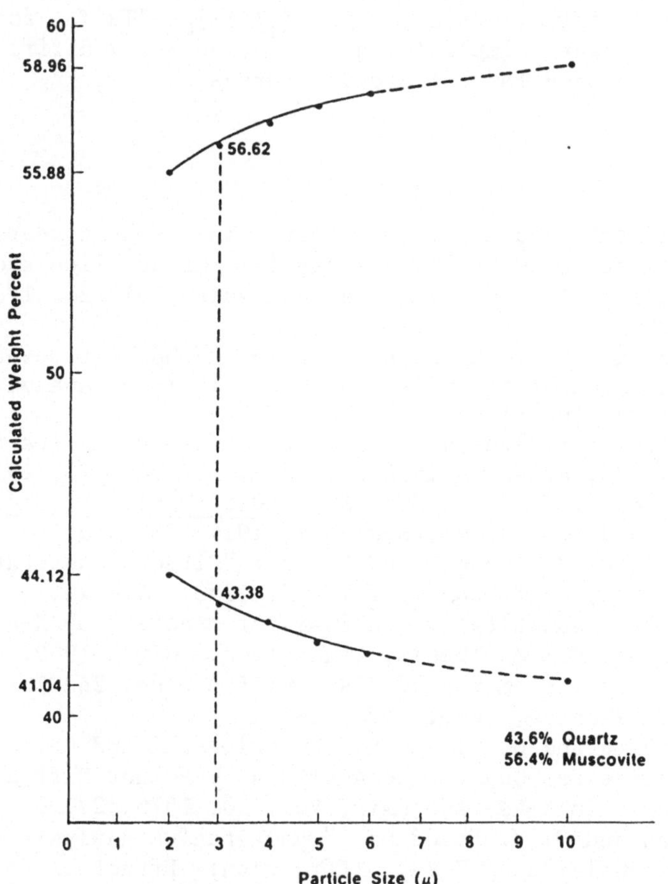

Fig. 1. Calculated weight fraction vs. particle size.

aluminum sample holder and examined with CuK_α radiation. It is important to note that the mixtures and their associated external standards were prepared by identical methods. Also, to eliminate preferred orientation in the top layer, a razor blade was scraped over the front of the aluminum holder to minutely disturb the top surface of the specimen. Particle sizing of the mixtures was provided by Coulter counter analysis. Table 1 shows the results of SCRIP's analysis of these mixtures.

CONCLUSION

A method of quantitatively analyzing core rock samples by X-ray diffraction utilizing external standards has been proposed. This software is not restrictive in its coding to rock analysis, but is applicable to any powder phase mixture lending itself to X-ray diffraction. The software is also written to be compatible with any computer system supporting FORTRAN IV. The computed mineral analyses of known mixtures have shown good correlation with the weight percentages to within 5% or less. The diffraction data measurements needed for this software are μ_j, ρ_j, $I_j/(I_i)_o$, 2THETA, and particle size. This software simplifies quantitative X-ray diffraction by eliminating the need for calibration curves.

REFERENCES

1. G. J. McCarthy and R. C. Gehringer, "Internal Standards for Quantitative X-Ray Phase Analysis: Crystalline and Solid Solution", Advances in X-Ray Analysis, Vol. 20, 1980, 253-264.
2. F. H. Chung, "A New X-Ray Diffraction Method for Quantitative Multicomponent Analysis:, Advances in X-Ray Analysis, Vol. 17, 1973, 106-115.
3. L. E. Alexander, "Forty Years of Quantitative Diffraction Analysis", Advances in X-Ray Analysis, Vol. 20, 1976, 1-13.
4. H. Klug and L. E. Alexander, X-Ray Diffraction Procedures, John Wiley and Sons, Inc., New York, 1974, 531-565.
5. J. Leroux, D. H. Lennox, and K. Kay, "Direct Quantitative X-Ray Analysis", Anal. Chem., Vol. 25, 1953, 740-743.
6. J. D. Yeko, "Quantitative Analysis of Minerals by X-Ray Powder Diffraction", MS Thesis, Texas Tech. Univ., 1980.
7. B. D. Cullity, Elements of X-Ray Diffraction, 2nd Edition, Addison-Wesley, Inc., 1978, 1-31.
8. C.R. Hubbard and D. K. Smith, "Experimental and Calculated Standards for Quantative Analysis by Powder Diffraction:, Advances in X-Ray Analysis, Vol. 30, 1976, 27-39.
9. E. Gavish and G. M. Friedman, "Quantitative Analysis of Calcite and Mg-Calcite by X-Ray Diffraction: Effect of Grinding of Peak Height and Peak Area,: Sed., Vol. 20, 437-444.

AN X-RAY DIFFRACTION STUDY OF $CaNi_5$ HYDRIDES USING IN SITU HYDRIDING AND PROFILE FITTING METHODS

G.J. Gainsford*, L.D. Calvert,
J.J. Murray and J.B. Taylor

Chemistry Division,
National Research Council of Canada
Ottawa, Canada. K1A 0R9

INTRODUCTION

The $CaNi_5$ system is unique among the AB_5-H systems in that it exhibits three distinct hydrides (β,γ,δ) at pressures below 65 atm. (Sandrock et. al., 1982). The present diffraction study was designed to characterise these phases, only one of which has been previously studied. Nowotny (1942) gave lattice parameters for a number of AB_5 (Haucke) phases including $CaNi_5$. Takéuchi et al. (1966) gave parameters for $CaNi_5$. Buschow (1974) reported on the entire Ca-Ni system. Oesterreicher et al. (1980) gave data on $CaNi_5$ and $CaNi_5H_{5.5}$. Enslen et al. (1981) indexed $CaNi_5H_{5.5}$ (γ-phase) as orthorhombic. Measurements made at NRC on one of the samples studied by Sandrock et. al (1982) and on three samples prepared at NRC are given in Table 1 together with the data from the literature. It is clear that there is a significant sample effect on the observed lattice parameters. It is well known that many AB_5 compounds have a range of composition. To avoid variations due to sample effects it was decided to characterise the hydride phases on a single specimen. In addition a special in situ hydriding apparatus was designed to avoid possible complications if "stabilized" samples were studied (Gualtieri et al., 1976). The use of in situ hydriding also made it possible to make a diffraction study of the activation process for which many explanations have been advanced (Schlapbach et al., 1980). After this work was completed Yoshikawa and Matsumoto (1982) reported results of in situ hydriding studies on $CaNi_5$. Their results agree generally with ours but the small scale graphs make numerical comparisons impractical.

* Guest Worker: Permanent address: DSIR, Petone, New Zealand

Table 1. $CaNi_5$ Lattice parameters

Year	Author or specimen	a,Å	c,Å
1942	Nowotny et al.	4.960	3.948[a]
1966	Takéuchi et al.	4.930	3.925[b]
1974	Buschow et al.	4.955[c]	3.941
1980	Oesterreicher et al.	4.950	3.936
1982	Yoshikawa et al.	4.943	3.942
*	T 80839[d]	4.9376(8)	3.9508(6)
*	F193	4.9468(5)	3.9433(5)
*	F163	4.9541(8)	3.9353(8)
*	F149	4.9516(6)	3.9373(5)

a Å from kX units
b Also quoted as 3.926
c Misprinted as 4.055
d Prepared by Ergenics Inc; used by Sandrock et al.
* This work, samples not activated

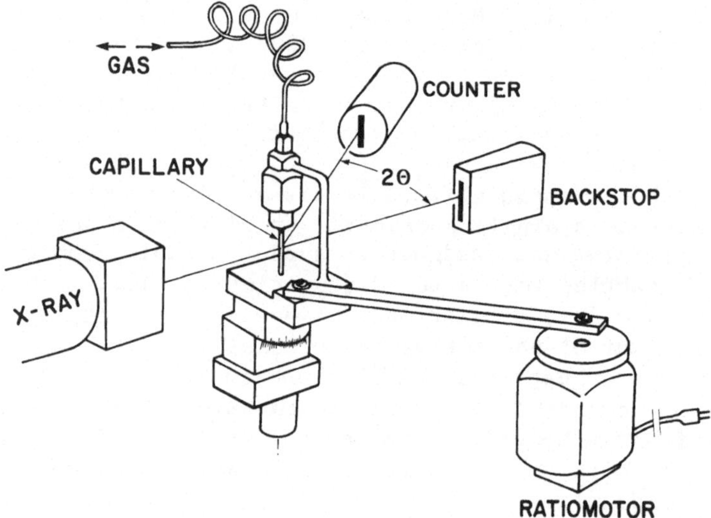

Fig. 1 The apparatus used for in situ hydriding studies.

EXPERIMENTAL

The samples were prepared by vapour phase reaction of 99.99% Ca and Ni in degassed stainless steel containers which were filled with a partial atmosphere of argon, welded shut and then annealed at temperatures up to 950°C. All sample handling was carried out in an argon-filled dry box with continuously monitored oxygen and water concentrations of less than 2 p.p.m. by volume. Samples were characterised by chemical analyses, spark-source mass spectrometry, optical metallography and X-ray diffraction. They had overall chemical purities of 99-99.9% with up to 1% free Ni in some samples. Alloys were crushed in the dry box and loaded into the special pressure cell which could be sealed and removed from the diffractometer. The apparatus is illustrated in Fig. 1. The pressure cell was comprised of a thin-walled glass capillary, which was glued into a steel sleeve, which in turn was held in a Swagelok fitting by nylon ferrules. This assembly, which could be sealed off with a vacuum-tight valve and then removed to the dry-box for sample loading, was connected to the fixed gas-handling system by a flexible spiral of stainless-steel capillary tubing. The gas handling system was similar to that of Murray et al. (1981) and the assembly has been tested to ~40 atm. The diffractometer was a horizontal one with a curved graphite monochromator in the incident beam; the beam was focussed on the detector slit, 130 mm from the monochromator. MoKα radiation was used and the data was collected and analysed using the profile-fitting algorithms of Sparks (1981) together with local programs for lattice parameter refinement. The data were first corrected for systematic errors by using an external calibration curve based on a mixed standard of Si, W and As_2O_3. Tests showed that lattice parameters could be measured to a precision of 1:30,000, angles to 0.006 - 0.013° and intensities to 2-6% ($\sigma \sim 0.04I$) for well crystallised materials; for the hydrides, lattice parameters were measurable from 1:1,000 to 1:10,000 and intensities to 5%.

Table 2. Lattice parameters β- $CaNi_5H_x$

x	a,Å	b,Å	c,Å	V,Å^3
.914	8.5883(18)	5.0818(13)	7.8467(30)	342.5(3)
1.058*	8.6088(19)	5.0930(11)	7.8466(22)	344.0(3)
1.058	8.6044(22)	5.0965(12)	7.8395(17)	343.8(2)
1.059	8.6077(17)	5.0954(9)	7.8427(18)	344.0(2)
1.100	8.6212(29)	5.1004(13)	7.8524(29)	345.3(3)

* deuteride

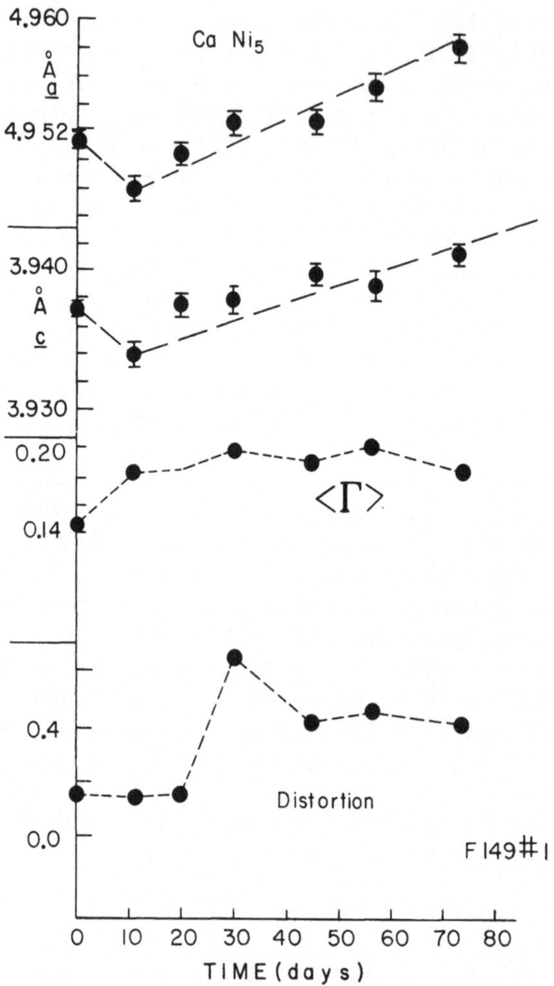

Fig. 2 The a and c parameters, the mean line-width and the "distortion" (see text) of $CaNi_5$ before activation and after each hydriding cycle. Where error bars are not given, the errors are less than the size of the plotted points.

THE $CaNi_5$ PHASE

The parameters of the starting material are given in Table 1 (Sample F 149). The specimen was activated by pressurising to 3.71 atmospheres of H_2 (γ-phase). After various measurements the sample was depressurised and pumped to remove H_2. The lattice parameters (Fig. 2) showed a distinct drop after this cycle followed by a slow rise to the value recorded after the 6th (final) cycle; the mean line width (Γ, FWHM) rose from 0.14°, after cycle 1, to 0.185° and stayed substantially constant (0.185(5)°). The other parameter, labelled "distortion", is the slope of the mean line obtained by plotting line-width (FWHM) agains $h^2 + k^2 + hk$ for each line of the $CaNi_5$ pattern; this is constant (~0.18) for two cycles and then rises to 0.5(1) after that. A similar or related effect has been reported by Achard et al. (1979) for $LaNi_5$. Several explanations are possible for this observation. It could be due to anisotropic particle shape, to varying a parameters (with c constant) or to a residual distortion in the cell, with a ≠ b. It is possible that these parameters are related to the activation process. Further studies are necessary to investigate these observations.

Table 3. Lattice parameters γ- $CaNi_5$ H_x

x	a,Å	b,Å	c,Å	V,Å^3
4.859	9.2437(26)	5.2609(14)	8.0802(26)	393.0(3)
5.142	9.2707(26)	5.2806(15)	8.0958(33)	396.3(4)
5.166	9.2792(23)	5.2846(23)	8.1036(29)	397.4(4)
5.166	9.2778(24)	5.2816(18)	8.1081(24)	397.3(4)
5.169	9.2750(40)	5.2801(38)	8.0954(42)	396.5(7)
*5.171	9.2731(33)	5.2808(22)	8.1353(57)	398.4(6)
5.52	9.3123(26)	5.3119(17)	8.1387(37)	402.6(4)

* deuteride

THE β-PHASE: $CaNi_5H_x$ (x ~ 1)

The β-phase was found to be orthorhombic, Im2m; parameters are given in Table 2. There is a range of parameters with composition (i.e. hydrogen pressure). The deuteride phase did not differ significantly from the hydride. The results of Yoshikawa and Matsumoto (1982) are in agreement with Table 2.

THE γ-PHASE: $CaNi_5H_x$ (x ~ 5)

This phase was also orthorhombic, Pmmm with a range of composition (Table 3). The deuteride has similar parameters although there is a possibility that the deuteride c- parameter differs. Yoshikawa and Matsumoto report a body-centred space-group.

THE δ-PHASE: $CaNi_5H_x$ (x ~ 6)

The δ-phase was indexed on a hexagonal cell with a range of parameters (Table 4). Our results are consistent with those of Yoshikawa and Matsumoto (1982). They report it to be trigonal; we have no evidence on this as yet.

Table 4. Lattice parameters δ- $CaNi_5H_x$(x~6)

p,atm.	a,Å	c,Å	V,Å^3
28.9	5.3424(11)	4.2478(7)	104.99(6)
28.9	5.3433(9)	4.2498(6)	105.08(5)
34.4	5.3502(9)	4.2547(9)	105.47(6)
34.5	5.3506(9)	4.2572(8)	105.55(6)
*13.6	5.3300(9)	4.2468(6)	104.48(5)
*13.6	5.3298(7)	4.2492(4)	104.54(4)
*28.5	5.3345(10)	4.2566(5)	104.92(8)

*deuteride

Table 5. Summary of data for $CaNi_5$-H Phases

Phase	$CaNi_5$	$CaNi_5H_{1.06}(\beta)$	$CaNi_5H_{5.17}(\gamma)$	$CaNi_5H_{6.2}(\delta)$
Space group or system	P 6/mmm	Im2m	Pmmm[a]	Hexagonal
a,Å	4.9516(6)	8.606(2)	9.277(2)	5.3504(9)
b,Å	-	5.096(1)	5.282(2)	-
c,Å	3.9373(5)	7.841(2)	8.102(6)	4.256(2)
V,Å^3	83.64(3)	343.9(2)	397.3(3)	105.50(6)
ΔV,%	-	2.8	18.8	26.1
ΔV/H, Å^3(b)	-	2.2	3.0	3.5

a Ensslen et al. (1981) gave a = 9.194, b = 5.250, c = 8.042 for the γ phase; Oesterreicher et al. (1980) gave the cell as hexagonal a = 5.305, c = 4.009Å.

b Yoshikawa (1982) gave ΔV/H as 2.3 (β), 2.5 (γ) and 3.0 (δ).

SUMMARY

The results for the four phases are compared in Table 5; average values have been rounded to the nearest significant digit and values from the literature are included. The most interesting observation is that the differential volume per hydrogen atom increases substantially with increasing hydrogen content, which is contrary to expectations. Yoshikawa and Matsumoto (1982) observe a similar change, although their values differ slightly from ours. The space groups for β and γ were assigned on the basis of model calculations. When these are completed they will be published elsewhere.

ACKNOWLEDGEMENTS

We thank G.J.G. Despault who prepared the samples.

REFERENCES

1. J.C. Achard, F. Givord, A. Percheron-Guegan, J.L. Soubeyroux and F. Tasset, (1979) J. de Physique 40: C5-218-219.

2. K.H.J. Buschow, (1974) J. Less-Common Metals 38: 95-98.

3. K. Ensslen, H. Oesterreicher and E. Bucher, (1981) J. Less-Common Metals 77: 287-9.

4. D.M. Gualtieri, K.S.V.L. Narasimhan, and T. Takeshita, (1976) J. Appl. Phys. 47: 3432-3438.

5. J.J. Murray, M.L. Post and J.B. Taylor, (1981) J. Less-Common Met. 80: 201-209.

6. H. Nowotny, (1942) Z. Metallkunde 34: 247-253.

7. H. Oesterreicher, K. Ensslen, A. Kerlin and E. Bucher, (1980) Mat. Res. Bull. 15: 275-283.

8. G.D. Sandrock, J.J. Murray, M.L. Post and J.B. Taylor (1982) Mat. Res. Bull. 17: 887-894.

9. L. Schlapbach, A. Seiler, F. Stucki and H.C. Siegmann, (1980) J. Less-Common Met. 73: 145-160.

10. R.A. Sparks (1981) Personal Communication, Nicolet Corp., 255 Fourier Ave.,Fremont, Calif. 94539, U.S.A.

11. Y. Takéuchi, K. Mochizuki, M. Watanabe and I. Obinata, (1966) Metall. 20: 2-8.

12. A. Yoshikawa & T. Matsumoto, (1982) J. Less-Common Metals 84: 263-271.

THE MEASUREMENT OF THERMALLY INDUCED STRUCTURAL CHANGES BY HIGH TEMPERATURE (900°C) GUINIER X-RAY POWDER DIFFRACTION TECHNIQUES

T. G. Fawcett, P. Moore Kirchhoff, R. A. Newman

The Dow Chemical Company
Midland, Michigan 48640

ABSTRACT

A new method for the collection and analysis of high temperature Guinier x-ray data has been devised at The Dow Chemical Co. This technique can be used to monitor various types of structural transformation and thermal expansions up to 900°C. The thermal expansions of α-Al_2O_3 and two TiO_2 structures, anatase and rutile, have been characterized for their use as high temperature internal standards.

INTRODUCTION

The use of microdensitometered, profile-fitted, calibrated Guinier films in obtaining high quality powder diffraction data has been discussed in several recent publications[1,2,3]. This method of analysis, when subjected to round robin studies, has provided reproducible results of high accuracy and precision[4,5]. This method is now interfaced to a high temperature (23-900°C) Guinier powder diffraction camera.

The ability to obtain both quantitative and qualitative data of high accuracy and precision at elevated temperatures depends on the development of high temperature standards. When used as internal standards these materials can be used to correct the data for many types of instrumental errors including sample displacement and film shrinkage errors. Since high temperature Guinier cameras are not in common use, it was necessary to measure the thermal expansion of several metal oxides for their future use as internal standards.

Lattice expansion measurements have been made on samples containing α-Al_2O_3 and the TiO_2 polymorphs, rutile and anatase.

A non-conventional data handling system was utilized in analyzing the data. This system will be discussed in detail.

EXPERIMENTAL

Samples of the TiO_2 polymorphs, rutile and anatase, and of α-Al_2O_3 were generated in-house at The Dow Chemical Company. These samples as analyzed by plasma emission spectroscopy were found to be 99.99% pure. The TiO_2 sample was a mixture of anatase and rutile while the α-Al_2O_3 sample was a pure phase of corundum. These samples had been previously characterized at room temperature by Guinier techniques described in this paper. NBS Si powder Standard Reference Material 640 was used as an internal standard. NBS SRM 640 Si was previously particle sized and a 6 micron fraction was used in the experiments. Particle sizing reduces orientation and microabsorption effects.

The oxide samples were ground to a fine powder in a mortar and pestle and subsequently blended with NBS Si. The mixture was then lightly pressed into a Pt-10% Rh 80 mesh gauze that was used as a sample holder. The 80 mesh Pt-10% Rh was obtained from Engelhard Industries and has a thickness of 76 microns. This sample preparation method has several advantages over conventional Pt loop sample holders in that the preparation is fast and binders do not have to be added to the sample matrix. In addition the Pt-10% Rh can be used as an internal standard to calibrate d-spacings and its thermally expanded d's can be used to verify the sample temperature. The sample and Pt-10% Rh gauze were then mounted on a goniometer for alignment in the Huber-Guinier high temperature camera.

The experimental setup and optical geometry are shown in Figure 1. A Huber-Guinier high temperature camera in the subtraction transmission mode was used in all the experiments. This camera is equipped with a curved focusing Ge monochromator. The monochromator separates CuK_{α_1} from CuK_{α_2} and CuK_{β} radiation emitted from a Philips x-ray generator equipped with a long fine focus Cu x-ray tube. When compared to conventional diffractometer and Debye-Scherrer techniques this optical geometry greatly reduces most instrumental effects which contribute to assymmetric peak line broadening[6]. Vertical divergence is reduced through the use of a focusing geometry, absorption and flat sample errors are reduced by using thin samples in a transmission geometry and the use of monochromatic radiation reduces most x-ray source errors. Utilization of small grained photographic films (CEA, Reflex 15), coupled with the use of internal standards reduces receiving slit width and misalignment errors commonly associated with electronic detection systems.

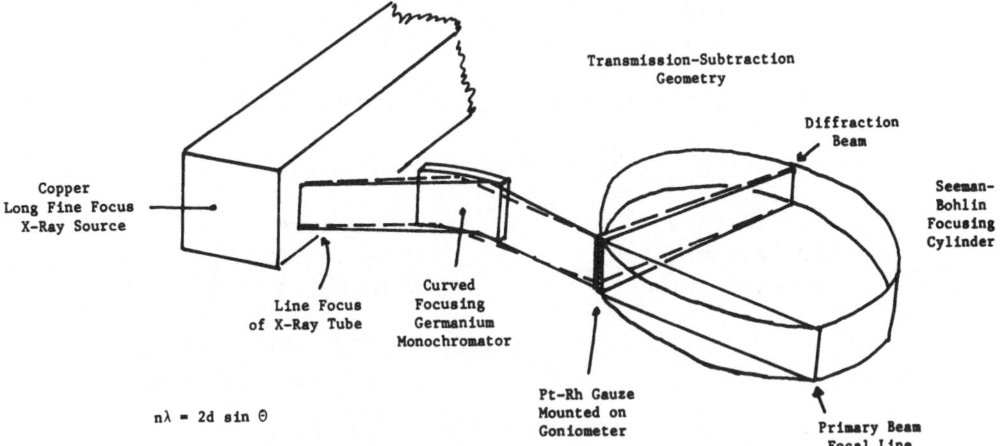

Figure 1. Huber-Guinier Diffraction System

The camera is microprocessor-controlled and records up to 16 exposures at any given temperature increment and time interval for a single sample. A scanning mode can also be used if desired. The camera is entirely self-contained and vacuum sealed. In the current experimental setup the experiment can be conducted in air, vacuum or N_2 atmosphere.

Figure 2 outlines the data reduction system used to process the high temperature data. This system has been the subject of several previous publications so the system will not be discussed in detail. The profile fitting program was developed by Edmonds and Brown[1]. The peak profiles are split and separately refined on either side of the peak centroid. Profiles are fitted to peak distributions which can be varied from Lorentzian to Gaussian by the following equation.

$$I_\theta = I_o \ (1 + k^2 \ x^2)^{-n}$$

k = scale factor related to halfwidth of the profile
I_θ = intensity at angle θ
I_o = peak maximum intensity
x = angular distance $\theta_o - \theta$ from peak position
n = variable (n = 1 Lorentzian, n = 24 Gaussian peak)

Since the samples had different crystallite sizes and chemical histories they did not fit the same peak profiles. In most cases the Pt-10% Rh, TiO_2 and Si profiles were best fitted with a modified Lorentzian shape (n = 2-4 in the previous equation). The α-Al_2O_3 peak profiles were best fitted with a more Gaussian peak shape.

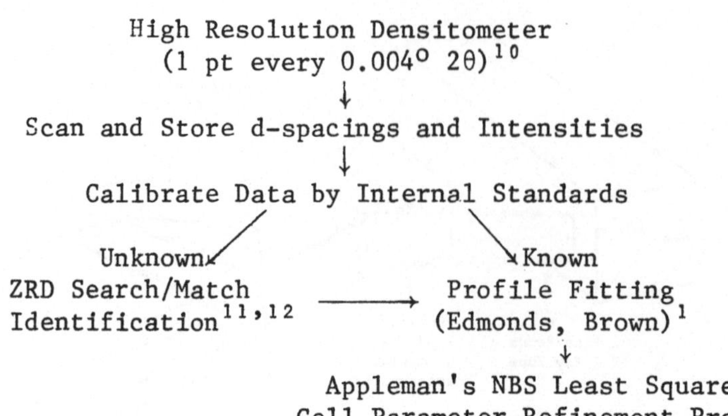

Figure 2. Data Reduction Flow Chart

Unweighted conventional R factors ranged from 7 to 0.5% for the individual integrated peaks when the peaks were at least three times the noise level. Typical peak widths and R factors are listed in Table I.

Cell dimensions were refined using Appleman's NBS Least Squares Cell Parameter Refinement program. The film cassette has a recording range of 0-60° 2θ when used in the subtraction-transmission mode with $CuK_{\alpha 1}$ radiation. Therefore, 6, 4, and 7 hkl reflections were used to refine the cell parameters of corundum, anatase, and rutile respectively. By comparing the high temperature data with a 0-60° 2θ range to room temperature data on a conventional Huber-Guinier system with a 0-160° 2θ range we found that the small number of reflections used in the high temperature experiments increased our cell edge error uncertainties by a factor of 10 in the case of α-Al_2O_3.

Table I. Full Peak Widths at Half Maximum Height (FWHM, Deg. 2θ) and Rf Values For Various Phases

Phase	Range of FWHM	Range of R Values
NBS Si	0.088-0.123	1.8-2.2
Pt-10% Rh	0.095-0.129	3.0-4.0
α-Al_2O_3	0.069-0.158	0.55-3.3
α-SiO_2	0.075-0.143	1.1-3.3
TiO_2-rutile	0.088-0.116	2.0-7.0
TiO_2-anatase	0.090-0.129	3.8-7.5

The room temperature (23°C) cell data for α-Al_2O_3 using 33 reflections over a 160° 2θ range are as follows: R3C, $a_o = b_o$ = 4.75942(3)Å, c_o = 12.9919(1)Å, vol = 254.87Å^3, Z = 6. When 6 α-Al_2O_3 reflections are used over a 60° 2θ range the cell parameters are as follows: $a_o = b_o$ = 4.7588(4)Å with c_o = 12.991(2)Å.

Experiments were run using a double exposure technique. Room temperature exposures of NBS Si were run on successive data tracks then the sample of interest was placed in the same sample position and heated. Sample positions before and after each experiment were monitored visually using a mounting cross-haired telescope which fixed the samples x and y direction. The z direction was fixed by aligning the sample with a small scored mark on the camera oven. The camera oven was not moved either during or between experiments.

Temperatures were calibrated by the use of a secondary thermocouple. The thermocouple can be mounted on a goniometer and aligned to the sample position by the aid of the mounting telescope. Temperatures were calibrated before and after each experiment. The secondary thermocouple was used to map temperature gradients around the oven and camera and used to monitor temperature fluctuation during heating and cooling cycles. These studies resulted in two practical changes, first, a 2 mm collimator was used to reduce the sample area irradiated for more accurate temperature control, secondly, 30 minutes were allowed between exposures to enable the sample to come to thermal equilibrium. When the camera unit is sealed, temperature stability is ∿±1° from 23 to 400°C and ∿±2°C from 400 to 900°C. Exposure times vary from 1-5 hours depending on sample crystallinity. Longer exposure times are needed in the high temperature experiments since the x-rays have to pass through a beryllium window on the camera and two coatings of plastic which surround the camera oven. The plastic prevents heating of the film cassette and motor drive during the experiments.

DISCUSSION

Initial experiments were run on the anatase-rutile TiO_2 system. These samples were chosen because high purity samples were readily available and there was a literature study on the thermal expansion of the polymorphs. Figures 3 and 4 show thermal expansion data taken from 23-700°C on both polymorphs. This data is compared to data taken by Rao, Naidu and Iyengar[7] on pure TiO_2 samples using a diffractometer with a back-reflection geometry. As shown, both polymorphs have anisotropic linear expansions over the cited temperature ranges. Even though the cell edges all expanded at different rates the volume expansions per TiO_2 molecule were identical for both polymorphs.

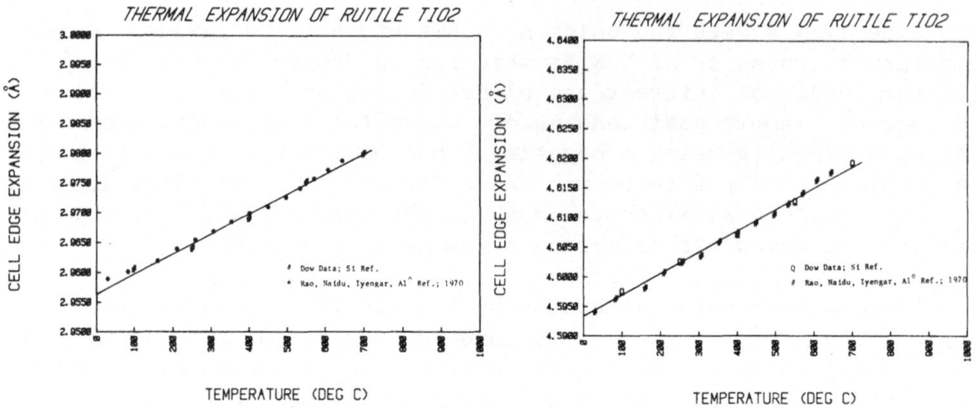

Figure 3. Thermal Expansion of the a and c Cell Edges of Rutile.

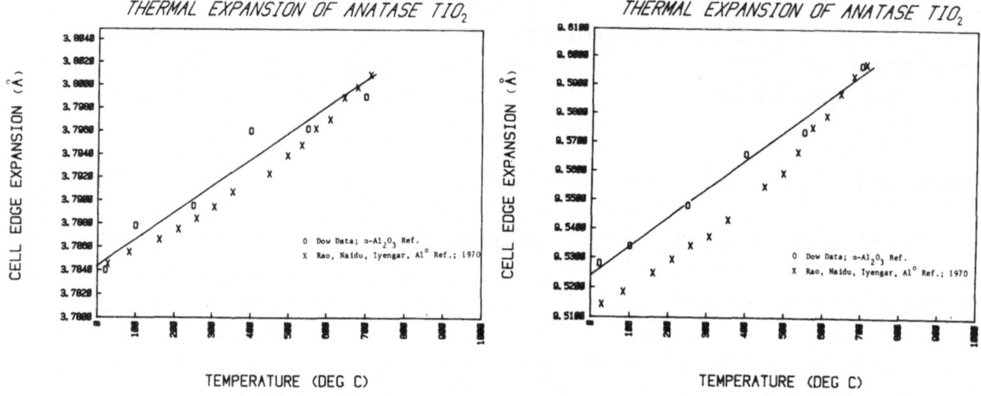

Figure 4. Thermal Expansion of the a and c Cell Edges of Anatase.

The data for rutile is nearly identical to the literature studies, however our data show a linear expansion for anatase while the previous literature exhibits a small non-linearity. Correlation coefficients for the linear expansions range from 96 (anatase) to 99.9% (rutile). Table II shows typical cell edge data for rutile and anatase. Average cell edge errors are in the range of 3-8 x 10^{-4}Å. A change in temperature of $\pm 1^{o}$ results in a fluctuation of 7 x 10^{-5}Å and 6 x 10^{-5}Å on the a and c cell edges of rutile, respectively. Thus the least squares cell edge errors are approaching the practical thermal cell edge errors imposed on the system by the temperature stability of the oven and surrounding environment. As previously mentioned, the least squares cell edge errors are slightly higher than usual due to the small number of reflections used in the limited 2θ of the experiment.

Table II. Refined Cell Parameters for Rutile and Anatase

TiO_2-Rutile	Temp. (^{o}C)	a_o(Å)	c_o(Å)
	22	4.5963(3)	2.9597(6)
	100	4.5975(5)	2.9612(8)
	250	4.6025(6)	2.965(1)
	400	4.6071(3)	2.970(1)
	550	4.6127(3)	2.9756(6)
	700	4.6192(9)	2.981(1)
TiO_2-Anatase	22	3.7831(4)	9.528(1)
	100	3.7859(7)	9.533(3)
	250	3.789(1)	9.544(4)
	400	3.796(1)	9.553(5)
	550	3.7963(1)	9.5721(1)
	700	3.7999(1)	9.5965(1)

The linear isotropic thermal expansion of the cubic phase Pt doped with 10% Rh was also measured. This alloy has a linear thermal expansion coefficient (α) of 10.6 x $10^{-6}/^{o}$C in the range of 22-800°C. Doping Rh into the Pt causes a shrinkage in the samples unit cell parameters since Rh has a smaller metallic radius than Pt (1.34Å vs 1.39Å)[8,9]. At room temperature the Pt-10% Rh alloy, which is used as a sample holder, has a unit cell dimension of 3.9025(5)Å in the cubic space group Fm3m. The hlk reflections 111 and 200, which expand linearly, can be used to either calibrate the other d-spacings at high temperature or be used as an internal temperature check on the sample temperature.

All the TiO_2 and Pt-10% Rh data sets were independently calibrated by both NBS Si and α-Al_2O_3. Only one standardized data set is reported in the Figures and Tables since the data from the two

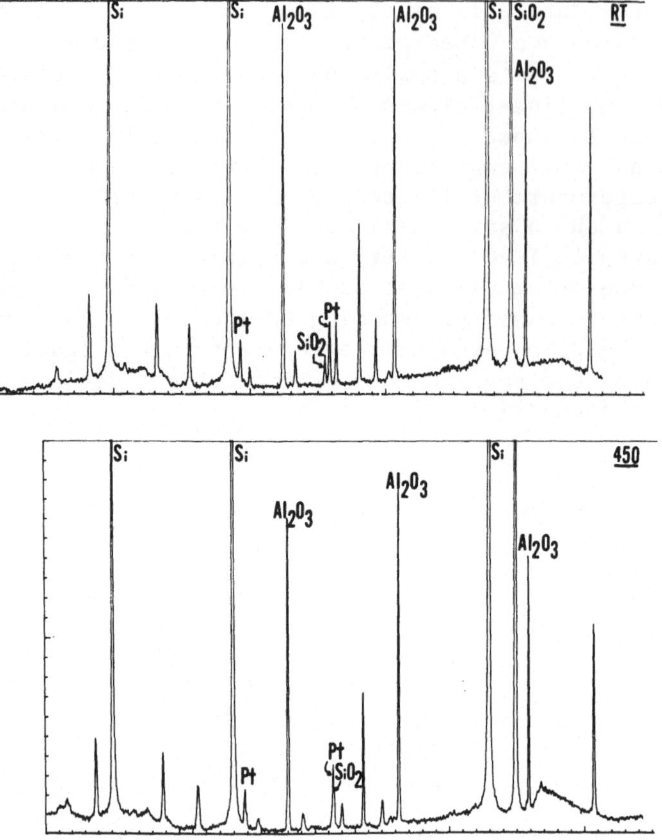

Figure 5. Densitometered data taken at 23°C (top) and 450°C (bottom).

data sets were nearly identical. Correlation coefficients calculated on an HP-85 statistical computer package were unable to differentiate between data sets standardized by the two internal standards on the basis of a linear regression fit on the thermal data from 22-700°C.

The thermal expansion of corundum (α-Al_2O_3) was measured in the range of 0-900°C. α-Al_2O_3 is hexagonal with room temperature cell parameters of a_o = 4.7588(4)Å and c_o = 12.992(2)Å. Thermal expansion data from 22-900°C were best fitted through the use of a second degree polynomial in the form

$$a_t = a_o + AT + BT^2$$

a_t = cell dimension at temperature T
a_o = cell dimension at 0°C
A,B = linear regression coefficients

α-Al_2O_3 expands anisotropically, for the a cell edge a_o = 4.7576Å with A = 1.20 x 10^{-5} and B = 3.03 x 10^{-8}, and for the c cell edge c_o = 12.991 with A = 8.81 x 10^{-6} and B = 1.43 x 10^{-7}. The non-linearity of the data may limit the utility of α-Al_2O_3 as an internal standard particularly at temperatures below 300°C. From 300-900°C α-Al_2O_3 expands close to linearity in both the a and c dimensions. The α-Al_2O_3 data was calibrated by NBS Si and experiments were run both in vacuum and N_2. Figure 5 shows densitometered data scans taken at room temperature (23°) and 450°C. The data have been scaled to emphasize the weaker reflections. The sharpness of the diffraction maxima is due to the removal of several instrumental aberrations through the use of a focusing subtraction transmission optical geometry. The figures also show the thermal expansion of α-SiO_2 (Minusil 5), however the data for this phase has not been processed at this time. The thermal expansion in Figure 5 can be observed by the merging of the α-SiO_2 d_{111} maxima with the room temperature Pt d_{200} diffraction maxima. At room temperature these reflections are distinct singlets with a Δd separation of 0.018Å and at 450° these peaks are a doublet with a Δd separation of 0.006Å.

CONCLUSIONS

We have interfaced a high temperature Guinier camera to a sophisticated data reduction package. With this system we have calibrated several materials which will be used as internal standards in high temperature experiments.

REFERENCES

1. A. Brown and J. W. Edmonds, The Fitting of Powder Diffraction Profiles to an Analytical Expression and the Influence of Line Broadening Factors, Adv. in X-Ray Analysis, 23:361 (1980).
2. J. W. Edmonds and W. W. Henslee, Application of Guinier Camera, Microcomputer Controlled Film Densitometry, and Pattern Search-Match Procedures to Rapid Routine X-Ray Powder Diffraction Analysis, Adv. in X-Ray Analysis, 22:143 (1979).
3. R. L. Snyder and J. W. Edmonds, Symposium on Powder Data Collection and Analysis, XII IUCr Congress, Ottawa, Canada (1981).
4. A. Brown, J. W. Edmonds and C. M. Foris, Reproducibility and Precision of Measurements of Guinier Powder Patterns Using Powdered Silicon Calibrant, Adv. in X-Ray Analysis, 24:111 (1981).

5. A. Brown, Symposium on Powder Data Collection and Analysis, XII IUCr Congress, Ottawa, Canada (1981), also see articles by A. Brown in Adv. in X-Ray Analysis, 26 (1983).
6. H. P. Klug and L. E. Alexander, Diffractometric Powder Technique: Profiles and Positions of Diffraction Maxima in: "X-Ray Diffraction Procedures", John Wiley and Sons, N.Y., N.Y. (1974).
7. K. V. K. Rao, S. V. N. Naidu and L. Iyengar, Thermal Expansion of Rutile and Anatase, J. Am. Ceramic Soc., 53:124 (1970).
8. W. B. Pearson, "A Handbook of Lattice Spacings and Structures of Metals and Alloys", Pergamon Press, N.Y., N.Y. (1958).
9. A. F. Wells, Metals and Alloys in: Structural Inorganic Chemistry", Clarendon Press, Oxford University, London, England (1962).
10. L. K. Frevel, Automated Measurement of Powder Diffraction Patterns, Anal. Chem., 38:1914 (1966).
11. L. K. Frevel, C. E. Adams and L. R. Ruhberg, A Fast Search-Match Program for Powder Diffraction Analysis, J. Appl. Cryst., 9:199 (1976).
12. J. W. Edmonds, Generalization of the Frevel ZRD-Search-Match Program for Powder Diffraction Analysis, Norelco Reporter, 27:22 (1980).

THE USE OF X-RAY DIFFRACTION AND INFRARED SPECTROSCOPY TO CHARACTERIZE HAZARDOUS WASTES

Douglas S. Kendall

Environmental Protection Agency
National Enforcement Investigations Center
Denver, Colorado

The National Enforcement Investigations Center of the EPA provides support services for the enforcement activities of the Agency. Recently, we have analyzed hazardous wastes as part of efforts to enforce the Resource Conservation and Recovery Act. Inorganic analysis often consists of elemental determinations. However, sometimes it is necessary to identify specific compounds. Some wastes are classified as hazardous by their origin, for instance, the wastes from the manufacture of several chromium pigments. On occasion, it is necessary to trace wastes to their source. In cleaning dumps and treating wastes, knowledge of the compounds present is useful.

Like x-ray diffraction (XRD) patterns, infrared (IR) spectra are highly characteristic of inorganic phases. This paper compares XRD and IR spectroscopy for the identification of inorganic compounds. IR spectroscopy has been effectively used for 30 years to analyze inorganic substances, including minerals. Recent advances have increased this effectiveness in ways which parallel those of XRD. The IR spectroscopist, like the diffractionist, can have an automated instrument which can collect high quality data and compare it with reference patterns on a disk.

The following examples are included as illustrations. It is not unlikely that an experienced analyst could have obtained much of the information from only one technique. The diffraction samples were scanned on a Philips APD 3600 and a proprietary search and match program was used to identify compounds. The IR spectra are of samples pressed into KBr pellets. In all cases, the identifications were aided by elemental analysis.

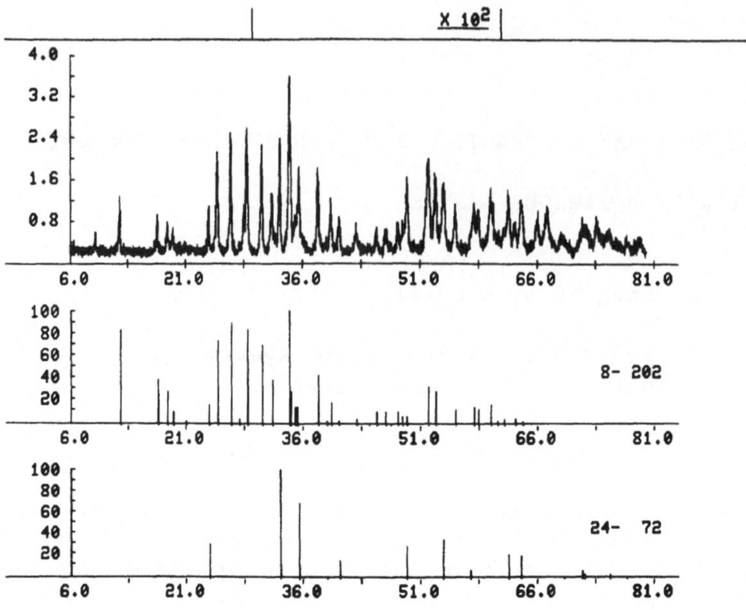

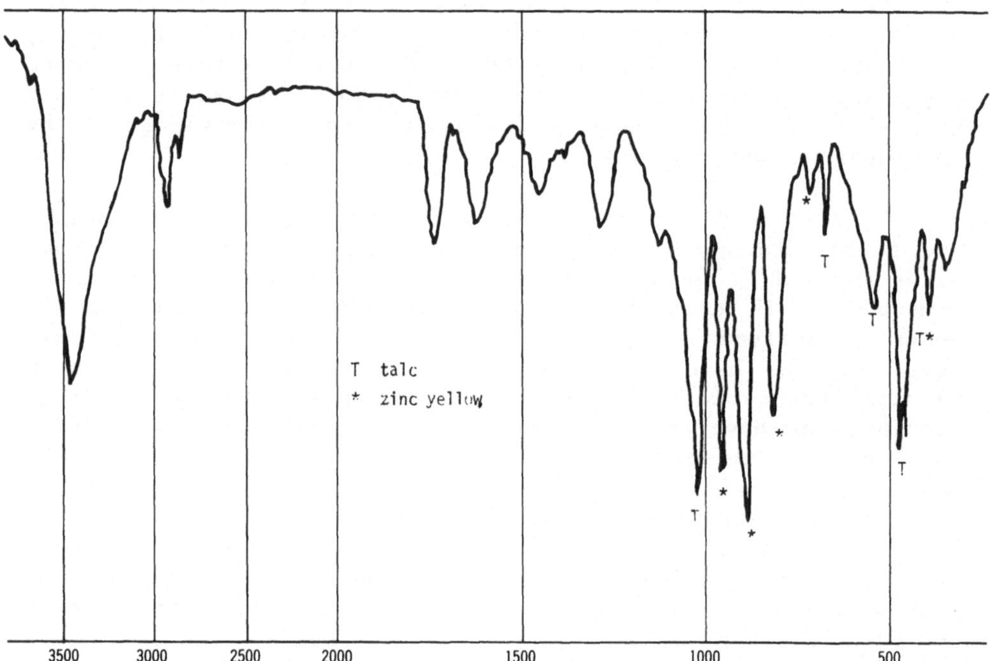

Fig. 1. (Top) Sample XRD pattern compared with zinc yellow (8-202) and ferric oxide (24-72). (Bottom) The IR spectrum of the same sample with the bands of zinc yellow and talc marked.

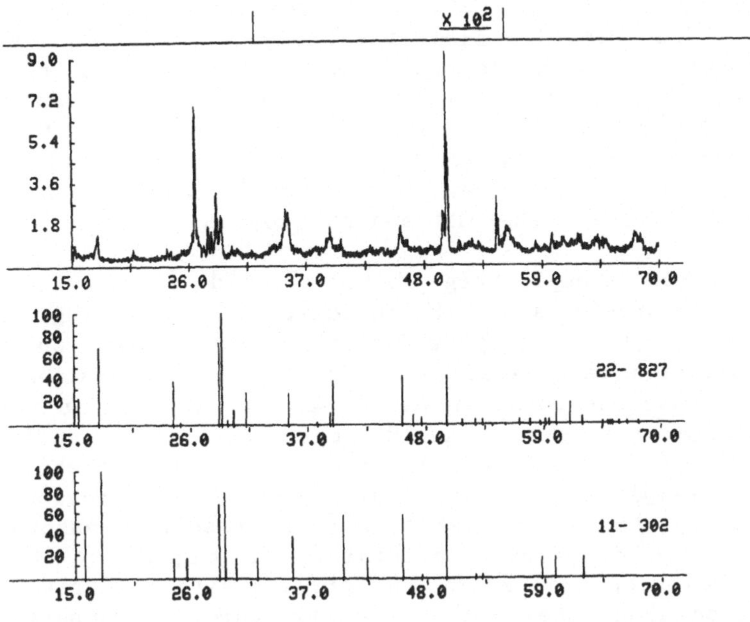

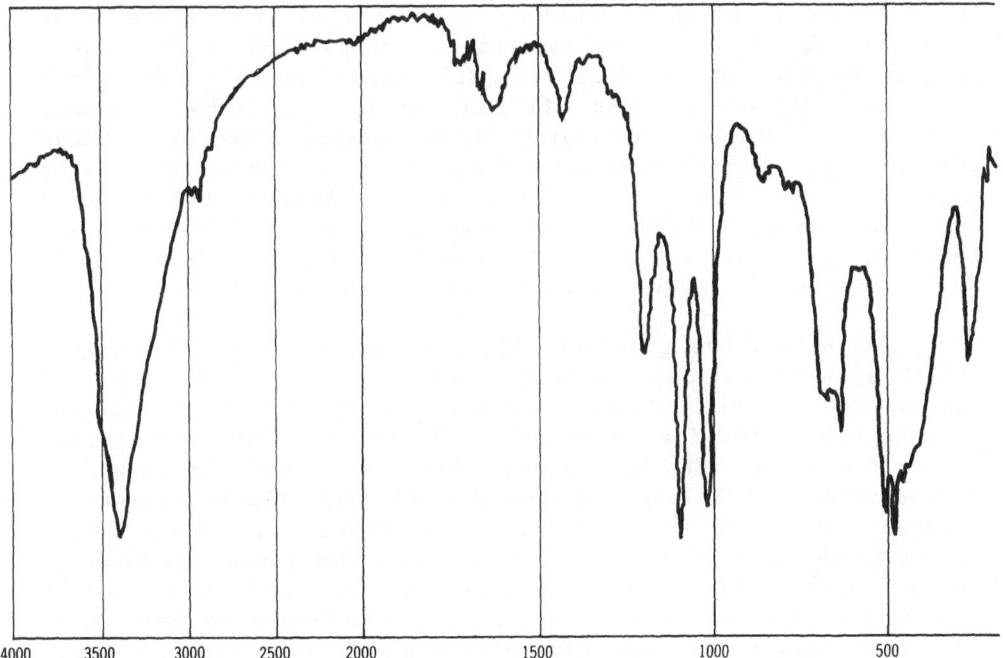

Fig. 2. (Top) The XRD pattern of the jarosite-like sample compared with jarosite (22-827) and natrojarosite (11-302). (Bottom) The IR spectrum of the same sample. The three sulfate stretching bands are near 1100 cm^{-1}.

The first example (Fig. 1) illustrates the complementary nature of XRD and IR. Elemental analysis showed that chromium was present. In both the powder diffraction pattern and the infrared spectrum the chromium pigment zinc yellow was easily identified. The sample also contained ferric oxide and talc. The ferric oxide was easily identified by XRD, but not by IR. In contrast, talc was quite evident in the IR spectrum, but not in the X-ray pattern. The XRD search program did not find talc.

The second example (Fig. 2) was a solid solution. IR helped to solve the problem since IR, in addition to providing a pattern for identification, also provides structural information, including information on crystal site symmetry. IR absorption is primarily by covalent bonds in polyatomic species, as in this sulfate containing example. The IR spectra of ordinary sulfates are characterized by a triply degenerate stretching vibration. However, the sample spectrum has three bands in this region. A detailed study[1] has shown that this type of sulfate absorption is characteristic of jarosite minerals. The degeneracy of the vibration has been removed by the crystal site. IR gives little information about the rest of the structure. A computer search of the XRD pattern identified jarosite, $KFe_3(SO_4)_2(OH)_6$, and a small amount of quartz. However, elemental analysis showed that less than 1% potassium was present. A comparison of the sample pattern to that of jarosite-like structures showed marked similarities. These compounds all had the $Fe_3(SO_4)_2(OH)_6$ grouping with various cations substituted for potassium. Several of these jarosite-type minerals were on the list of about 25 likely matches generated by the computer search. Without the infrared evidence, identification would have been much more difficult. The principal phase was a solid solution similar to jarosite with Cu, Al, Na and Zn substituted for most of the potassium.

Many substances give good XRD patterns but poor IR spectra, including many halides and many oxides. On the other hand, IR spectra can be obtained from amorphous substances. There are inorganic substances for which infrared provides more information. An example is provided by a sample which was predominantly sodium ferrocyanide. A strong absorption band, very characteristic of cyanides, was easily identified. Additional bands show that a ferrocyanide was present. In contrast, the powder pattern of sodium ferrocyanide has only two strong lines and was missed by the search program. It is important for analysts to know that, for some samples, such as hazardous wastes, which are completely unknown, infrared data can be an important addition to diffraction studies.

REFERENCES

1 *H. H. Adler, P.F. Kerr, Amer. Min.,* 50 *(1965) 132-147.*

COMPARISON OF X-RAY POWDER DIFFRACTION TECHNIQUES

K. Das Gupta

Radiation Research Laboratory

Texas Tech University, Lubbock, Texas 79409

Following the discovery of powder diffraction by Debye and Scherrer, a major improvement of the intensity of diffraction patterns was attained by introducing focusing cameras. The focusing principle is exploited in the Seemann-Bohlin camera and with improved general background in the Guinier focusing camera. Guinier used a bent crystal to focus Kα lines from the target radiation. Das Gupta, Schnopper and Metzger reported[1] a modification of the Seemann-Bohlin camera by replacing the slit with a target positioned on the focusing circle. This was further improved by using a microfocus target, 30 x 100 micron, positioned accurately on the focusing circle, Fig. 1 (a). The arrangement was very successful for determining thermal expansion coefficients of elements and a few superconducting alloys at different temperature regions. Results of thermal expansion coefficients of Nb, Sn and Nb_3Sn taken between room temperature and 200°C will be reported elsewhere along with measurements at the liquid nitrogen temperature region.

A significant improvement in intensity sacrificing the general background has been obtained by using a convergent Soller slit system of tantalum of thickness 0.007 inches. The target radiation is focused to a sharp line onto the focusing circle by using six concentric channels of a tantalum Soller slit system. The experimental setup is shown in Fig. 1 (b). We have also set up in symmetric position convergent Soller slit focusing diffraction cameras of diameters 5 cm and 15.4 cm. The diffraction pattern of amorphous and glassy materials, fibers, polymers, liquids have been obtained using films and an exposure time of 30 minutes when a copper tube is operated at 35 Kv, 20 mA.

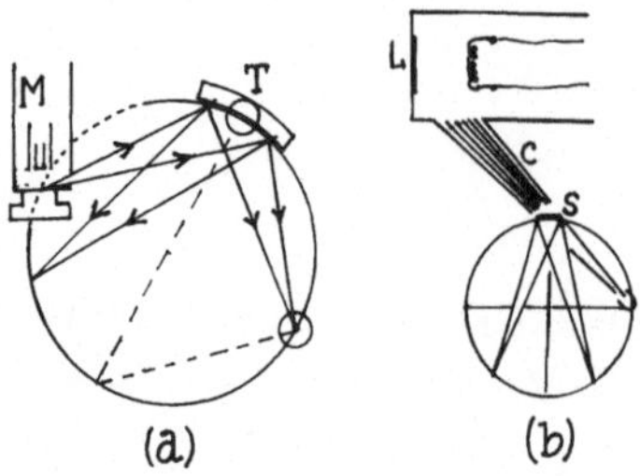

Figure 1 (a) Diffractometer with microfocus M on the focusing circle. The sample at T for measurements of thermal expansion coefficients at high and low temperatures.

(b) Focusing diffraction camera using the convergent Soller slit system C to focus the target radiation from L that passes through sample S.

We have set 2 flat crystals of LiF (200) in the dispersive (1,+1) position using a collimated beam of copper target radiation to obtain a sharp peak for CuKα_1. We have also set 2 flat crystals of Si(111) in the (1,+1) position to obtain CuKα_1 peak. The setting has been done on either side of the sealed off x-ray tubes. Contact prints of quartz powder with the LiF crystals and Si crystals are shown in Fig. 2 (a) and 2 (b) respectively.

Figure 2 (a) Diffraction of quartz powder with LiF(1,+1).

(b) Diffraction of quartz powder with Si(1,+1).

The $CuK\alpha_1$ diffraction pattern using double monochromatizing by two bent crystals of mica in succession, and in a different set of experiments with two bent crystals of silicon in succession, have been reported earlier.[2] The intensity of the diffraction pattern with two flat crystals is about three times stronger than that with two curved crystals. This is a rough estimation, made by knowing the time of exposure required to obtain the same intensity of the diffraction pattern using the same sample.

The work is supported by the Robert A. Welch Foundation.

REFERENCES

1. K. Das Gupta, H. Schnopper and A. Metzger, Advances in X-Ray Analysis, Volume 9, (Plenum Press, New York, N.Y., 1966) p. 221-241, U.S. Patent number 3,440,419, patented approved April 22, 1969.
2. K. Das Gupta, U.S. Patent number 3,379,876, patented approved April 23, 1968.

THE USE OF MULTI-SCAN DIFFRACTION IN PHASE IDENTIFICATION

Gordon S. Smith, M. C. Nichols

Chemistry Department,Lawrence Livermore National Laboratory, Livermore, CA 94550; Materials Science Department, Sandia National Laboratory, Livermore, CA 94550

ABSTRACT

Phase identification by X-ray diffraction techniques in a complex mixture would be greatly simplified if the component phases could be physically separated. As opposed to current computer search–match algorithms for phase identification, which presuppose a single diffraction scan on a carefully prepared sample, we propose multi-scan data-taking on a not-so-carefully prepared sample so as to exploit certain aberrations in the diffracted intensities. The result can effectively be a physical separation by diffraction. Examples include exploitation of samples having a preferentially oriented component as well as samples with components having differing crystallite sizes. The techniques can involve diffractometer as well as film techniques.

INTRODUCTION

It is well recognized that phase identification by X-ray diffraction would be greatly simplified if the component phases could be separated or at least differentiated from one another. Nichols and Johnson (1980) and Nichols, Smith and Johnson (1982) have proposed novel differentiation strategies which would provide phase characterization by making multiple scans of the same sample. Frevel (1982) has suggested possible restructuring of the powder diffraction data file so that groups of compounds can be rapidly differentiated.

We propose to extend the multi-scan strategy so as to deliberately exploit some of the aberrations that affect diffracted intensities, and that are usually minimized by careful and sometimes lengthy sample preparation. The result of the extensions proposed can result in practical separations of phases via diffraction.

BACKGROUND

The traditional method of data-taking in powder diffractometry involves a single scan over scattering angles using symmetric reflection geometry and no rotation of the sample within its plane. In addition, the diffraction data are typically recorded on a strip-chart and are hand reduced to d's and I's. This method presents several deficiencies; preferred orientation effects can drastically alter the intensities, and reproducible intensity measurements are impossible to obtain without a very finely powdered sample.

Most of the thinking going into current computer search–match algorithms is being influenced by the methodology described above. For example, most search–match algorithms tacitly assume only a single scan of the sample was made, and that preferred orientation and crystallite-size effects have been eliminated by the diffractionist.

With the advent of automation, digitized patterns are both more easily obtained and can be subsequently massaged by computer processing. Also, manufacturers are presently coming out with more versatile diffractometers that are capable of 1) transmission as well as reflection geometry, 2) sample rotation done routinely if desired, and 3) θ and 2θ driven separately.

With these advances, we believe it is possible to come up with new methods involving multi-scans of the same sample. These methods deliberately exploit intensity aberrations present in the traditional method, and will be of value in the phase identification of mixtures.

NEW METHODS FOR PHASE IDENTIFICATION

Example 1. Exploitation of preferred orientation in a loose powder by multi-scan diffractometry.

Figure 1 shows three powder patterns of a platy material (MoO_3) under different diffraction geometries. Preferred orientation effects are apparent. Suppose we had a mixture of two components, one having a platy habit and the other having a random arrangement of grains. We could take two diffraction scans on the same sample, one using transmission geometry and the other in reflection. Computer subtraction of the two digitized patterns would show the lines due to the platy material deviate from zero with both plus and minus excursions, whereas the lines due to the randomly oriented phase would tend toward zero. Thus we would have achieved a separation of the phases by diffraction. (Strictly speaking, the two patterns would also differ because of sample absorption; computer applied corrections for this difference could easily be made.)

It should also be noted that even though the phases were separated in this example, we could not readily identify the platy material (MoO_3) through use of today's search-match algorithms. This is because most such algorithms implicitly assume the absence of marked orientation. When the two diffraction patterns were added (transmission and reflection), a positive identification was made.

Example 2. Exploitation of preferred orientation in pressed samples.

Figure 2a diagrammatically shows what happens when a mixture of a platy material and a polyhedral material are uniaxially pressed. The platy material will tend to orient during compaction; the other phase will not. If we carefully grind or cut a flat on the cylinder wall (Figure 2b), two diffraction scans in reflection geometry can be made on each surface. The two digitized patterns are then subtracted as in Example 1. Diffraction lines due to the platy material will show excursions from zero; those due to randomized material will tend toward zero. Each separate set of lines are sent through a

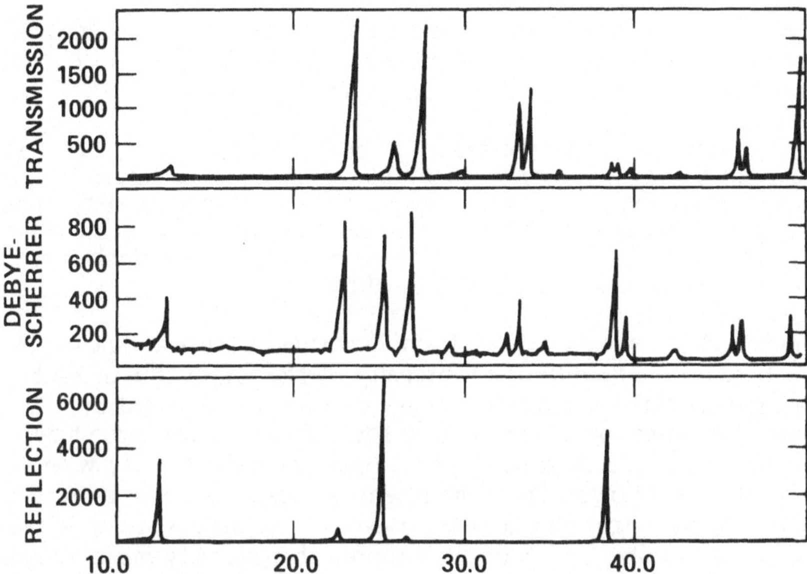

Fig. 1. Powder diffraction patterns for a material having a plate-like habit (MoO_3). Top and bottom spectra are diffractometry scans using transmission and reflection geometries, respectively; the middle spectrum is a trace of a Debye-Scherrer pattern.

Fig. 2. Preferred orientation effects in a uniaxially pressed sample. Crystallites of a platy phase, indicated by –, will orient during compaction; those due to a phase having a polyhedral habit, indicated by *, will not. Diffraction scans on the two surfaces in Figure 2b can be taken and compared.

computer search-match program for phase identification and a separation of phases by diffraction has been achieved. (Differences in the patterns due to dissimilar areas scanned can be computer-adjusted; it is simpler, however, to place a mask over each surface to achieve equal areas of diffraction.)

Example 3. Exploitation of grain-size differences by means of multi-scan diffractometry.

Suppose we had a mixture of two components; one coarse grained and the other fine grained (a frequent occurrence in actual practice). The sample could be a loose or a compacted powder sample.

Two diffraction scans are taken in reflection geometry, one with no sample spinning and the other with sample spinning. The intensities due to the coarse-grained component will vary greatly as a function of the position of the sample in the sample spinner, whereas the intensities due to the fine-grained phase will not. Thus we know which lines belong to which component and we can easily carry out a phase identification.

Figure 3 shows such a pattern from an actual mixture of quartz (44–74μ) and α-Al_2O_3 (0.5μ). The diffracted intensity of the major line of each phase was measured with the sample stationary in steps of 60° rotation. The corresponding intensities for a continuously rotated sample are also shown. The intensity due to the coarse-grained phase (quartz) varies greatly with sample orientation within its own plane; the intensity due to the fine-grained phase (alumina) does not. Figure 3 also illustrates that our proposed method should not be followed blindly. For example, suppose the intensity measurements of the non-rotated sample had been made at $\phi \approx 120°$ and were compared with the continuously rotated sample. Since the diffracted intensities for that particular line are so similar in the stationary and continuously rotated sample, one would incorrectly conclude that the line belongs to a finely divided phase. For this reason we recommend that the stationary intensity measurements be made at a number of rotation angles.

In practice it will of course be necessary to investigate the intensity fluctuations or lack thereof for a number of the major lines present in the composite pattern. With the ease of data-taking that is rapidly becoming available, this will be possible using computer control, although the film methods discussed next may turn out to be a preferred technique.

Example 4. Exploitation of grain-size differences by multi-diffraction film-methods.

Debye-Scherrer patterns were taken of a portion of the same mixture of quartz and alumina as in Example 3. Figure 4a shows a normal pattern from a capillary sample that has been rotated; Figure 4b shows a pattern when the capillary is not rotated.

As can be seen, diffraction lines from the fine-grained phase are smooth and continuous in both films. However, those from the coarse-grained material have "disappeared" in the pattern for which the sample was not rotated. The now-separated diffraction lines could have been presented to a computer search–match program for an easy phase identification of both substances.

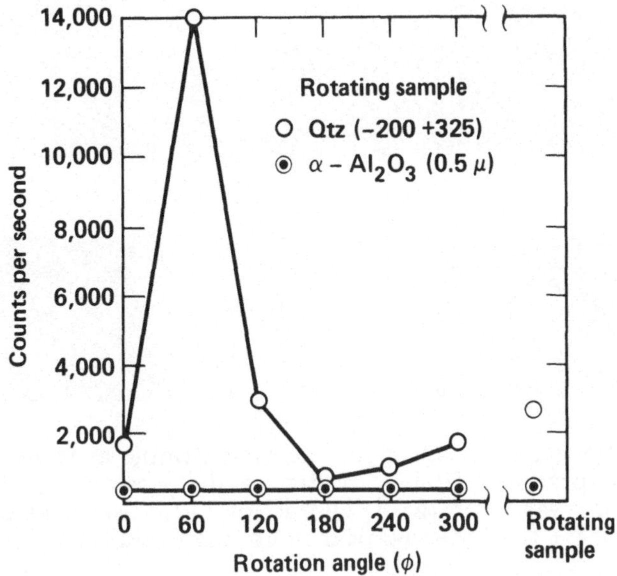

Fig. 3. Diffracted intensities vs. sample rotation angle, ϕ, for a coarse-grained component (quartz) and a fine-grained component (alumina). Intensities for the continuously rotated sample are also shown.

We showed these patterns informally to a colleague without divulging the nature of the sample. He immediately and unhesitatingly identified Figure 4b as α–alumina. This personal pattern recognition method did not work for the α–alumina in Figure 4a where confusion was created by the presence of lines from a second phase.

In Figures 5a and 5b we display the case of Debye-Scherrer patterns from materials having differing ranges of crystallite sizes — minus 325 mesh (–44μ) quartz and 0.5μ alumina. As in Figure 4, the capillary is first rotated normally (Figure 5a), and then not rotated (Figure 5b).

Again, diffraction lines due to the smaller grained material, α–alumina, are smooth and continuous in both patterns; lines due to quartz, the component with the larger crystallite sizes, tend to disappear or become spotty, just as was the case in Figure 4b where the quartz crystallites were even larger.

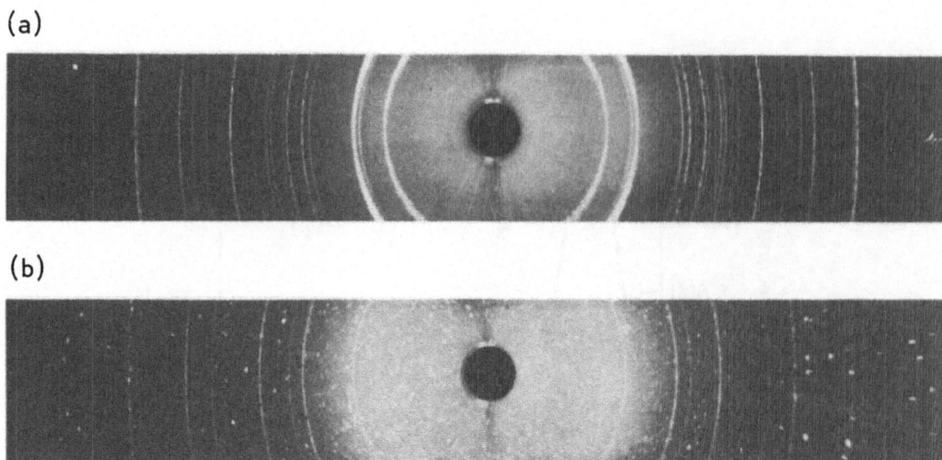

Fig. 4. Debye-Scherrer patterns for the same sample as in Figure 3. Figure 4a was produced from a rotated capillary sample; in Figure 4b the capillary was not rotated. Diffraction lines due to the coarse-grained component tend to disappear in 4b due to its larger size.

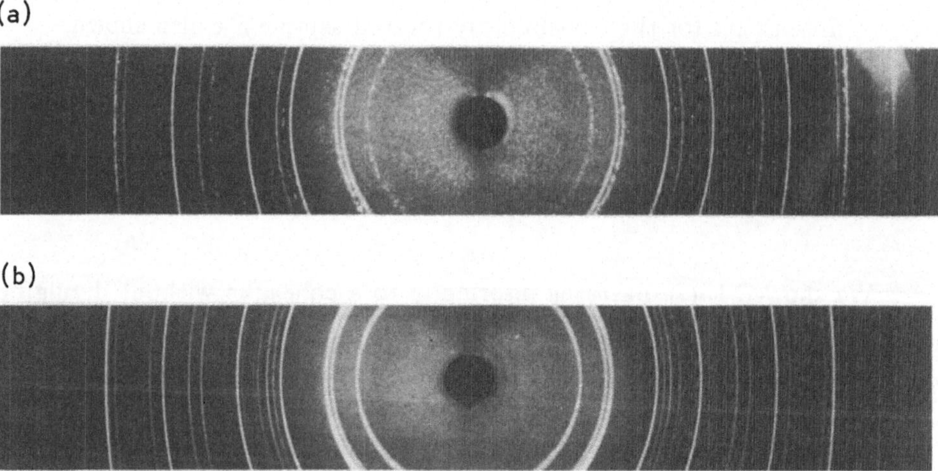

Fig. 5. Debye-Scherrer patterns for finely divided phases having differing ranges of crystallite sizes — -44μ quartz and 0.5μ alumina. The sample that produced Figure 5a was rotated; in Figure 5b the sample was stationary. The diffraction lines from the phase having the larger crystallite size and larger range of sizes (quartz) still tends to disappear or become spotty in Figure 5b in spite of the small particle sizes for both of the phases.

This is particularly interesting because conventional wisdom in sample preparation for X-ray powder diffraction is to grind until the material passes through a 325 mesh (-44μ) screen. However, as is readily seen in Figures 5a and 5b, these two phases can be readily separated by their diffraction patterns taken under different conditions of rotation.

It is somewhat ironic that graininess effects are so obvious in Debye–Scherrer photographs and are difficult to bring out in diffractometer geometry. On the other hand, the converse is true for preferred orientation effects.

OTHER POSSIBLE METHODS TO AID IN PHASE IDENTIFICATION

Once the possibility of phase differentiation is admitted, a number of other methods come to mind:

1) Multiple scans in Debye-Scherrer geometry employing coarse and fine collimators to bring out differences in graininess.

2) Multiple scans using different θ-values on a diffractometer having independent θ-drive to show up preferred orientation effects.

3) Multiple scans before and after cold-working of the sample could differentiate phases subject to cold-working from those which do not.

4) Physical and chemical separation methods: density gradient, magnetic separation, chemical leaching or selective dissolving, and of course, as indicated in the present examples, sieving and sizing of the as-received sample.

SUMMARY AND CONCLUSIONS

The standard practice for single-scan Debye-Scherrer analyses and especially for diffractometer analyses has been to grind the as-received sample to a fine powder to minimize errors due to preferred orientation and coarse crystallite size. This has required in many instances a careful and lengthy sample preparation.

However, by making multiple diffraction scans, we can sometimes exploit these effects to provide differentiation among component phases in a mixture, and can therefore obtain information that makes the interpretation of results much simpler.

The extent to which differences in preferred orientation and crystallite size can be detected and be useful for phase identification is not really known. The effects were quite obvious in our experimental examples, but these examples may not be typical. We do note, however, that Parrish and Huang (1982) have reported seeing crystallite size effects even at the 5–10μ size level.

It should be emphasized that with today's automation and greater versatility of diffractometer design, multi-scan powder data is relatively easy to obtain and manipulation of multiple patterns is easily done by computer.

We also believe multi-scan diffraction could have a positive impact on future developments in computer search–match algorithms. Instead of having a single figure-of-merit scheme for computer search–match, it will probably be useful in the case of multiple patterns of the same sample to have more than one figure-of-merit available at the phase identification stage. For example, a standard figure-of-merit could be used for a component having randomly oriented grains and a different one that downplays the importance of intensities for a phase showing non-random orientation.

ACKNOWLEDGEMENT

We wish to thank B. G. Nichols and D. R. Boehme for their help during the course of this work.

REFERENCES

Frevel, L. K., 1982, Structure-Sensitive Search–Match Procedure for Powder Diffraction, Anal. Chem., **54**, pp 691–697

Nichols, M. C. and Johnson, Quintin, 1980, The Search–Match Problem, in: "Advances In X-ray Analysis", Plenum Press, NY, Vol 23, p273

Nichols, M. C., Smith, D. K. and Johnson, Quintin, 1982, Differential X-ray Diffraction by Wavelength Variation: A Theoretical Basis, Submitted to J. Appl. Cryst.

Parrish, W. and Huang, T. C., 1982, Accuracy and Precision in X-ray Polycrystalline Diffraction, in: "Advances In X-ray Analysis", Plenum Press, NY, Vol 25.

AN AUTOMATED X-RAY DIFFRACTOMETER FOR DETECTION AND IDENTIFICATION OF MINOR PHASES

John C. Russ and Thomas M. Hare

School of Engineering
North Carolina State University
P.O. Box 5995, Raleigh, NC 27650

In the preceding volume in this series, we described a stepping motor automation package for XRD data acquisition, based on the Apple microcomputer with an interface card that produces the stepping motor pulses and counts X-ray pulses from the normal X-ray detector and amplifier. We are presently using the system to automate a General Electric XRD-5 diffractometer mechanism, by replacing the original synchronous motor with the stepping motor, and installing a 10:1 reduction gear instead of the chain drive to drive the double worm gears. This gives 2500 steps of the motor per degree two theta. Motor pulses are counted by a small machine language routine which also handles accelerating the mechanism when slewing over long angular ranges.

X-ray Diffraction spectra are collected either by step scanning over a single long scan (perhaps 10-15,000 points) or a series of shorter step scans over the segments of the entire two-theta range which are of interest. In either case, as the data are acquired, they are stored (as dead time corrected counts per second) on floppy disk. They are also displayed on the computer's video monitor. Once recorded, the displays can be reexamined or printed out on a hard copy graphics printer, in the form of strip chart recordings with scale markings and peak indentification.

Peaks can be located in the spectrum either by manual marking or automatic peak search. In the former case, the user views the displayed spectrum and positions a cursor using joystick or potentiometer knob. A multipoint parabola can be fitted to the data to locate peak centroids more accurately than the operator's eye.

Alternately, the automatic peak search tracks changes in derivative to find peaks, and then locates centroids in the same way. The user enters a minimum peak width (defined as a number of steps), and a minimum peak to background ratio, to select peaks for storage. Up to 100 peaks are then written into a disk file as a list of d-spacings and normalised intensities for subsequent analysis and identification.

This system is being used to study phase development in SYNROC, a multiphase ceramic intended for use as a host for radioactive waste from power reactors, to immobilize the many (30 or more) ionic species for geological periods of time. The major phases in the host, before waste is added, are Hollandite ($BaAl_2Ti_6O_{16}$), Perovskite ($CaTiO_3$) and Zirconolite ($CaZrTi_2O_7$). By careful balance of the raw materials, it is possible to control the production of these phases without other minor phases appearing in the diffraction patterns when no waste is added. However, the addition of 10-15 weight percent of the waste ions affects the crystal structure. Many of the waste ions occupy either interstitial or substitutional sites in the crystal structure of the three phases (indeed, this is the reason for their selection). However, this produces some change in the lattice spacings and a corresponding changes in the measured diffraction patterns. The nonuniform distribution of the elements produces broadening of peaks as well as shifts.

Of great concern for the intended use of this material is the possibility that additional phases might be formed due to the presence of the waste. If these phases were less stable than the desired ones, they might be attacked (for instance by water leaching out the soluble ions) and cause release of the ions into the biosphere. Because the distribution of waste ions into the various phases is far from equilibrium, even a small amount of such a phase could be associated with significant release of the hazardous ions. By determining the presence and amount of such unwanted phases, the raw material proportions, fabrication and sintering processes can be tailored to eliminate them.

Patterns (sets of d and I values) for all of the likely or possible oxide forms of the various cations were stored in disk files for comparison to the measured unknown patterns. Many were taken from standard published sources, including the JCPDS files, and others were measured on carefully prepared single phase materials produced in our own laboratory. Up to 100 such patterns can be stored on a 5" floppy disk. A search match program compares the measured spectrum to the stored patterns to determine which of them may be present in the unknown.

To perform a search match, the operator enters the number of peaks to compare and the minimum number which must be present in the unknown for the compound to be considered further. Figure 1 shows the results when a search is made using the three most intense peaks and requiring that all three be present. The phases found include the three expected ones (Hollandite, $CaZrTi_2O_7$, and $CaTiO_3$). The Al_2O_3 and and Na_2TiO_3 also appear possible based on visual comparison of the spectra. This is shown in Figure 2, comparing each of the phases to the unknown.

In the original search, a wide tolerance window for d-spacing mismatch was used. This is an operator parameter, which multiplies a tolerance curve that varies with d-spacing. The nominal acceptable error is +-.02 A at d=2.0, and in this example the multiplier was 1.5. Narrowing the window results in the rejection of more peaks. Figure 3 shows a plot of the figure of merit (which will be defined shortly) as a function of this multiplier. As the spectrum is shifted slightly in two theta (to compensate for misalignment in the spectrometer or change in the lattice parameter), the smallest error window which does not reduce the fitting figure of merit changes. For the zirconolite, a shift of -0.2 degrees permits the window multiplier to be reduced to 0.75. Figure 4 shows that a similar shift, probably due to spectrometer misalignment, can be used to improve the results for most of the candidate minor phases.

Using this shift, the search match was repeated using the most intense 18 peaks for each standard pattern. The results (Figure 5) show the same candidate phases and a few others. The figure of merit shown is the sum of the fractional intensities of the peaks which match. A value of 1 indicates that all peaks were found. Values are reduced from one when a line is missing from the pattern, but the amount of reduction is greater if a major line is missing than if a minor line is absent. Notice that the first three phases, with figures of merit greater than 0.9, are the three expected host phases, followed by the same two possible candidate phases.

Phases found	# Peaks	F. M.
HOLLANDITE	3	1.0
CAZRTI2O7	3	1.0
CATIO3	3	1.0
AL2O3	3	1.0
NA2TIO3	3	1.0
TIZRO4	3	1.0
ZRO2:CAO(85/15)	3	1.0

Figure 1: Results from search match requiring 3/3 most intense peaks to match.

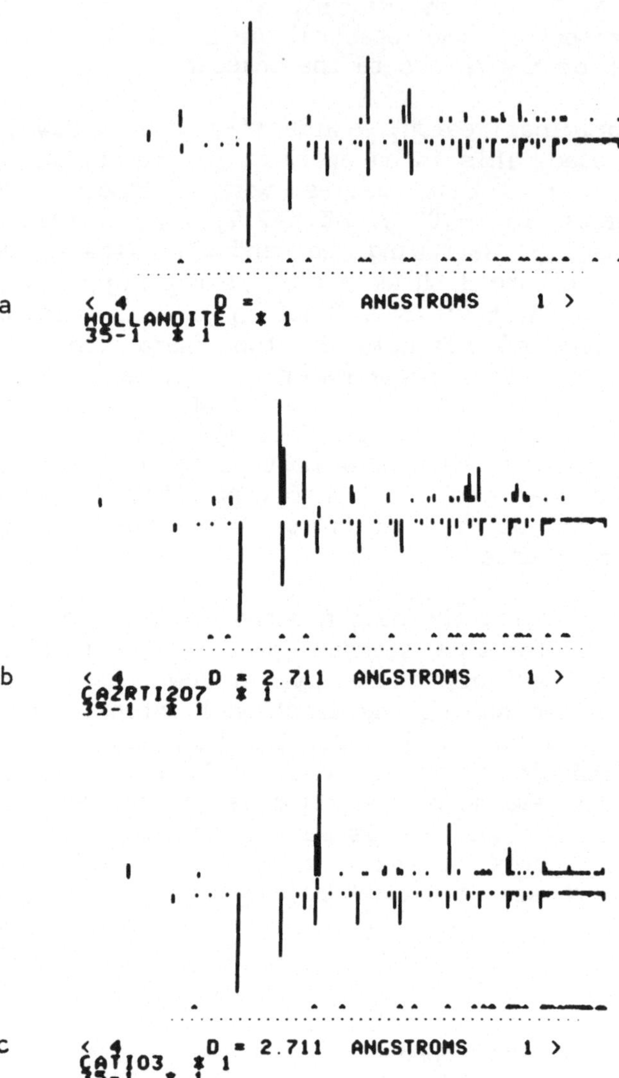

Figure 2: Comparison of measured spectrum (d,I data shown at bottom of each figure) with stored pattern for each matched compound. a) Hollandite, b) Zirconolite, c) Perovskite, d) Alumina, e) Sodium Titanate.

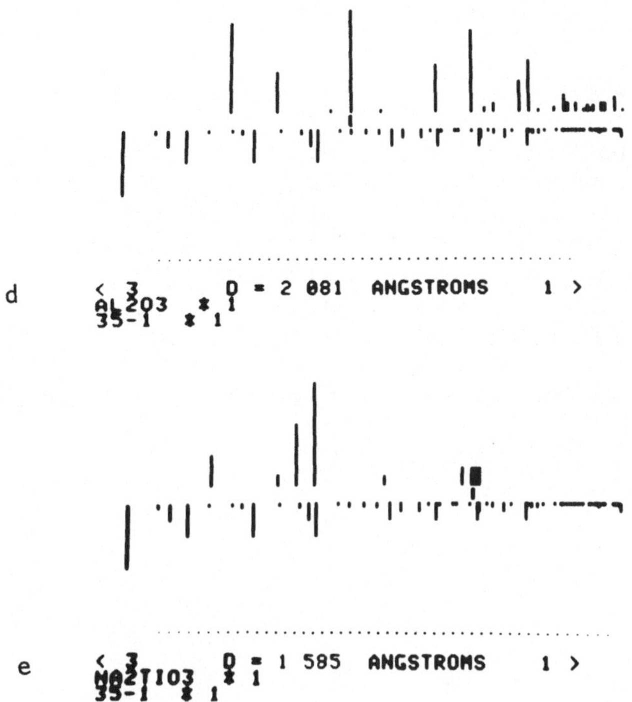

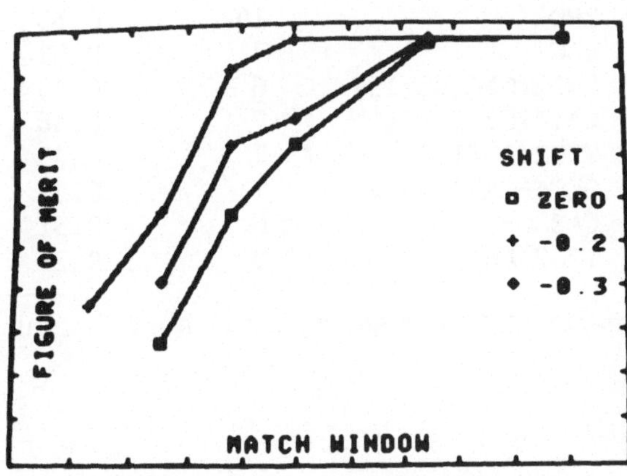

Figure 3: Variation of Figure of Merit with Window width multiplier for Zirconolite, as function of shift in pattern (angle in degrees two theta).

Phase	Offset	Multiplier
Hollandite	-0.1	0.5
CAZRTI207	-0.2	0.75
CATIO3	-0.2	0.5
AL2O3	-0.2	0.5
NA2TIO3	-0.2	0.75
TIZRO4	0	1.25
ZRO2:CAO(85/15)	-0.2	1.0

Figure 4: Optimum shift for each of the matched phases.

Phases found	# Peaks	F. M.
HOLLANDITE	17	0.97
CAZRTI207	14	0.90
CATIO3	13	0.90
AL2O3	14	0.79
TIZRO4	10	0.74
NA2TIO3	7	0.67
BATIO3	8	0.55
ZRO2	10	0.55
BAO	7	0.52
ZRO2:CAO(85/15)	6	0.48
CATI2O5	7	0.46
TIO2(RUT)	7	0.43
BETA-CA3(PO4)2	7	0.35
CAZRO3	6	0.31
TIO2(ANA)	9	0.28

Figure 5: Search Match results using most intense 18 peaks.

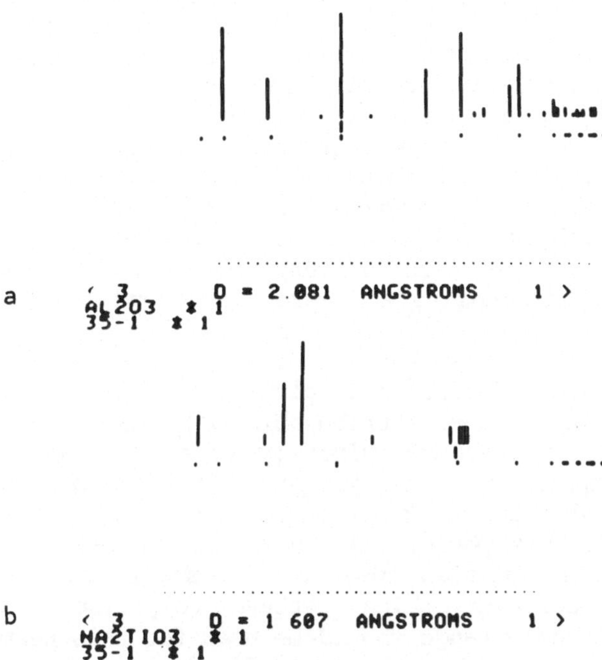

Figure 6: Comparison of candidate trace phase patterns to residual after stripping away of three major phases. a) Alumina, b) Sodium Titanate

Phase	Peak Ratio
HOLLANDITE	32/64
CAZRTI2O7	16/40
CATIO3	16/50
AL2O3	4/100
NA2TIO3	4/70
TIZRO4	8/15
ZRO2:CAO(85/15)	8/45

Figure 7: Intensity ratio of uniquely matched peaks for phases found by search match procedure.

Figure 6 shows the comparison of the patterns from these phases with the residue after stripping away the peaks from the three major host phases. Note that several of the major peaks are missing due to removal along with the overlapping peaks from the major phases. The amount of intensity to remove from a line is scaled by the ratio of intensity in the unknown to that in the standard for the major peaks. A somewhat better indication of the possible amount of the phases present is shown in the table in Figure 7, which gives the ratio of intensity of the peaks in the unknown to those in the standard for only the uniquely matched peaks (ones which only a single standard can account for). The numbers are consistent with the expected 50:25:25 proportion for the Hollandite, Zirconolite and Perovskite, and a trace amount of the Alumina.

The unequivocal detection of alumina is dependent on measuring a number of unique peaks attributable only to the major alumina peaks and distinct from the minor peaks of the identified major phases. By comparison, no unique peaks from the sodium containing phase were found with the full range scan. This did not rule out the presence of this phase, but rather identified the requirement for a second scan at slow speed over a small two-theta range to resolve the uncertainty. The pattern comparison in Figure 6 suggested which angle range should be used for this purpose, and in the scan no evidence of the phase was found.

For this material, the presence of about 1% of alumina (confirmed by quantitative metallographic techniques and point counting in the Scanning Electron Microscope) does not deleteriously affect the ceramic's properties. Another similar specimen showed the presence of a small amount of rutile (TiO_2), also using this method. The presence of the sodium titanate phase would have been extremely important because it is readily attacked by leachants.

This paper illustrates the utility of a rather simple, low cost computer control system for X-ray diffraction data acquisition and analysis, including search-match comparison of complex measured patterns to stored phases. Positive identification of minor phases can be made by combining the proper choice of simple search-match parameters with additional slow scan measurements over ranges selected with the aid of the analysis system.

TIME SHARE COMPUTER CAPABILITY FOR PHASE IDENTIFICATION BY X-RAY DIFFRACTION

Andrew M. Wims

Analytical Chemistry Department
General Motors Research Laboratories
Warren, MI

INTRODUCTION

In the past, we were limited to manual methods for the identification process. Identification was done manually[1,2] with a data base of reference spectra (in card and search manual format) supplied by the Joint Committee on Powder Diffraction Standards (JCPDS). Most single substances can be identified with relative ease using this approach. But when the number of components gets much beyond two, the measured pattern is a complex composite of sub-patterns which makes the identification a tedious and uncertain process taking 60 to 90 minutes per pattern in most simple cases.

The Johnson/Vand (J/V) search/match program and the magnetic tape data base are designed for use on a large computer in the batch mode. The computer identification process involves both a search and match process[3,4,5] of the unknown sub-patterns in the unknown sample pattern. In addition to the data base for the full file on magnetic tape, JCPDS also supplies the J/V search/match program which is designed for use on large computers, e.g., IBM 370, in the batch mode[6,7,8]. The analyst had to supply all the input information key punched on computer cards and the data base on magnetic tape to the computer center for processing. This procedure was found to be cumbersome and a hindrance to the frequent use of the program in our laboratory.

EXPERIMENTAL

For general x-ray diffraction analysis, a Siemens, D-500, diffractometer system equipped with a carbon monochromator was used. A microprocessor automatically controls the operation of the diffractometer, and a PDP 11/03 computer was used to locate the peaks and to calculate the d and I data from the diffraction peaks.

PREPARATION OF SEARCH/MATCH DATA FILE (IBM 370/3033)

To use the J/V search/match program for identifying an unknown mixture, one must set up a data file containing a large number of input parameters including chemistry in addition to a list of d and I data from the diffraction peaks. Several lines of input are required which must agree with an exact format structure as defined in the J/V program. On punched cards and in the TSO mode of operation, the setup of the data file by the analyst proved to be inconvenient leading to input errors because each line of input required a specific, different format. When data for several unknown mixtures had to be typed into one data file for processing, the inconvenience and difficulties increased significantly. Even when many of the input lines for subsequent sets of data required minimal or no changes, each line had to be typed following the specified format. Because of these factors, the advantages of the computerized system were often not realized.

To solve these problems, a program was designed and written that provides a conversational mode of communication between the computer and the analyst for setting up the data file from a remote terminal in the laboratory. The program is operational on the IBM 370/3033 in the TSO mode. The user is prompted for information without concern for the format structure except for the simple case of entering the chemical symbols. Preselected default values are automatically set into the data file unless the analyst overrides them when prompted by the computer. Because the analyst almost always wants to change some of the chemistry defaults, the group of chemistry options is separated from the other default options in the program.

At the point in the program where input of the d and I data is requested, the analyst selects an option to either type in the data at the terminal or to have the data picked up from a file stored on the IBM 370. This IBM file contains data generated on the Siemens XRD system and transferred over a data set link to the IBM 370.

After all of the required information is supplied for one unknown mixture, the program queries the analyst regarding data for another sample. If an answer of YES is given, the analyst is asked for information about the next sample. The default parameters now in effect would be those set for the previous sample; as before, the default parameters can be overridden. An answer of NO terminates the questions, and the final, properly formatted data file ready for the J/V search/match program is generated. A typical dialog from the programs is shown in Table 1. A copy of the program is available from the author.

Other than the J/V search/match program, two additional useful programs are provided on the magnetic tape from JCPDS. Both programs

Table 1. Dialog from File Set Up Programs (XRDNEW.CNTL & XRDNEW.FORT)

```
• EXEC 'XRDNEW.FORT' LIST
• DELETE OUTPUT SUMMARY FILE (Y OR N): Y
  DEL JCPDS.OUTPUT.DATA
  COPY XRDNEW.FORT XRDTEMP.FORT
  ALLOC FI(FT05F001)DA(*)
  ALLOC FI(FT06F001)DA(*)
  ALLOC FI(FT07F001)DA(JCPDS.INPUT, DATA)
• ENTER NO. SAMPLES FOR ID:1
  XRD SPACING AND INTENSITY DATA FROM
• 1-TERMINAL, 2-IBMFILE.SRC, 3-IBMFILE.ADR:2
• INPUT IBM FILE NAME: C6545.SRC
  ALLOC FI (FT08F001) DA(C6545.SRC)
  E XRDTEMP.FORT
  220  KPPT = 1-0
  225  NF = 3
  SAVE
  END
  RUN XRDTEMP.FORT
  SAMPLE IDENTIFICATION?
• TEST AIR SENSOR SAMPLE
• CHEMISTRY INFO (Y/N)?Y
  SELECT OPTION: 1-MAJOR, 2-MINOR, 3-TRACE, 4-UNKNOWN
  5-ABSENT, 6-FUNC. GRP., 7-NOMORE, 8-GUESSES
• SELECT OPTION?1
  MAJOR ELEMENTS
• ZNMG O
• SELECT OPTION?7
-----------------------------------------------------------------------
• CHANGE DEFAULT PARAMETERS(Y/N)?
  SELECT OPTION:1-ERROR WINDOW, 2-DATA TYPE
  3-RADIATION, 4-SEARCH FILE, 5-SIZE FILE, 6-PRINTS, 7-NOMORE
• SELECT OPTION?5
  FILE SIZE: O-MAXI, 1-MINI?O
• SELECT OPTION?7
  ALLOC FI(FT07F001) DA(JCPDS.INPUT.DATA) MOD REU
  FREE FI(FT08F001)
  DEL XRDTEMP.FORT
  FREE FI(FT05F001, FT06F001, FT07F001)
  END
  READY
• SUBMIT JCPDS.SEARCH.CNTL
  READY
-----------------------------------------------------------------------
```

are designed to assist the analyst in setting up special subfiles from the full files. One program (# 6.2) selects compounds from the full file based on chemical input, e.g., a list of the noble metals. The second program (# 6.5) builds the specific subfile after editorial changes by the user of the output of # 6.2. The smaller files can be searched faster with wider tolerances permitted on both d and I than the full file, and a listing of impossible phase matches will be shortened. Two special files have been set up containing a list of common phases: the Dow file (∽ 300 standards) and the frequently encountered phase file (∽ 2500 standards).

ACKNOWLEDGMENTS

I want to thank Prof. Gerald G. Johnson, Jr. of the Pennsylvania State University for his useful advice and help in implementing his search/match programs. I also want to acknowledge Dr. Jack Johnson and Mr. Sarmad Hermiz of the General Motors Research Laboratories for their assistance.

REFERENCES

1. W. F. McClume, Managing Editor, Powder Diffraction File, JCPDS, International Center for Diffraction Data, Swarthmore, PA. (1981).
2. J. D. Hanawalt, S. Rinn, and L. K. Frevel, Chemical Analysis by X-Ray Diffraction, Ind. Eng. Chem. Anal. Ed., 30:457 (1938).
3. M. C. Nichols and Q. Johnson, The Search-Match Problem, "Advances in X-ray Analysis," 23:273, J. R. Rhodes et al, eds., Plenum Press, New York (1980).
4. T. C. Huang and W. Parrish, A New Computer Algorithm for Qualitative X-Ray Powder Diffraction Analysis, "Advances in X-ray Analysis," 25:213, J. C. Russ et al, eds., Plenum Press, New York (1982).
5. R. L. Snyder, A Hanawalt Type Phase Identification Procedure for a Minicomputer, "Advances in X-ray Analysis," 24:83, D. K. Smith et al, eds., Plenum Press, New York (1981).
6. G. J. McCarthy and G. G. Johnson, Jr., Identification of Multiphase Unknowns by Computer Methods: Role of Chemical Information, the Quality of X-ray Powder Data and Subfiles, "Advances in X-Ray Analysis," 22:109, G. J. McCarthy et al, eds., Plenum Press, New York (1979).
7. P. F. Dismore, Computer Searching in JCPDS Powder Diffraction File, "Advances in X-ray Analysis," 20:113, H. F. McMurdie et al., eds., Plenum Press, New York (1977).
8. G. G. Johnson, Jr., User Guide Data Base and Search Program, JCPDS, Swarthmore, PA (1975).

THE USE OF Mn-Kα X-RAYS AND A NEW MODEL OF PSPC IN STRESS ANALYSIS OF STAINLESS STEEL

Yasuo Yoshioka

Musashi Institute of Technology
1 Tamazutsumi, Setagaya, Tokyo 158, Japan

Ken-ichi Hasegawa and Koh-ichi Mochiki

Dept. of Nuclear Engg., University of Tokyo
7 Hongoh, Bunkyo, Tokyo 113, Japan

INTRODUCTION

The authors previously reported[1] stress measurement in stainless steel by the use of monochromatic Cr-Kβ X-rays and a position sensitive proportional counter. Results indicated that a stress value can be obtained with high precision on account of the subtraction of background and the elimination of αFe(211) peak by Cr-Kα X-rays. The major disadvantage of this method, however, is that the intensity of Kβ X-rays monochromatized is essentially weak and it is complicated to eliminate Kα X-rays for practical use.

The use of Mn-Kα X-rays is an interesting problem for stress measurement in austenitic stainless steel because a fine profile with low background would be obtained from the γFe(311) crystal plane. The wavelength of Mn-Kα is 0.21021 nm and the lattice constant of γFe is 0.3571 nm. Thus the γFe(311) diffraction line of 154.92° in 2θ is just suitable line for the stress measurement. This γFe(311) line is also free from interference by other diffraction lines. An X-ray constant $K(=E.\cot\theta_0/2(1+\nu))$ is calculated as -282 MPa/deg. with Young's modulus of 193 GPa and Poisson's ratio of 0.3. This value is smaller than the value of -366 MPa/deg. used in measurements with Cr-Kβ X-rays, and the stress value will be obtained precisely in consort with the fine profile.

In the present study, we attempted to measure residual stresses in austenitic stainless steel by the use of Mn-Kα X-rays. A new position sensitive proportional counter made in our laboratory was used as an X-ray detector.

DESIGN AND STRUCTURE OF PSPC

We had already reported a charge division type PSPC consisting of a wire cathode and an anode.[2,3] However, since it is difficult to make the resistive cathode because a fine Ni-Cr wire has to be wound in order to get a higher resistance value, we tried to make a cathode consisting of printed-circuit boards and resistors.[4] With this type it is much easier to make the cathode with any pattern. Fig. 1 shows the pattern of PC board used. In a PSPC for stress measurement, a uniform angular resolution is required over an angle range of about 30° in 2θ. For this purpose, fifty radial lines on a 0.6° pitch were printed. These radial lines intersect at a point at a distance 200 mm apart from the lower edge in this figure and this point agrees with the irradiated position of the X-ray beam on the specimen surface. A schematic structure of the PSPC and a block diagram of the signal processing circuit are shown in Fig. 2. Two PC boards are placed at intervals of 10 mm, and a couple of cathode lines are positioned between the upper and lower boards. Fifty resistors of 1 KΩ are connected in series between pairs of cathode lines.

Four gold-plated tungsten wires are stretched at 2 mm intervals in the middle of both PC boards as an anode. Such structure is in

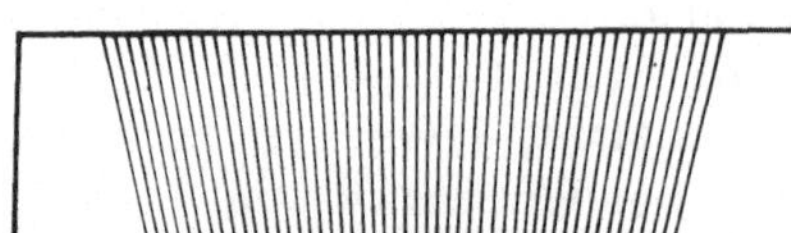

Fig.1. Printed pattern of cathode.

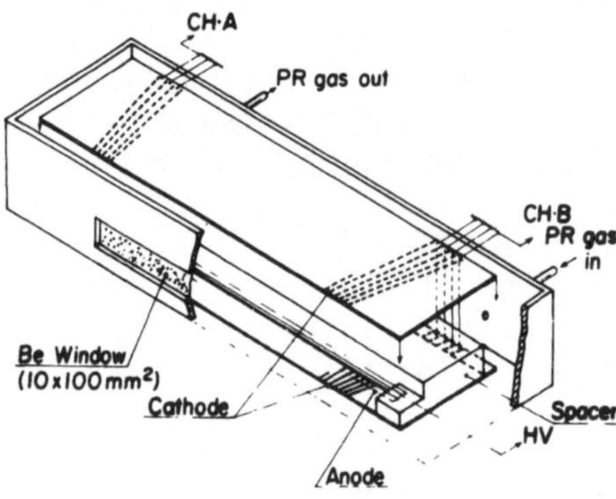

Fig.3. Schematic bird view of PSPC.

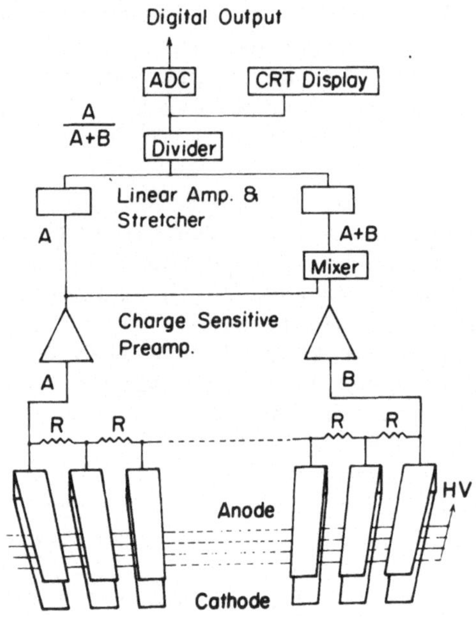

Fig.2. Schematic structure of PSPC and block diagram of signal processing circuit.

principle the same as the PSPC reported in the past paper.[2] A combination of the multiwire anode and the fan-shape cathode decreases the deterioration of angular resolution with increase of incident angle of the X-ray beam to the PSPC.

A bird's-eye view of the PSPC is shown in Fig. 3. The active depth is 10 mm and the entrance window has the dimensions of 10x100 mm^2 and it is covered with a Be foil of 0.3 mm thickness. The gas filling is a mixture of argon 90% and methane 10% of 1 atm.

Each end of the series resistors is connected to a charge sensitive preamplifier. Signal A and the mixed one (A+B) are amplified and shaped by linear amplifiers and stretchers, and the A is divided by the (A+B) with the help of an analogue dividor. The digital output pulses are processed by a 512 channel pulse height analyzer, and these data are analysed by a microcomputer.

PSPC PERFORMANCE

The linear response and the angular resolution with respect to the path angle α have been determined. The X-ray beam collimated by a 50 μm wide slit was directed into the PSPC with the angle of α from the irradiated position mentioned. In Fig. 4, closed circles show the linear response. A plot is obtained that indicates good linearity in the range of α=-10° to +10°. Open circles in this figure show the angular resolution. The minimum value is about 0.14° in 2θ (FWHM) at α=0° and it slightly increases with path angle α, but it is a maximum of 0.2° in 2θ at α=10°. This resolution is sufficient for stress measurement.

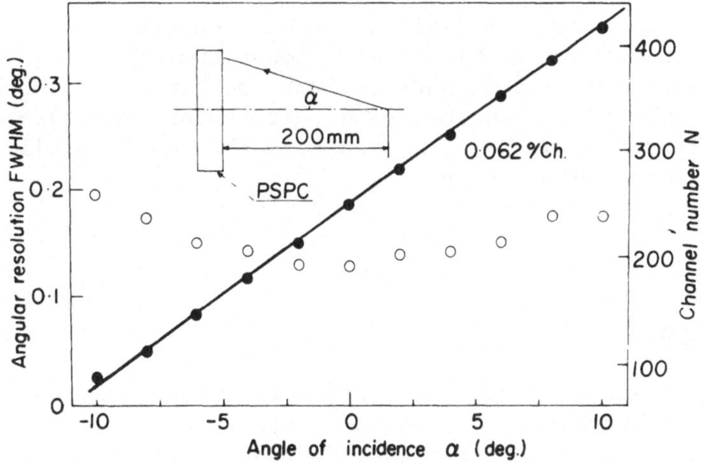

Fig.4. Linearity and angular resolution along channel number.

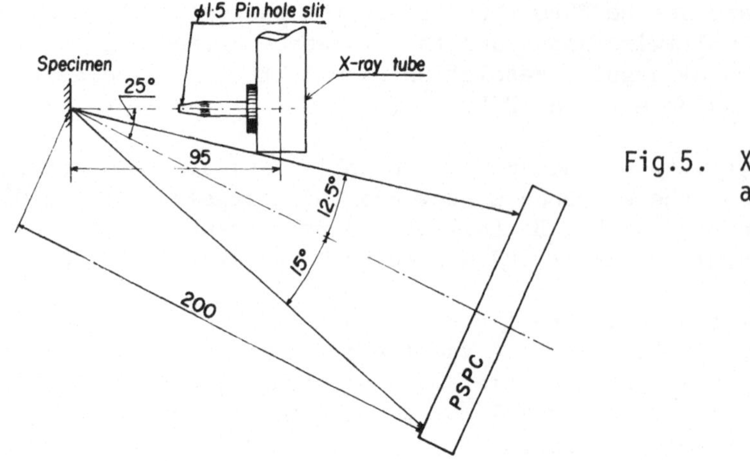

Fig.5. X-ray optical alighment.

METHOD, EQUIPMENT AND SPECIMENS FOR STRESS MEASUREMENT

The basic configuration of X-ray source, PSPC and specimen is shown in Fig. 5. When diffraction by Mn-Kα X-rays is going to be carried out, an important problem is to get a stable X-ray source with high power. As far as we know, only Philips Co. produces a Mn target X-ray tube in answer to a user's request. As we could not obtain it at a stage when we had begun this experiment, we used a microbeam X-ray generator with a prototype Mn target made by Rigaku Co. The X-ray beam irradiated the specimen surface through a 1.5 mm diameter single pin hole. The distance between PSPC and specimen was 200 mm, and the PSPC was placed at an angle of 25° to the X-ray beam. The output power of the X-ray tube was 40 KVP and 0.5 mA.

Two sorts of type 304 stainless steel specimens were prepared. One is the specimen for residual stress measurements, which is made in the form of a small block with dimensions of 50x40x8 mm^3. After being solution-annealed, surfaces were machined or ground under various conditions. Another is a rectangular beam type specimen for a calibration of X-ray stress by mechanically applied stress. Surfaces were finish ground.

RESULTS

Observation of Diffraction Profile

Fig. 6 shows examples of diffraction profiles; one is from an as-solution-annealed specimen and another is from one that was surface machined by a sintered carbide tool. The ratio of peak intensity to background is found to be 3.3 and 2.1 respectively. These values are a little inferior to those measured by monochromatic

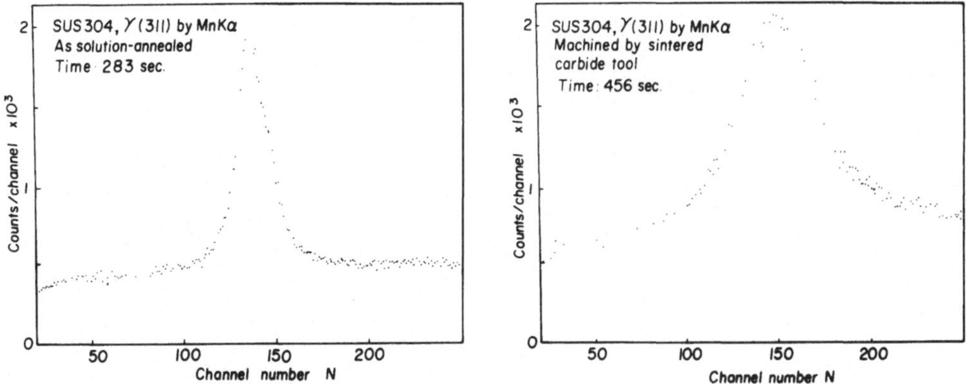

Fig.6. Examples of diffraction profile. a) As solution-annealed specimen, b) machined by a sintered carbide tool.

Cr-Kβ X-rays[1] because of non-use of a Cr foil filter for eliminating Kβ X-rays. However, the αFe(211) line, which appears in the neighborhood of the γFe(311) line by Cr-Kβ X-rays accompanied by Cr-Kα ones, is naturally not observed even on the specimen in which the strain-induced transformation has occurred.

The time required for data accumulation is about 5-15 minutes when the peak count was preset to be 2048. It is long because the X-ray power is very limited for fear of the contamination, but it can be very much reduced by the use of a high power X-ray source.

Residual Stress Measurements

Residual stresses were measured on various specimens of type 304 stainless steel. An example of a $\mathrm{Sin}^2\psi$ diagram is shown in Fig. 7; $\Delta\sigma$ in this figure means a 95% confidence limit of stress σ, and it is calculated from the following equation:

$$\Delta\sigma = K\cdot t(\alpha,n-2)\sqrt{\frac{\Sigma\{2\theta i-(2\theta_0+M\cdot \mathrm{Sin}^2\psi i)\}^2}{(n-2)\Sigma(\mathrm{Sin}^2\psi i-\overline{\mathrm{Sin}^2\psi})^2}} \tag{1}$$

where $t(\alpha,n-2)$ is the t-distribution under the degree of freedom of $(n-2)$ and the confidence limit of $(1-\alpha)$, n the sample size, M the gradient of $\mathrm{Sin}^2\psi$ diagram and $\overline{\mathrm{Sin}^2\psi}$ is the mean value.

This $\Delta\sigma$ is generally being used in Japan as a parameter on the precision of stress value σ by the $\mathrm{Sin}^2\psi$ method. A good linear relation is found to hold on this diagram. Stress values measured in this experiment are almost equal to those measured by monochromatic Cr-Kβ X-rays.

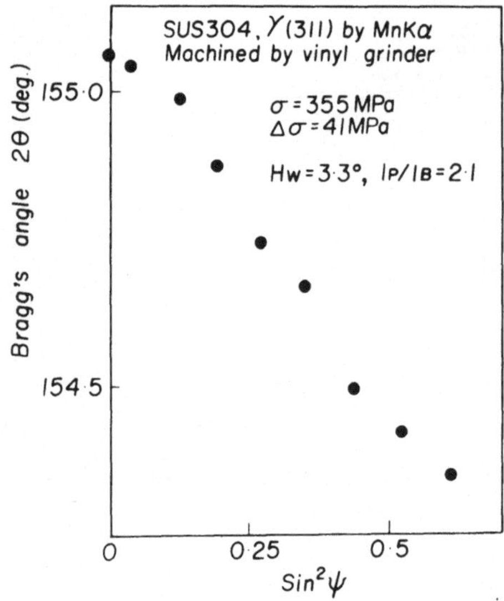

Fig.7. An example of $\mathrm{Sin}^2\psi$ diagram. Surfaces were machined by a vinyl grinder.

$\Delta\sigma$ is a statistical error of stress; it depends on the shape of the diffraction profile. The precision of the diffraction peak measured is affected by line broadening, peak intensity, background intensity and other factors. An error parameter is derived as follows[5] from the statistical variation of the profile,

$$\Delta\sigma \propto \frac{Hw}{\sqrt{IP}} \sqrt{(1+2IB/IP)^2 \Delta 2\theta} \qquad (2)$$

where Hw is half the value breadth of the profile, IP the peak intensity, IB the background intensity, and $\Delta 2\theta$ the sampling interval. This parameter was calculated on all stress measurements, and the 95% confidence limit by eq.(1) was plotted against this parameter. Fig. 8 shows the result. As apparent in this figure, $\Delta\sigma$ increases with increase of parameter. The result indicates the

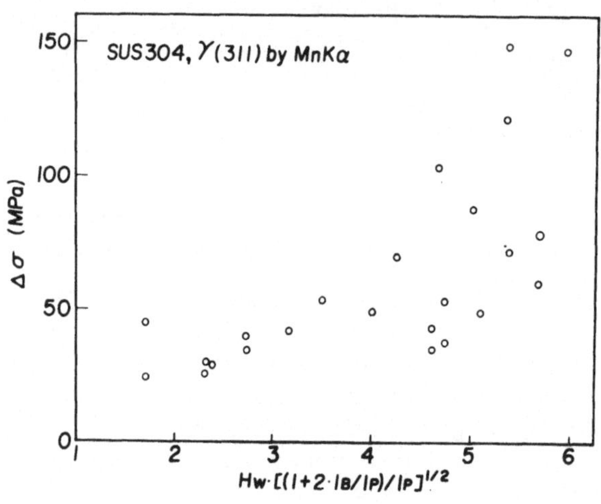

Fig.8. Relation between 95% confidence limit and error parameter.

peak intensity should increase on the profile with a large half value breadth when a precise value of stress is demanded. It is very easy to increase the peak intensity in the case of the PSPC method. This relation is similar to that for the monochromatic Cr-Kβ X-rays. But it is obviously good in comparison with that for non-monochromatic Cr-Kβ X-rays.

Calibration of X-ray Stress by Applied Stress

Several bending stresses were applied to a specimen with a rectangular shape,and the X-ray stress at each load was measured. In Fig. 9, values indicated as open circles were measured by the use of Mn-Kα X-rays, while values indicated by closed circles were done by the use of the monochromatic Cr-Kβ X-rays in the previous study.[1] A relation between applied stress and stress measured by Mn-Ka X-rays indicates a nearly straight line which has a slope of about 1 as well as the result of using the monochromatic Cr-Kβ X-rays.

CONCLUSIONS

A prototype PSPC with a resistive cathode consisting of printed circuit boards and resistors was made and residual stresses in several specimens of type 304 stainless steel were measured by the

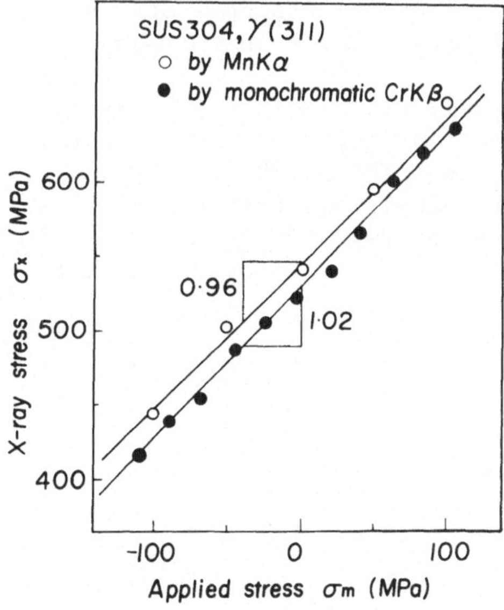

Fig.9. Relation between applied stress and X-ray stress.

use of this PSPC with Mn-Kα X-rays. The following conclusions can be drawn from the successful results.

1) The printed-circuit board type cathode can be made as easily as a resistance wire cathode.
2) Both the linearity and the angular resolution of this PSPC are sufficiently good for the stress measurement.
3) The use of Mn-Kα X-rays is suitable for stress analysis in type 304 stainless steel. When the stresses measured were compared with those measured with monochromatic Cr-Kβ X-rays, the precision of stress values is almost similar; the setting of apparatus is apparently easy and simple.
4) In the present study, we had to use a small power X-ray source. Thus the time required for data accumulation was long despite the use of the PSPC method. However, this problem will be solved if a high power X-ray source with Mn target would be developed in the future.

REFERENCES

1. Y. Yoshioka, K. Hasegawa and K. Mochiki, "Stress Measurement in Stainless Steel by Use of Monochromatic Cr-Kβ X-Rays and a Position Sensitive Detector," *Advances in X-Ray Analysis* 24:167 (1981).
2. Y. Yoshioka, K. Hasegawa and K. Mochiki, "Study on X-Ray Stress Analysis Using a New Position-Sensitive Proportional Counter," *Advances in X-Ray Analysis* 22:233 (1979).
3. Y. Yoshioka, K. Hasegawa and K. Mochiki, "A Position-Sensitive Proportional Counter for Residual Stress Measurement by Means of Microbeam X-Rays," *Advances in X-Ray Analysis* 23:325 (1980).
4. E. Gatti, A. Longoni, R. A. Boie and V. Radeka, "Analysis of the Position Resolution in Centroid Measurements in MWPC," *Nuclear Inst. and Methods* 188:322 (1981).
5. T. Shiraiwa and Y. Sakamoto, "Error of Peak Position Measurement in X-Ray Stress Analysis," *Standard for X-Ray Stress Measurement* The Society of Material Science, Japan, 63 (1973).

MEASUREMENT OF STRESS GRADIENTS

BY X-RAY DIFFRACTION

J.M. Sprauel, M. Barral and S. Torbaty

Ecole Nationale Supérieure d'Arts et Métiers
151 Boulevard de l'Hôpital
75013 Paris, France

INTRODUCTION

Among the limitations of the classical methods for measuring stresses by X-ray diffraction, the existence of stress gradients constitutes a particularly delicate problem. On the theoretical plane, this problem has been tackled by Dölle, Hauk and Cohen.[1,2] These authors showed that whilst the gradients of the shear stresses σ_{13} and σ_{23}† are relatively easy to bring to the fore, as their presence is reflected in an opening out of the curves for $2\theta_{\phi\psi}=f(\sin^2\psi)$, those of the direct stresses σ_{11}, σ_{22} and σ_{33} are very difficult to detect. In the latter case it is demonstrated that even when the gradients attain very high values, the curves of $2\theta_{\phi\psi}=f(\sin^2\psi)$ remain practically linear. This may invite one to apply the classical $\sin^2\psi$ law, with the result that the stress values so determined do not correspond with the real mechanical state of the surface of the specimen.

On the other hand, experimental measurements of stress gradients are practically non-existent. The work of Evenshor[3] concerning the determination of the gradients of the shear stress σ_{13} in a test piece of ground steel★ is an exception. The procedure involved varying the X-ray depth of penetration by utilizing three different wavelengths; viz. CrKα, CoKα and MoKα. These measurements, however, can only be regarded as approximate since no variation of the depth of X-ray penetration with the angle ψ (Fig. 1) was taken into account.

†Direction 3 being normal to the surface of the specimen (Fig. 1(a))
★Grinding being in direction 1

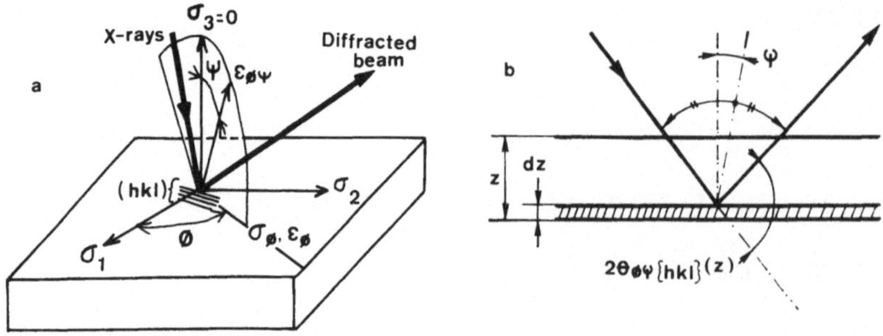

Fig. 1. Basic geometry and coordinate system

In the study reported in this paper,we undertook to adapt the theoretical formulations of Dölle, Hauk and Cohen to the experimental determination of the stress gradients on the surface of a ground specimen of alloy steel. Like Evenshor, we used radiation of different wavelengths. However, rather than concentrate on the measurement of the gradient of just one stress component, we endeavoured to measure all the six components of the stress tensor together with their first and second derivatives. Furthermore, we have been able in this work to allow for the variation of the depth of X-ray penetration with the angle ψ.

PRINCIPLE OF THE METHOD

In order to achieve different penetration depths on steel, four different types of radiation may be used, as shown in Table 1. Often, however, measurements with MoKα radiation are not possible on account of the fluorescence of the material and the very small intensity of the diffracted beam. So we are left with the first three types giving penetration depths between 5μm and 12μm. Yet this choice implies working on different families of crystallographic planes and thus it becomes necessary in this method to account for two different factors: (i)the effect of X-ray penetration and (ii)the effect of X-ray mechanical anisotropy of the material.

Effect of X-ray Penetration[2,4]

The X-ray diffraction integrates the different layers dz (Fig. 1 (b)) of the specimen, where each layer may be defined by its depth z below the surface and its diffraction angle $2\theta_{\phi\psi\{hkl\}}(z)$ for a specific family $\{hkl\}$ of crystallographic planes.

Now, the position of the diffraction peak is the weighted mean of all the values of $2\theta_{\phi\psi\{hkl\}}(z)$ over the thickness of the specimen, assumed infinite, thus:

Table 1. Different types of radiation which may be used for the measurement of stress gradients on steel

RADIATION	{hkl}	2θ(deg.)	PENETRATION DEPTH (μm)
Cr Kα	{211}	156.3	5.9
Fe Kα	{220}	145.7	9.6
Co Kα	{310}	161.7	12.1
Mo Kα	{651}	154.8	16.6

$$2\theta_{\phi\psi\{hkl\}}(z) = \frac{1}{\tau_{\{hkl\}}(\psi)}\int_{o}^{\infty}\{2\theta_{\phi\psi\{hkl\}}(z).\exp(\frac{-z}{\tau_{\{hkl\}}(\psi)})\}dz \quad (1)$$

Equation (1) introduces an exponential absorption factor which varies with the mean depth $\tau_{\{hkl\}}(\psi)$ penetrated by the X-rays at a given ψ angle. This depth, in turn, varies with the wavelength of the radiation used and with the angle ψ (Fig. 2); its value may be calculated from one of the following expressions, depending on the goniometer used:

$$\tau_{\{hkl\}}(\psi) = \left.\begin{array}{ll} \dfrac{\sin^2(\theta_{\phi\psi\{hkl\}})-\sin^2\psi}{2\mu\sin(\theta_{\phi\psi\{hkl\}})\cos\psi} & \Omega\text{-goniometer} \\ \sin(\theta_{\phi\psi\{hkl\}})\cos\psi/2\mu & \psi\text{-goniometer} \end{array}\right\} \quad (2)$$

In Eqn. (2), μ is the linear absorption coefficient.

The position of the diffraction peak is now expressed as a function of its derivatives and of the depth $\tau_{\{hkl\}}(\psi)$, by Maclaurin's theorem:

$$2\theta_{\phi\psi\{hkl\}} = 2\theta_{\phi\psi\{hkl\}}(0) + \tau_{\{hkl\}}(\psi)\frac{d}{dz}\{2\theta_{\phi\psi\{hkl\}}(0)\} + \tau^2_{\{hkl\}}(\psi)\frac{d^2}{dz^2}\{2\theta_{\phi\psi\{hkl\}}(0)\} + \ldots \quad (3)$$

Effect of X-ray Mechanical Anisotropy

In order to understand the effect of the mechanical anisotropy of crystallites on the X-ray method it is necessary to distinguish between the values of strain measured by X-ray diffraction $\varepsilon_{ij\{hkl\}}$ and the macroscopic values ε_{ij}. The X-ray value of strain $\varepsilon_{\phi\psi\{hkl\}}$ is

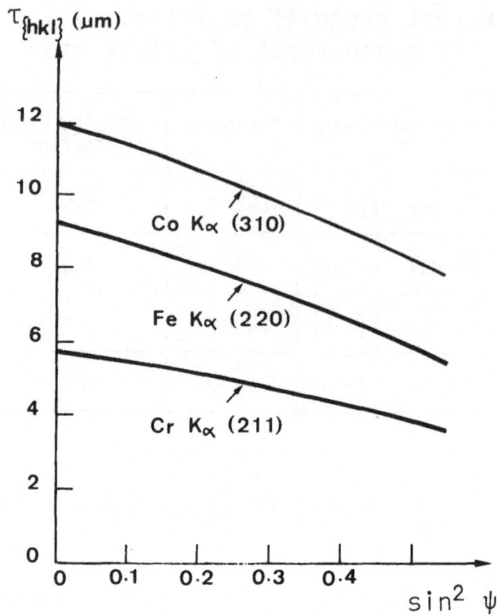

Fig. 2. Variation of the X-ray penetration depth with ψ

related to the macroscopic stresses σ_{ij} by the X-ray elastic constants $\frac{1}{2}\,S_{2\{hkl\}}$ and $S_{1\{hkl\}}$, as follows:

$$\varepsilon_{ij\{hkl\}} = \frac{1}{2}\,S_{2\{hkl\}}\cdot\sigma_{ij} + \delta_{ij}S_{1\{hkl\}}\cdot\sigma_{kk} \tag{4}$$

The macroscopic value $\varepsilon_{\phi\psi}$, on the other hand, is related to the macroscopic stresses σ_{ij} by the macroscopic (or mechanical) elastic constants $\frac{1}{2}S_2$ and S_1:

$$\varepsilon_{ij} = \frac{1}{2}\,S_2\sigma_{ij} + \delta_{ij}S_1\sigma_{kk} \tag{5}$$

Finally, the X-ray strain $\varepsilon_{\phi\psi\{hkl\}}$ is linked to the macroscopic value $\varepsilon_{\phi\psi}$ by two anisotropy factors $A_{1\{hkl\}}$ and $A_{2\{hkl\}}$:

$$\varepsilon_{ij\{hkl\}} = A_{2\{hkl\}}\,\varepsilon_{ij} + \delta_{ij}A_{1\{hkl\}}\,\varepsilon_{kk} \tag{6}$$

These latter factors depend only on the indices $\{hkl\}$ of the family of diffracting planes and on the X-ray anisotropy factor for the material, A_{rx}.[5] Thus,

$$\left.\begin{aligned}
&A_{1\{hkl\}} = -(0.2 - r_{\{hkl\}}).\Delta \\
&A_{2\{hkl\}} = 1 - 3A_{1\{hkl\}} \\
\text{where}\quad &r_{\{hkl\}} = \frac{h^2k^2 + h^2l^2 + k^2l^2}{(h^2 + k^2 + l^2)^2} \\
&\Delta = 5(A_{rx}-1)/(3 + 2A_{rx}) \\
&A_{rx} = \tfrac{1}{2}S_{2\{h00\}}/\tfrac{1}{2}S_{2\{hhh\}}
\end{aligned}\right\} \qquad (7)$$

Equations (7) are general. In the particular direction of measurement, they are replaced by the following:

$$\left.\begin{aligned}
&\varepsilon_{\phi\psi\{hkl\}} = \{l_i l_j A_{2\{hkl\}} + \delta_{ij} A_{1\{hkl\}}\}\varepsilon_{ij} \\
\text{where}\quad &l_1 = \cos\phi\sin\psi \\
&l_2 = \sin\phi\sin\psi \\
&l_3 = \cos\psi
\end{aligned}\right\} \qquad (8)$$

Determination of the Stress Tensor and Its Derivatives

So far, we have been dealing with average values of $\varepsilon_{ij\{hkl\}}$ which vary with the depth of X-ray penetration. We must now take this variation into account. Thus, using Eqns. (6) and (8), together with the Maclaurin's expansion for ε_{ij} (similar to the one for $2\theta_{\phi\psi\{hkl\}}$ of Eqn. (3)), we deduce the following expression for $\varepsilon_{\phi\psi\{hkl\}}$:

$$\begin{aligned}
\varepsilon_{\phi\psi\{hkl\}} = {} & \{l_i l_j A_{2\{hkl\}} + \delta_{ij} A_{1\{hkl\}}\}\times \\
& \{\varepsilon_{ij}(0) + \tau_{\{hkl\}}(\psi)\frac{d}{dz}\varepsilon_{ij}(0) \\
& + \tau^2_{\{hkl\}}(\psi)\frac{d^2}{dz^2}\varepsilon_{ij}(0) + \ldots\}
\end{aligned} \qquad (9)$$

It is now possible to fit the experimental results to this form of expression by using a least squares method, thus giving the components of the strain tensor and their derivatives. In order to do this, the values of $\varepsilon_{\phi\psi\{hkl\}}$ have to be calculated. This, however, requires knowledge of the Bragg angle $2\theta_{o\{hkl\}}$ for the unstressed specimen:

$$\left.\begin{aligned} \varepsilon_{\phi\psi\{hkl\}} &= K(2\theta_{\phi\psi\{hkl\}} - 2\theta_{\circ\{hkl\}}) \\ \text{where} \qquad K &= -\cot(\theta_{\phi\psi\{hkl\}})\cdot\frac{\pi}{360} \end{aligned}\right\} \qquad (10)$$

The angle $2\theta_{\circ\{hkl\}}$ is difficult to measure; it may also vary with the depth $\tau_{\{hkl\}}$ penetrated by the X-rays. However, equilibrium of a surface element in direction 3 (Fig. 1) requires that:

$$\left.\begin{aligned} &\sigma_{33}(z=0) = 0 \\ &\frac{\partial}{\partial x}\sigma_{13} + \frac{\partial}{\partial y}\sigma_{23} + \frac{\partial}{\partial z}\sigma_{33} = 0 \end{aligned}\right\} \qquad (11)$$

Also, measurements at different points (x,y) on the surface of the specimen indicate σ_{13} and σ_{23} to be practically constant, if not zero. Consequently, their gradients have to be negligible. It follows from Eqns. (11) that $\sigma_{33}(z)$ and all its gradients should be negligible too. Using this supplementary condition, it is possible to arrive, iteratively, at a good estimate of the angle $2\theta_{\circ\{hkl\}}$.

The value of the stresses and their gradients may now be calculated using the classical relations of elasticity.

EXPERIMENTAL RESULTS

In our study, measurements were carried out on a ground specimen of alloy steel (0.35% C; 1.00% Cr; 0.6% Mo) with CrKα, FeKα and CoKα radiation. For each wavelength, 8 values of the angle ϕ and 9 values of the angle ψ were used. Moreover, by using an unstressed powder as

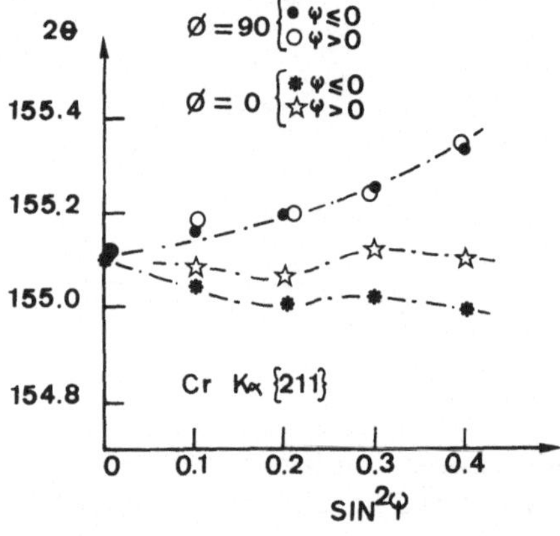

Fig. 3. Example of curves $2\theta=f(\sin^2\psi)$ obtained with ground alloy steel

reference, any systematic error from the equipment was eliminated. All the measurements were carried out using an automatic goniometer[6] with a position sensitive detector "Elphyse". The time for one peak registration was about 60 seconds.

Figure 3 gives an example of the type of curves $2\theta=f(\sin^2\psi)$ obtained in the direction of grinding ($\phi=0$), while in Table 2 are shown the components of the complete stress tensor and its first and second order derivatives obtained while keeping the value of σ_{33} and its derivatives fixed at zero. The results show large values of the direct stresses and their gradients in the direction of grinding (the 11 elements) and perpendicular to it (the 22 elements), while the values of most of the remaining elements of the tensors are one order of magnitude smaller. The results also show that the actual state of stress is almost biaxial.

Knowledge of the components of the stress tensor and its derivatives makes it possible to calculate the variation of the various stresses with depth (Fig. 4). While σ_{12} and σ_{13} are almost zero throughout the depth of the specimen, σ_{11} varies rapidly across a thin surface layer: from a compressive value on the surface of the specimen it becomes tensile at a depth of about 3μm, and exhibits a maxima in tension at a depth of about 12μm.

Table 2. Stress tensor and its first and second derivatives

STRESS TENSOR [MPa]

-523 ± 61	107 ± 43	17 ± 14
107 ± 43	-276 ± 61	- 4 ± 14
17 ± 14	- 4 ± 14	0 ± 43

FIRST DERIVATIVES [MPa·μm^{-1}]

197 ± 12	-39 ± 8	- 12 ± 2
- 39 ± 8	46 ± 12	1 ± 2
- 12 ± 2	1 ± 2	0 ± 8

SECOND DERIVATIVES [MPa·μm^{-2}]

- 15 ± 1	3 ± 1	1 ± 0
3 ± 1	-4 ± 1	-0 ± 0
1 ± 0	-0 ± 0	0 ± 1

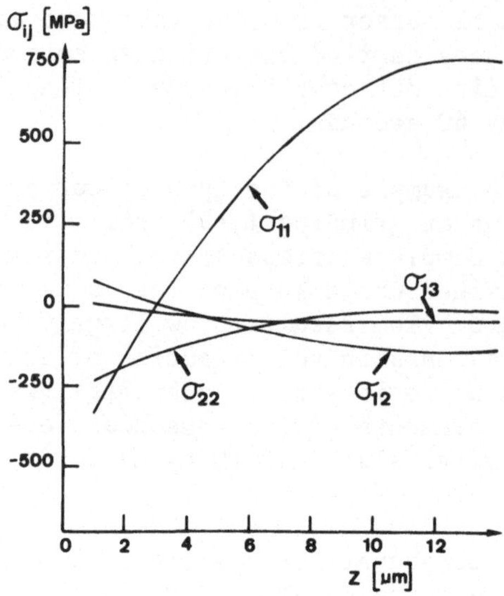

Fig. 4. Variation of stresses with depth ($\sigma_{23}=\sigma_{33}=/0/$)

CONCLUSIONS

In conclusion, we may sum up the main points of our study as follows:

It is possible to measure stress gradients by X-ray diffraction while taking account of the effects of penetration of X-rays and of the mechanical anisotropy of crystallites.

Using equilibrium conditions, it is possible to do without knowledge of the angle $2\theta_o$ for an unstressed specimen.

Acknowledgement

The authors much appreciate the contribution of Dr. P.H. Markho, Associate Professor at ENSAM, to the writing and presentation of this paper.

REFERENCES

1. Dölle, H. and Cohen, J.B.,Met. Trans.,Vol.11A,1980,pp.159-164
2. Dölle, H. and Hauk, V.,HTM,Vol.34,1979,pp.272-277
3. Evenshor, P.D.,Z.Metallkde,Vol.73,1982,pp.387-389
4. Shiraiwa, T. and Sakamoto, Y.,The Sumimoto Search 7,1972,pp.109-135
5. Hauk, V. and Kockelmann, H.,Materialpruf.,Vol.21,1979,pp.201-205
6. Lebrun, J.L.,Sprauel, J.M. and Maeder, G.,Adv.in X-ray Anal., Vol.24,1981,pp.143-148

A METHOD FOR X-RAY STRESS ANALYSIS OF THERMOCHEMICALLY TREATED MATERIALS

Rolf A. Prümmer and H.W. Pfeiffer-Vollmar

Fraunhofer-Institut für Werkstoffmechanik

Freiburg, W-Germany

INTRODUCTION

X-ray stress analysis is a nondestructive method enabling one to measure residual and loading stresses in polycrystalline materials.[1-3] Known as a difficult method for several decades, the introduction of microprocessors allowed the automation of the measuring procedure and the subsequent storage and computation of diffraction data. Also the availability of position sensitive counters increased the efficiency of the method. Nowadays a residual stress or loading stress determination can be a problem of a few minutes if the time required for installation and alignment of the equipment is neglected. As the penetration of the X-ray beam into the investigated surface of polycrystalline materials is low and only a few micrometers the obtained information is that of a surface stress state. A further specialty of the method is the selective nature of X-ray stress analysis: lattice strains are measured in certain crystallographic directions. Therefore the elastic anisotropy of the single crystal has to be taken into account. If second phases are present in the investigated sample, also the effect of heterogeneity contributes to the stress analysis. Therefore, the "X-ray elastic constants" had to be introduced. They can deviate from the corresponding mechanical values as much as 40%.[4] Also plastic deformation and severe lattice distortions, such as those introduced by hardening of steel, can influence large deviations in the X-ray elastic constants.

The assumption of a plane-stress state in the surface layers of polycrystalline samples was the subject of an investigation of Vasilev.[5] It was found that the principal stress acting in the direction of the surface normal indeed has its effect upon the lattice strain distribution that is obtained. The theory of elasticity requires a straight distribution of lattice strains vs. $\sin^2\psi$ (with ψ = angle between measured direction and surface normal). Deviations from linearity can occur when the principal stresses are inclined with respect to the surface geometry or when stress gradients occur in the sample's surface within the depth of the penetration of the X-ray radiation.[6]

The purpose of this paper is to introduce a factor which is important when gradients of composition exist in the examined surface in a direction parallel to the normal. A method is presented to account for this effect in X-ray stress analysis of thermochemically treated materials. Experiments with chromized and manganized iron are performed.

CALCULATION OF EFFECT OF GRADIENT

The penetration of X-ray radiation during X-ray stress measurement is dependent on the diffraction angle θ and the inclination angle ψ. Fig. 1 shows the path of the impinging and reflected beam in a ω-type diffractometer. After passage of the depth z the reflected beam is attenuated due to absorption. The depth τ which leads to an intensity of the reflected beam of $I = I_0/e$, where e is the natural number, is the effective penetration depth. In the case of an ω-type diffractometer it is given by

$$\tau = \frac{\sin^2\theta - \sin^2\psi}{2\mu \sin\theta \cdot \cos\psi} \qquad (1)$$

where μ is the mass absorption coefficient. For instance in the case

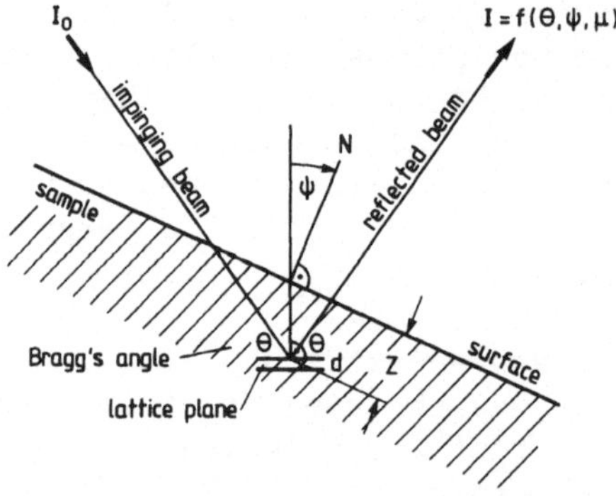

Fig. 1. Geometry of impinging and reflected X-ray beam in X-ray stress analysis

of Cr-Kα-radiation and iron, where X-ray stress measurements usually are performed on {211}-planes, τ is 5.5 µm at $\psi = 0$ and 3.8 µm at $\psi = 45°$.

As long as the lattice constant, a, of the polycrystalline sample does not vary with depth z the reflected beam occurs at one angle, given by Bragg's equation. However, if the lattice spacing, a, is any function of z, there occur different diffracted beams from different layers at different depths z. In the measurement a mean value of the lattice constant is therefore registered. As the top layers of the sample contribute to a greater extent to the measurement compared to deeper layers, a weighted mean value of a(z) has to be taken in order to obtain the effective lattice constant $\langle a \rangle$:

$$\langle a \rangle = \frac{\int_0^T a(z)\, e^{-z/\tau} dz}{\int_0^T e^{-z/\tau} dz} = a(\psi) \qquad (2)$$

The integration depth T has to be chosen such that all diffracting layers are covered, and therefore the thickness of the sample is taken. With $\tau = \tau(\psi)$, $\langle a \rangle$ becomes a function of ψ. This means that even in an annealed and stress free sample, with a gradient of composition in the surface layers, a lattice strain distribution $\langle a \rangle = f(\psi)$ with nonzero slope is obtained, indicating a fictitious residual stress σ_{grad}. This case was experimentally demonstrated with a 6% chromium steel coated with thin layers of pure iron of 1 and 3 µm thickness.[7] Although the steel and the layers were stress free annealed separately, fictitious residual stresses σ_{grad} of -26 and -80 N/mm^2 were measured, respectively. These were in good agreement with the calculated ones. In the following, measurements of chromized and manganized samples are described and the correction for σ_{grad} is applied.

EXPERIMENTAL PROCEDURE AND RESULT

Iron samples with the dimension 20 x 10 x 3 mm were packed into a container together with a mixture of Cr- or Mn-powder with additives of NH_4Cl and Al_2O_3-powder and exposed to a temperature of 1000°C for 6 hours. A subsequent slow furnace cooling was applied in order to prevent thermal gradients in the sample during cooling. Fig. 2 shows the micrographs of the surface layers of the two samples. The concentration profiles of Cr and Mn in Fig. 2 were obtained by scanning electron microscopy. They were used in order to establish the relationship between the lattice parameter, a, and depth, z, given at the lower part of Fig. 2 for the chromized and manganized sample. Experimental data describing the Vegard's relationship between the lattice parameter and the alloying content in the α-Cr Fe[8] and the γ-Mn Fe and α-Mn Fe solid solutions were taken from literature.

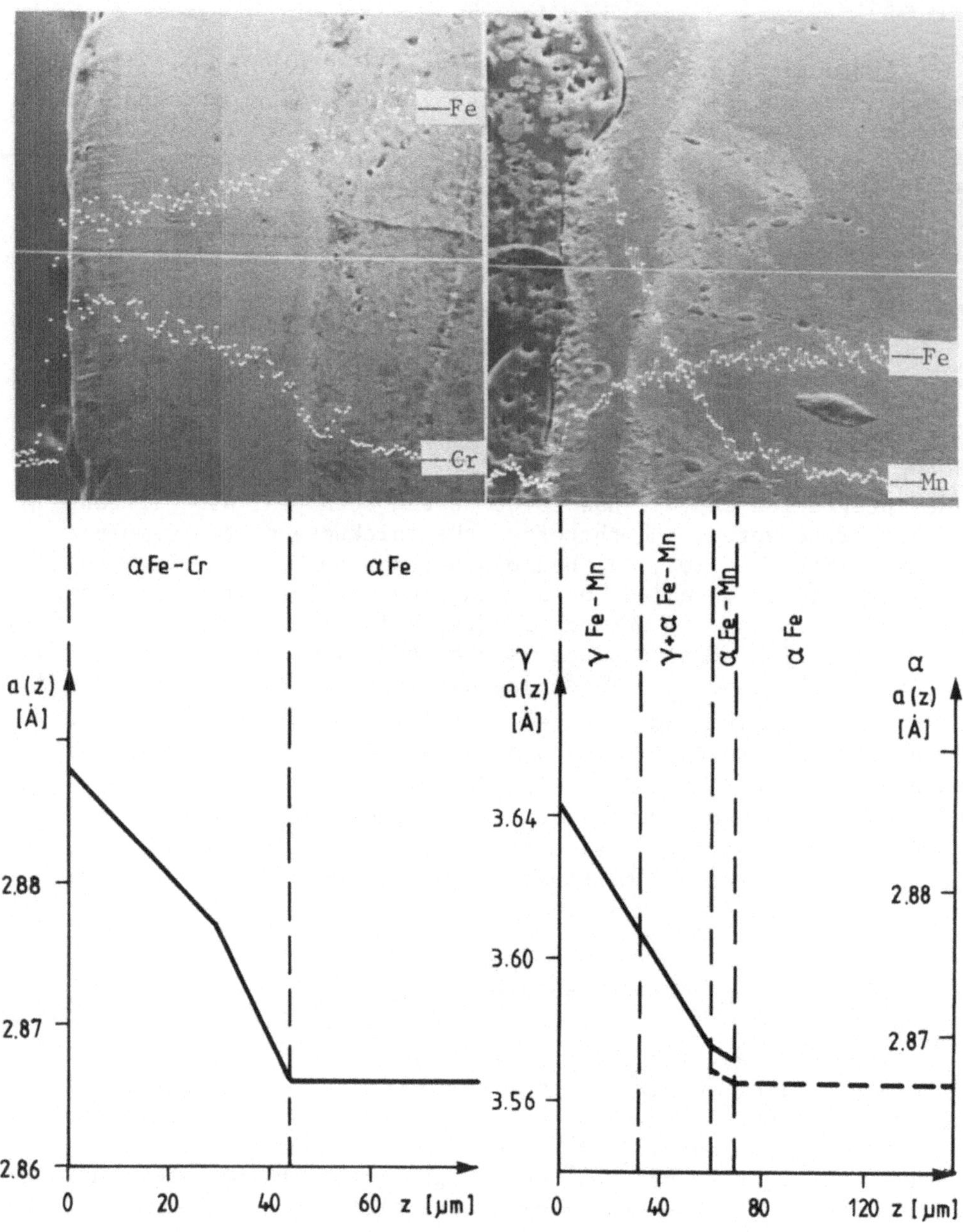

Fig. 2. SEM of chromized (left) and manganized (right) iron and lattice parameters as a function of depth under the surface (lower parts)

X-ray measurements were performed with a Siemens D500-diffractometer and a position sensitive counter at {211} lattice planes of the α-iron with Cr Kα-radiation and {311} lattice planes of the γ-iron with Cr Kβ-radiation.[10] After measurement of the residual stress at the immediate surface, thin layers were successively electrolytically polished away and the X-ray residual stress measurement repeated.

Fig. 3 gives the residual stress distribution for both the chromized and manganized sample as a function of depth z. The surface layers of the chromized sample reveal a compressive stress of $\sigma_E = -180\ N/mm^2$. The correction for the effect of the chrome's gradient in the surface layers after eq. (2) and eq. (1) is leading to a much larger value of the compressive stress in the surface layer: $\sigma_E = -410\ N/mm^2$. Residual stress determinations performed by mechanical methods at a similarly chromized sample by Dubinin et al.[11] lead to a surface stress of $-360\ N/mm^2$. The calculation of the residual stresses as a result of different thermal expansions in successive layers during cooling by means of finite element methods on the other hand gives a value of the surface residual stress of $-450\ N/mm^2$ and is also in fairly good agreement with the value obtained by X-ray stress analysis. The manganized sample shows a tensile residual stress of $660\ N/mm^2$. The correction for the Mn-content gradient reduces this value to $340\ N/mm^2$.

In deeper layers the residual stresses reduce in amount, and at a depth of ∿ 40 μm in the chromized sample and ∿ 60 μm in the manganized sample they become practically zero after some changes in compressive/tensile character.

DISCUSSION OF RESULTS

In cases of thermochemically treated iron it is shown that in X-ray residual or loading stress determinations a correction is necessary. It accounts for the effect of a gradient of chemical composition with depth under the surface. If the correction is neglected even a stress-free annealed specimen could indicate fictitious residual stresses either of compressive or tensile character, depending on the effect of the alloying element on the crystalline parameters of iron.

The origin of the residual stresses obtained in chromized and manganized iron after slow furnace cooling from chromizing/manganizing temperatures are differences in mechanical properties (Young's modulus and Poisson's ratio) and thermal expansion coefficients of the surface layers and deeper areas. Finite element methods applied to the investigated surface areas of the chromized sample lead to surface stresses which are in agreement with the measured ones.

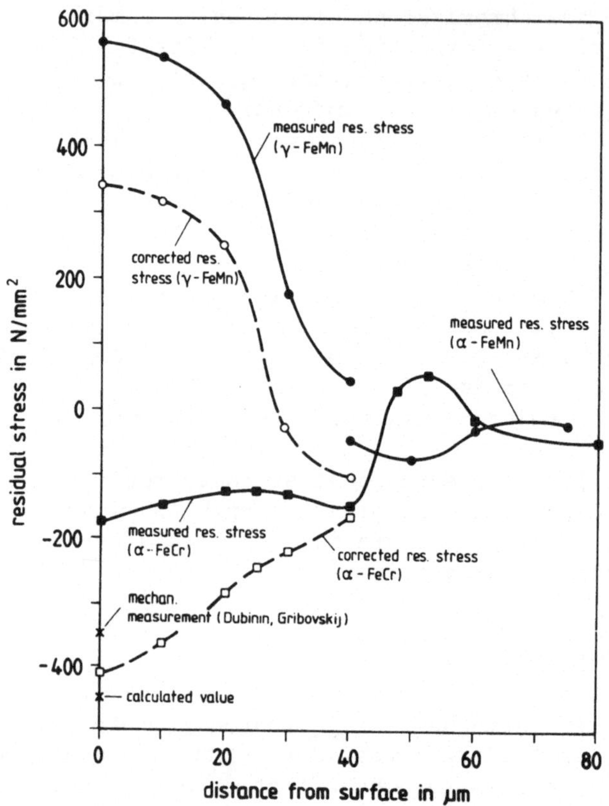

Fig. 3. Residual stresses in chromized and manganized sample vs. depth under surface.

There are, however, limitations of the applied method. The correction for concentration gradients after eq. (1) and (2) is only valid for the case of a homogeneous solid solution. If the alloying element precipitates or forms new phases, the determination of the alloying content in the surface layers with scanning electron microscopy does not represent the content which would be effective for the crystalline parameters following the Vegard relationship. Furthermore it is known that not only the mechanical but also the X-ray elastic constants are altered by alloying of iron, as was previously shown in the case of chrome-iron alloys.[12]

The effect of concentration gradients of alloying content in the surface layers of thermochemically treated materials on X-ray stress analysis is the greater the steeper the gradients and the deeper the penetration of X-rays. For a correct interpretation of X-ray stress analysis the application of the described method of correction is necessary.

BIBLIOGRAPHY

1. R. Glocker, Materialprüfung mit Röntgenstrahlen, Springer-Verlag, Bln, Hdbg., New York (1971).
2. C. S. Barrett, Structure of Metals, New York, McGraw Hill (1952).
3. E. Macherauch, u. P. Müller, Z. angew. Phys., 13 (1961) 305.
4. R. Prümmer, Kerntechnik, Isotopentechnik und Chemie 13 (1971) H.2, 68-77.
5. D. M. Vasilev, Sovj. Phys. Techn. Phys. 3 (1958) 2315.
6. H. Dölle a. J. B. Cohen, Metallurg. Transact. A, 11A (1980) 159.
7. R. Prümmer u. H. W. Pfeiffer-Vollmar, Z. Werkstofftechn. 12 (1981) 282.
8. L. Zwell, G. R. Speich a. W. C. Leslie, Metallurg. Transact. 4 (1973) 1990.
9. E. Öhmann, Z. phys. Chem. B 8 (1930) H.1/2, 81.
10. R. Prümmer u. H. W. Pfeiffer-Vollmar, Siemens ATM 293 (1982).
11. G. N. Dubinin a. L. Gribowski, Izv. Fiz. - cernaja metallurgija 11 (1962), 170.
12. R. Prümmer, unpubl. results.

APPLICATION OF A POSITION SENSITIVE SCINTILLATION DETECTOR FOR NONDESTRUCTIVE RESIDUAL STRESS MEASUREMENTS INSIDE STAINLESS STEEL PIPING

C.O. Ruud, P.S. DiMascio, and D.M. Melcher

Materials Research Laboratory
The Pennsylvania State University
University Park, PA 16802

INTRODUCTION

As early as 1974 cracking was observed in the austenitic stainless steel piping systems of several Boiling Water Reactors [1,2]. Failure analysis indicated that the cracks developed through intergranular stress-corrosion cracking and an active interest in residual stress measurement methodologies developed. This paper describes the procedures and demonstration testing employed to provide absolute residual stress measurement, nondestructively, on the inside surface of pipe specimens. A Ruud-Barrett position sensitive detector (PSSD)* was used to build an EPRI pipe stress analyzer which was developed for these residual stress measurements [3,4].

OBJECTIVE

The Materials Research Laboratory of The Pennsylvania State University (PSU) was contracted to develop operational procedures for the application of an EPRI pipe stress analyzer to residual stress measurements on the inside surface of stainless steel piping. Also, PSU was to demonstrate the viability of the procedures through proof testing of zero stress powders and welded pipe specimens. However, as the PSU program progressed, considerable effort had to be expended in redesigning and rebuilding the instrumentation supplied by a previous EPRI contractor. This effort was necessary in order to provide the electronic reliability and stability necessary for XRD residual

*Presently available from Denver X-Ray Instruments, Northglenn, CO.

stress measurements. Thus the objectives of the program developed to be:

1. Redesign and Rebuild the EPRI Pipe Stress Analyzer.
2. Test Calibration Procedures for Absolute Stress Measurement on Austenitic Stainless Steel.
3. Develop and Test Procedures for Data Refinement and Stress Calculation.
4. Demonstrate Stress Measurement Capabilities on Zero Stress and Pipe Stress Specimens.

This paper will primarily describe the procedures and results concerned with the last two objectives.

INSTRUMENTATION

The basic concept of the EPRI pipe stress analyzer is essentially the PSSD described in previously referenced papers [3,4]; interfaced with a PDP 11/03 computer with 32K memory. Peripherals include a teleprinter terminal, dual floppy discs, and CRT display. Data processing programs are supplied on floppy disc and programming runs under RT 11 control. The x-ray stress head, including the PSSD, is mounted upon a cantilevered support which allows for its insertion into a horizontally positioned pipe specimen, up to thirty-six inches. The collimated incident beam from the standard Amprex x-ray tube strikes the surface of the pipe at an angle (beta) of twenty-five degrees to provide residual stress measurement in the axial (longitudinal) direction of the pipe. It had previously been determined that Cr K-beta radiation filtered by Cr foil and diffracted at a Bragg angle of 73.9° from the (311) planes of the austenite grains provided the best combination of measurement accuracy and compactness of the x-ray optics [5].

PROCEDURES AND RESULTS

The four major objectives of the program undertaken by PSU are listed in a previous section. This paper will not be concerned with the first two items listed, except to state that the analyzer as rebuilt by PSU was eventually demonstrated as reliable and stable and that the previously developed instrument calibration procedures* were shown to provide absolute stress readings without the necessity of calibration on a zero stress standard.

*A method patent is pending on the procedure.

Data Refinement and Stress Calculation

The software employed in the EPRI pipe stress analyzer uses a number of algorithms to ameliorate electronic noise and artifacts as well as to correct for mechanical misalignments and x-ray focusing errors. This paper will not describe in detail the derivation and/or basis for the algorithms, but the following is a brief description of the sequence of corrections as displayed on the CRT during data refinement.

Upon command, the instrument will initiate x-ray data collection; this normally requires about 30 seconds at 40 KV and 10 Ma using a full wave rectified x-ray power supply. At the conclusion of the collection period the x-ray raw data is displayed upon the CRT, Fig. 1A. At this point the operator gives the command to calculate stress if the displayed data is satisfactory. The raw data is then corrected for electronic noise and gain inconsistencies between the individual silicon diode arrays, Fig. 1B. Next, a group of data points on both sides of each peak is used to provide a linear least square fit for x-ray background correction, Fig. 1C. Finally, the calculated background is subtracted point by point across all of the collected data and a parabola is fit by a least square routine to the upper half of the x-ray peak, Fig. 1D. The position of the apex of the parabolic fits are then located mathematically and these two values are used to calculate the stress using the following equation.

$$\sigma = K \frac{L - R + f(R_o)}{R_o}$$

where K = x-ray stress constant from four-point bend tests (see next section); R_o = the calculated distance from the specimen to the PSSD surfaces; L = the position of the left x-ray peak apex; and R = the position of the right x-ray peak apex.

Four-Point Bend Tests

Engineers and scientists experienced in XRD stress measurements are in general agreement that experimental derivation of the stress constant, i.e., x-ray peak shift versus applied stress, is necessary for accurate absolute residual stress readings [6]. The XRD stress constant was obtained for the experimental condition described above (see Instrumentation) by the four-point bend technique described by Prevey [7]. The constant was 0.00808° 2θ/KSI; this compares well to the 0.00795° 2θ/KSI value as determined from previous data [5]. The difference between these values is less than two percent.

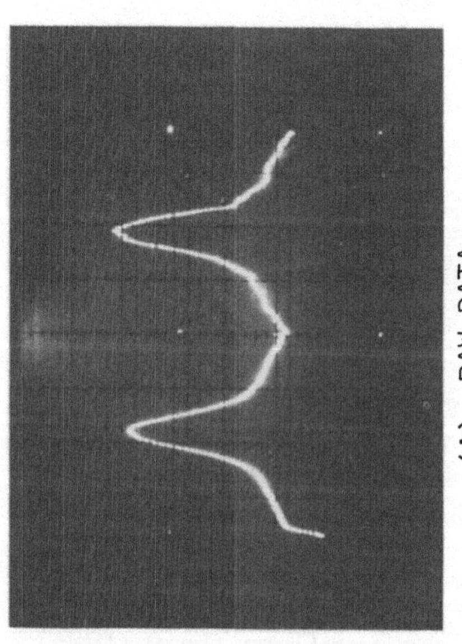
(A) RAW DATA

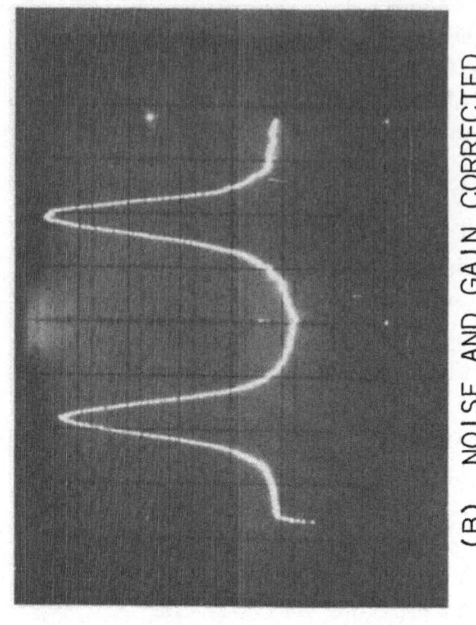
(B) NOISE AND GAIN CORRECTED

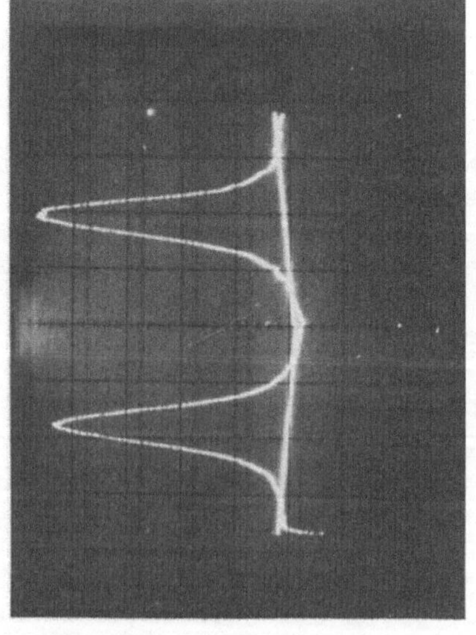
(C) BACKGROUND FIT

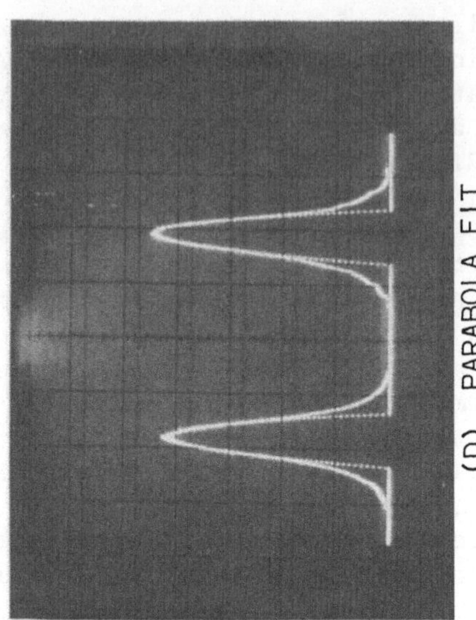
(D) PARABOLA FIT

Fig. 1

Zero Stress Conformation Testing

In order to establish the accuracy and precision of the EPRI pipe stress analyzer for absolute residual stress measurement, a zero stress specimen was obtained. This was done using a -400 mesh, loose stainless steel powder which, because of lack of bonding between the individual powder grains, will not support residual stress. Such a loose powder then is a reliable zero stress specimen and has been accepted as such by ASTM. The 304 austenitic stainless steel powder used here was calibrated against NBS SRM-640 using a state of the art APD-3600 x-ray diffractometer.

The zero stress tests were performed at β=25° and R_o = 40±0.9 mm. The R_o range was larger than for the subsequent pipe tests in order to illustrate the effect of R_o changes, as well as to provide data for accuracy estimation under conditions where both the left and right peak apexes were significantly displaced from the center of the PSSD detection surfaces.

The accuracy of the zero stress loose powder absolute stress measurements was shown as a mean of 0.1 KSI and a standard deviation of 3.4 KSI.

Comparison Testing

A small weldment coupon in which the residual stresses had been measured on several occasions by R. Chrenko at The General Electric Research and Development Center in Schnectady, New York, using a Rigaku Strainflex XRD stress analyzer was obtained. The coupon was approximately 2 x 1 x 0.5 inches (50 x 25 x 12 mm) in size and had been removed from a ten-inch diameter 304 stainless steel pipe with a 0.5 inch (12 mm) wall thickness. It included a 1-inch (25 mm) length of the weld across the long dimension and at its center. The long dimension, 2 inches (50 mm), was parallel to the pipe axis. On one side of the specimen, i.e., on one side of the weld, areas of the coupon representing the inside surface of the original pipe weldment had been marked into designated sections. These sections were approximately 0.08 x 1.0 inches (2 x 25 mm) with the long dimension parallel to the weld. These areas were used as indexing locations to establish the location of measurements made at G.E. using the Rigaku Strainflex instrument.

Several stress measurements were made at varying distances from the weld fusion line on that side of the weld marked into sections by G.E. These measurements were made over a period of five weeks by various operators at β=25° and R_o = 40±0.4 mm. The location of the measurements were indexed with respect to the weld fusion line.

Figure 2 shows measurements obtained from the EPRI pipe stress analyzer from March to late April. These readings are not corrected for beta angle error caused by the up-slope of the pipe along the axial traverse within 0.31 inches (4 mm) of the weld fusion line (WFL). Such an error is on the order of about 3 KSI at 0.16 and 0.31 inches (4 and 8 mm) from the WFL. The tolerance given is the estimated standard deviation (SD) of the accuracy obtained by summing the SD of the three or more readings at each location and the SD of the accuracy of the loose powder (i.e., 3.4 KSI). The SD of the accuracy of the G.E. readings was estimated. The plotted data includes bars representing the SD of the accuracy and the uncertainty in location of the readings. The SD of accuracy and uncertainty in location from the G.E. specimen were estimates based upon a letter and telephone conversation with R. Chrenko.

The data plotted in Fig. 2 shows good consistency between the various types of data especially when the error bars are considered as well as other factors such as the lack of LPA correction in the G.E. data and β angle correction in the PSU data. Furthermore, there is a steep stress gradient in this specimen beginning at the WFL and extending well beyond 0.10 inches (3 mm) from the WFL. Therefore, slight errors in measurement location can markedly effect the measured stress.

Pipe Testing

Demonstration of the pipe stress measurements were made using four pipe specimens identified as HT8013N, HSW5G-105, HSW5G-104 and HSW5G-103, and supplied by the Electric Power Research Institute. A minimum of three axial stress measurements were made at each location on traverses parallel to the pipe axis at four azimuths, 0, 90, 180, and 270 degrees, on both sides of the weld for a minimum of 72 stress measurements on each pipe. Approximately one half of a day was required to obtain the data from each pipe. The three or more readings at each location were reported as a mean stress and their standard deviation was incorporated into a standard deviation of the accuracy estimate.

Figure 3 shows all the data from all four azimuths from pipe HT8013N plotted. Azimuths 0, 90 and 180 on Side A are remarkably consistent; however, the data from those azimuths on Side B are not as uniform. The data indicates that Side B experienced additional nonuniform heating before, during, or after welding. Azimuth 270 is very different than the rest due to a repair weld that had been made in that region. Comparing the readings at 0.53 inches taken on Side A show that the counter bore machining produced very uniform residual stress (about -160 KSI); however, on Side B only the 90° azimuth seemed to show the same -160 KSI residual stress.

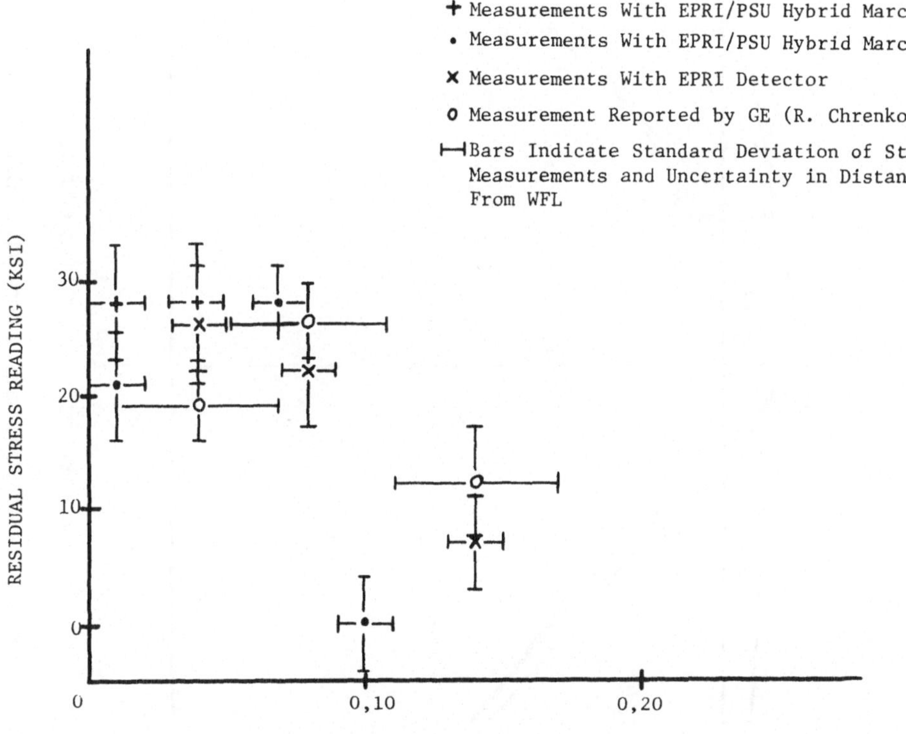

Fig. 2. Instrument comparison on G.E. pipe weldment "calibration" coupon

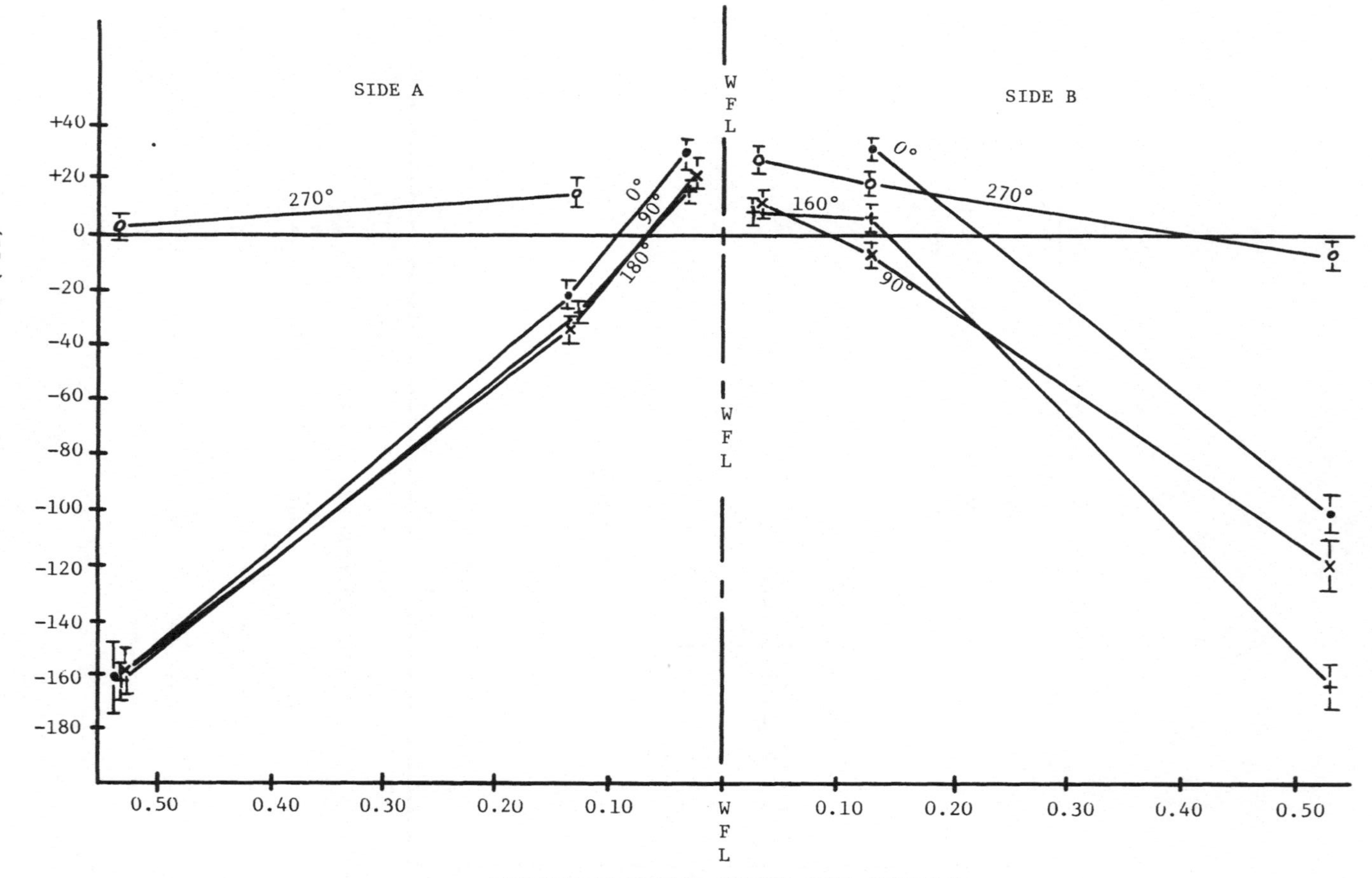

Fig. 3 XRD stress readings from pipe HT8013N.

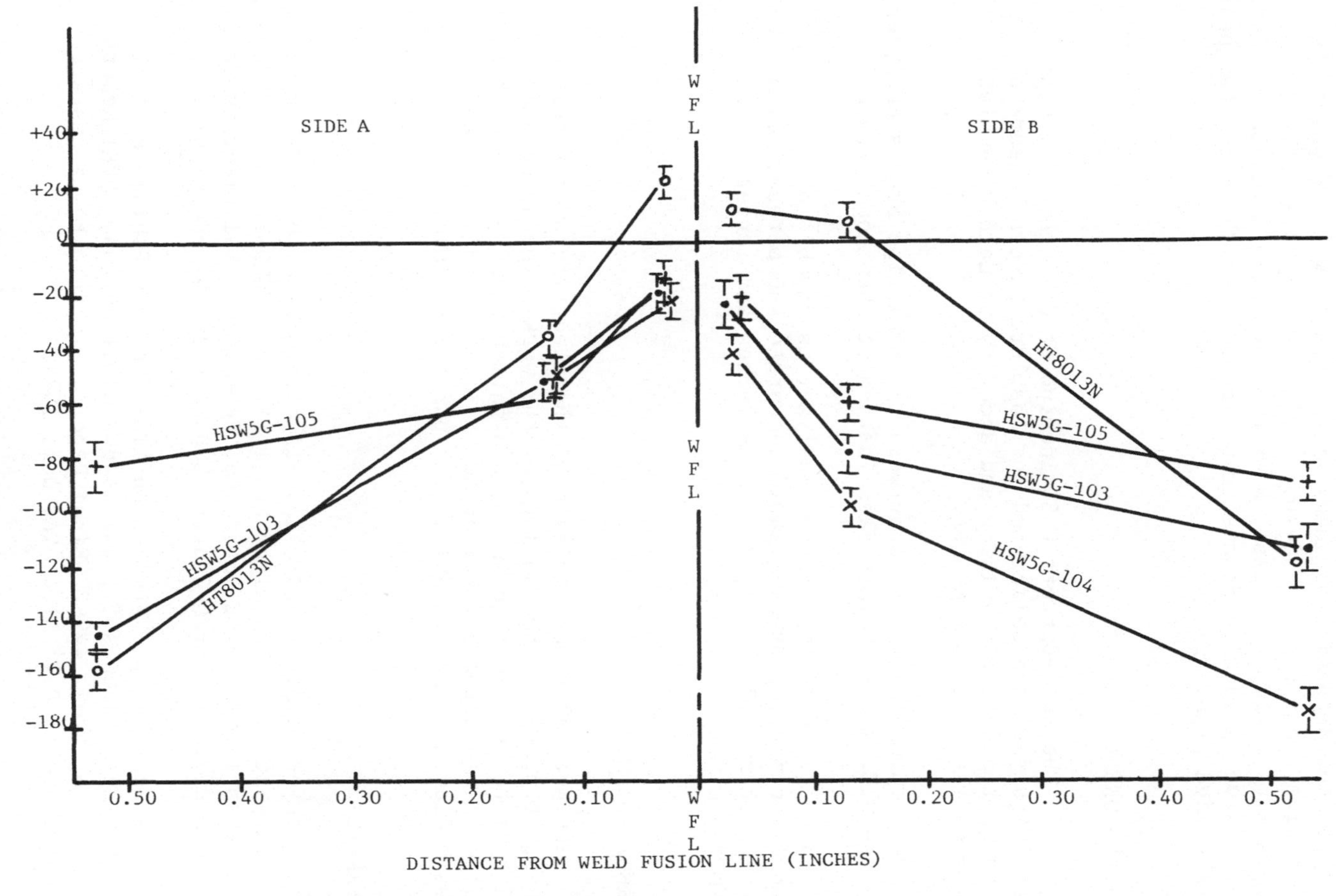

Fig. 4 XRD stress readings from four pipe - 90° azimuth.

Figure 4 shows the data from the 90 degree azimuth, from all four pipe plotted together. Near the weld (i.e., 0.03 inches) the HT pipe showed tensile residual stress due to welding while the other pipe specimens (HSW5G-103, -104, -105) consistently showed compressive residual stress. However, HSW5G-103 often showed lower compressive residual stress at 0.03 inches than the other two heat sink welded pipe.

CONCLUSIONS

1. The Ruud-Barrett PSSD incorporated in the EPRI Pipe Stress Analyzer and Calibrated for Absolute Stress Measurement shows unprecedented precision and accuracy at data accumulation times of less than one minute.

2. Comparative tests between the EPRI Pipe Stress Analyzer and the G.E. Rigaku Strain Flex instrument show good agreement.

3. The EPRI Pipe Stress Analyzer is capable of nondestructively providing rapid, accurate XRD stress readings on the inside surface of pipes of nine inches and larger inside diameter.

ACKNOWLEDGEMENTS

The Nuclear Power Division of the Electric Power Research Institute provided the funding for the work. The authors are indebted to Dr. James Quinn for his patience and advice on this project. Also, the efforts of Ms. Lois Annechini-Moore and Miss Kelley Ruud in preparing this paper were invaluable.

REFERENCES

1. Klepfer, H.H. et al., "Investigation of Cause of Cracking in Austenitic Stainless Steel Piping," NEDO-21000-1, 75NED35, Class 1, General Electric Co. Report, General Electric Co., Corporate Research Center, Schenectady, NY, July 1975.

2. Gianuzzi, A.J., "Studies on AISI Type-304 Stainless Steel Piping Weldments for Use in BWR Applications," EPRI Report NP-944, Proj. 449-2, Final Report, Electric Power Research Institute, Palo Alto, CA, Dec. 1978.

3. Steffen, D.A. and Ruud, C.O., "A Versatile Position Sensitive X-Ray Detector," Adv. in X-Ray Anal., Vol. 21, pp. 309-315, Plenum Press, NY, 1978.

4. Ruud, C.O., "Feasibility of Determining Stress in BWR Pipes With the DRI X-Ray Stress Analyzer," EPRI Report NP-914, Proj. 823-1, Final Report, Electric Power Research Institute, Palo Alto, CA, Oct. 1978.

5. Ruud, C.O. and Barrett, C.S., "Use of Cr K-Beta X-Rays and a Position Sensitive Detector for Residual Stress Measurement in Stainless Steel Pipe," Adv. in X-Ray Anal., Vol. 22, pp. 247-249, Plenum Press, NY, 1979.

6. SAE, "Residual Stress Measurement by X-Ray Diffraction - SAE J784a," Society of Automotive Engineers Inc., 1971.

7. Prevey, P.S., "A Method of Determining the Elastic Properties of Alloys in Selected Crystallographic Directions for X-Ray Diffraction Residual Stress Measurement," Adv. in X-Ray Anal., Vol. 20, Plenum Press, 1977, pp. 345-354.

ON THE X-RAY DIFFRACTION METHOD OF MEASUREMENT OF TRIAXIAL STRESSES WITH PARTICULAR REFERENCE TO THE ANGLE $2\theta_o$

S. Torbaty, J.M. Sprauel, G. Maeder and P.H. Markho

Ecole Nationale Supérieure d'Arts et Métiers
151 Boulevard de l'Hôpital
75013 Paris, France

ABSTRACT

Certain machining processes (e.g. turning, grinding, ...) lead to a splitting up and an opening out of the curve of $2\theta_{\phi\psi} = f(\sin^2\psi)$. In order to determine the complete stress tensor, it is necessary in this case to use the Dölle-Cohen method but this requires the knowledge of the diffraction angle $2\theta_o$ for the unstressed material. Such a material, however, is extremely difficult to obtain in practice and the method usually employed uses annealed powder for this measurement. In our study, we show that it is nevertheless possible to determine the angle $2\theta_o$ on a specimen for which the curve of $2\theta_{\phi\psi} = f(\sin^2\psi)$ does not open out. Applying the Dölle-Cohen method to turning of a medium carbon steel, we have been able to bring to the fore: (i) the presence of an opening out of the curve $2\theta_{\phi\psi} = f(\sin^2\psi)$, for measurements made in the direction of lay, which is attributed to a shear stress σ_{23}, and (ii) shear stress σ_{12} which causes a rotation of the stress distribution diagram $\sigma^\phi = f(\phi)$ in the plane of the specimen.

INTRODUCTION

When analysing stresses by X-ray diffraction, it is now well known that in the case of those processes that introduce shear stresses parallel to the surface of the material (e.g. turning, grinding, ...), the curves $2\theta_{\phi\psi}=f(\sin^2\psi)$ may split up in a manner corresponding to positive or negative values of ψ. This splitting up and opening out of these curves, which varies with the angle ϕ, may readily be explained by considering a state of triaxial stresses.[1-3]

In fact, in a reference coordinate system linked to the specimen (Fig. 1), the angle of diffraction may be expressed in terms of stresses according to the following relation:[1]

$$2\theta_{\phi\psi} = 2\theta_o + \frac{1}{K_1}(\sigma_{11}^{\phi}\sin^2\psi + \sigma_{33}^{\phi}\cos^2\psi + \sigma_{13}^{\phi}\sin 2\psi) + \frac{1}{K_2}(\sigma_{11}^{\phi} + \sigma_{22}^{\phi} + \sigma_{33}^{\phi}) \quad (1)$$

In Eqn. (1), K_1 and K_2 are constants and σ_{ij}^{ϕ} are the stresses in the measurement coordinate system X_i^{ϕ}. This equation may be considered as a generalization of the $\sin^2\psi$ law which will be considered later. The term in $\sigma_{13}\sin 2\psi$ readily explains the opening out of the curve mentioned earlier if σ_{13} is non-zero.

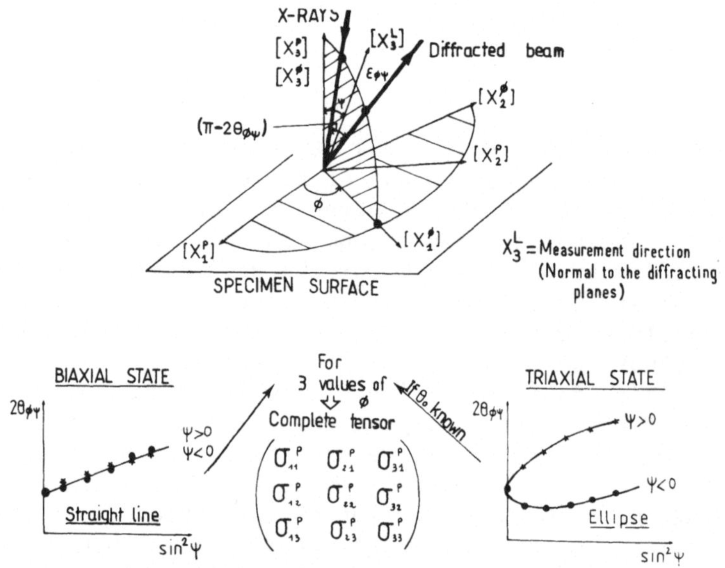

Fig. 1 Scheme for analysing a state of triaxial stresses by X-ray diffraction. (X_1^p, X_1^{ϕ} = specimen, measurement coordinate system respectively)

Meanwhile, if we want to determine the complete stress tensor, it is necessary to know the value of $2\theta_o$ which is the angle of diffraction for the same family of planes studied and for the same material as that on which the stress measurements are made, but in the unstressed state. Now, to obtain a stress-free material is very difficult, so the method usually employed consists of taking an annealed powder on which to do this measurement.[2,4]

In the course of a study of the stresses introduced by turning of a medium carbon steel (0.38% C) in the annealed or quenched-and-tempered state,[5] we demonstrate that it is possible to determine the value of $2\theta_o$ using a real specimen which is identical to the as-machined specimens but for which the curve $2\theta_{\phi\psi} = f(\sin^2\psi)$ does not open out.

EXPERIMENTAL CONDITIONS

The measurements by X-ray diffraction were carried out with an experimental set-up equipped with a position-sensitive detector (PSD)[6] and entirely computer-controlled (Fig. 2). A special mounting allows for measurements to be made in any direction ϕ, from 0° to 360°.[7]

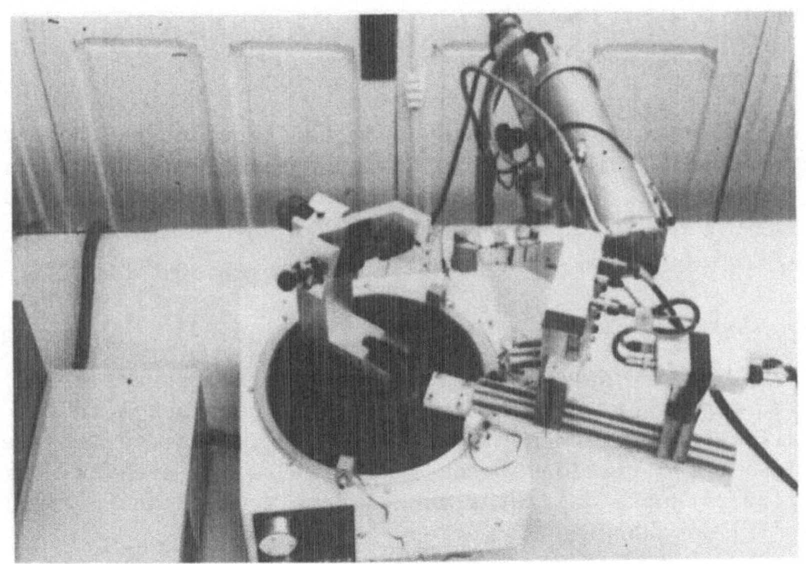

Fig. 2 View of the measurement system

The measurement conditions were as follows:

- Incident radiation: λK_α Cr (30 kV; 20 mA)
- Vanadium filter in front of the detector
- Diffraction plane {211}
- Irradiated area: circle of diameter 1.5 mm
- 13 values of ψ such that $\sin^2\psi = 0, 0.1, 0.2, \ldots, 0.6$
- Measurement time for one value of ϕ: 25 min
- Specimens: cylindrical (ϕ50 mm × 50 mm long), turned under different conditions of cutting.

CLASSICAL DETERMINATION OF $2\theta_o$

In our method we start with a classical determination of $2\theta_o$. Under a system of biaxial stresses, the $\sin^2\psi$ law is written as follows:

$$2\theta_{\phi\psi} - 2\theta_\circ = \frac{1}{K_1}\sigma_{11}^{\phi}\sin^2\psi + \frac{1}{K_2}(\sigma_{11}^{\phi} + \sigma_{22}^{\phi}) \qquad (2)$$

Thus the slope of the straight line $2\theta_{\phi\psi} = f(\sin^2\psi)$ gives σ_{11}^{ϕ}. Furthermore, if $\psi = 0$, then $2\theta_\circ = 2\theta_{\phi,\psi=0} - \frac{1}{K_2}(\sigma_{11}^{\phi} + \sigma_{22}^{\phi})$. Now, with measurements made at $\phi = 0$ and $\phi = 90°$, we may be able to determine $\sigma_{11}^{\phi} + \sigma_{22}^{\phi}$ since for $\phi = 0$, $\sigma_{11}^{\phi=0} = \sigma_{11}^{p}$ and for $\phi = 90°$, $\sigma_{11}^{\phi=90°} = \sigma_{22}^{p}$. (Here, p refers to the specimen coordinate system (Fig. 1)). Thus,

$$\sigma_{11}^{\phi} + \sigma_{22}^{\phi} = \sigma_{11}^{p} + \sigma_{22}^{p}$$

and the value of $2\theta_\circ$ may now be calculated.

By utilizing a specimen taken from the interior of the bar and polished electrolytically, the following results were obtained:

$\phi = 0$: $\sigma_{11}^{p} = -15 \pm 11$ MPa; $2\theta_{\phi=\psi=0} = 156.301° \pm 0.02°$

$\phi = 90°$: $\sigma_{22}^{p} = -9 \pm 16$ MPa; $2\theta_{\phi=90°,\psi=0} = 156.305° \pm 0.02°$

Hence, $\sigma_{11}^{p} + \sigma_{22}^{p} = -24$ MPa and $(2\theta_{\phi,\psi=0})_{mean} = 156.303°$. By using this value of $2\theta_\circ$, we are able to analyse the triaxial stress state in relation to depth for test pieces which have been subjected to different metallurgical treatments (annealed or quenched-and-tempered). Fig. 3 shows the variation of the shear stress σ_{23} and the normal stress σ_{33} with depth (z) for a turned sample which had been: (a) annealed, (b) quenched and tempered.

We observe in Fig. 3 that the shear stress σ_{23} is effectively down to zero at a depth of about 100 μm in (a) and 80 μm in (b). At about the same values of depth, the normal stress σ_{33} has almost levelled out to a constant value of about -20 MPa in (a) and -150 MPa in (b).

Now, static equilibrium of the material in the direction X_3^{ϕ} imposes the following relation:

$$\frac{\partial\sigma_{13}}{\partial X_1} + \frac{\partial\sigma_{23}}{\partial X_2} + \frac{\partial\sigma_{33}}{\partial X_3} = 0 \qquad (3)$$

When measurements at several points of the plane X_1X_2 indicate σ_{13} and σ_{23} to be zero (as no opening out of the curves in $\sin^2\psi$ is observed), their gradients too (in the direction of X_1 and X_2 respectively) must be zero. It follows from the previous equation that the stress σ_{33} should be constant over the layer penetrated by the X-rays. However, the fact that σ_{33} does not appear to reduce to zero at the same time as σ_{23} must be due to an incorrect value of $2\theta_\circ$.

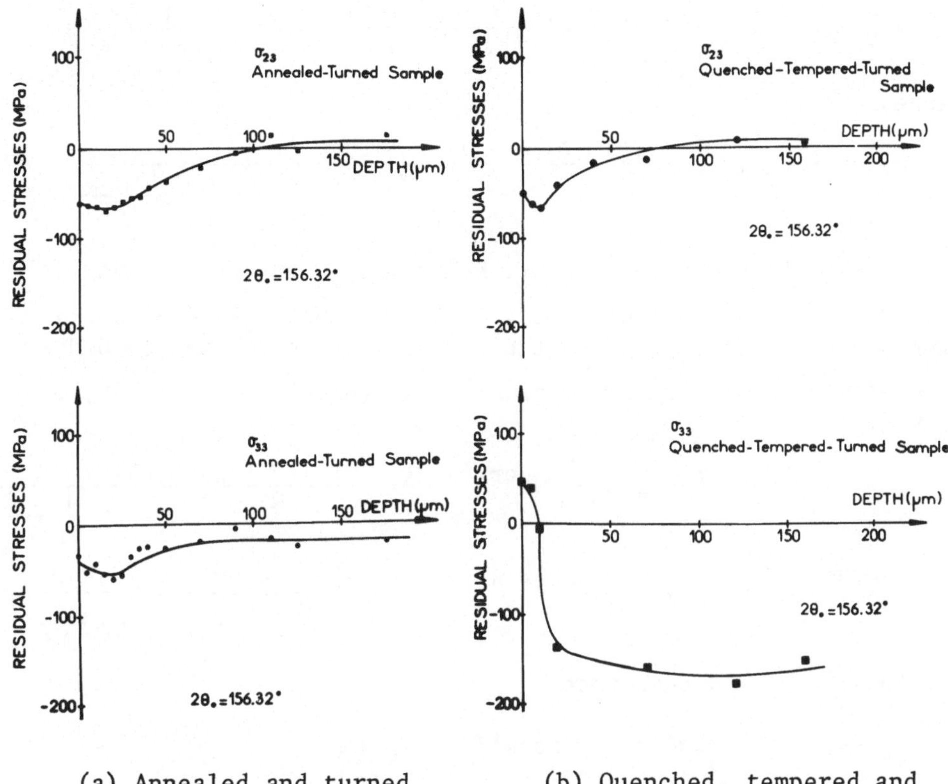

(a) Annealed and turned sample

(b) Quenched, tempered and turned sample

Fig. 3 Variation of the shear stress σ_{23} and the normal stress σ_{33} with depth (z)

CORRECTED VALUE OF $2\theta_\circ$

For the case of $\phi = 0$, Eqn. (1) may be stated as follows:

$$2\theta_{\phi=0,\psi} = 2\theta_\circ + \frac{1}{K_1}(\sigma_{11}\sin^2\psi + \sigma_{33}\cos^2\psi + \sigma_{13}\sin 2\psi) + \frac{1}{K_2}(\sigma_{11} + \sigma_{22} + \sigma_{33}) \tag{4}$$

With ψ=0, Eqn. (4) reduces to

$$2\theta_{\phi=\psi=0} = 2\theta_\circ + \frac{\sigma_{11} + \sigma_{22}}{K_2} + \sigma_{33}\left(\frac{1}{K_1} + \frac{1}{K_2}\right)$$

This equation may be rearranged as follows:

$$2\theta_{\phi=\psi=o} = 2\theta_o + \frac{(\sigma_{11} - \sigma_{33}) + (\sigma_{22} - \sigma_{33})}{K_2} + \sigma_{33}\left(\frac{1}{K_1} + \frac{3}{K_2}\right)$$

Whence,

$$\sigma_{33} = \left\{2\theta_{\phi=\psi=o} - 2\theta_o - \frac{(\sigma_{11} - \sigma_{33}) + (\sigma_{22} - \sigma_{33})}{K_2}\right\} \times \frac{1}{\frac{1}{K_1} + \frac{3}{K_2}} \quad (5)$$

In Eqn. 5, the values of K_1, K_2, $2\theta_{\phi=\psi=o}$, $(\sigma_{11} - \sigma_{33})$ and $(\sigma_{22} - \sigma_{33})$ are independent of $2\theta_o$. If we denote $\sigma_{33}^{corr'd}$ as the value of the stress σ_{33} after the correction (here equal to 0 MPa) and $\sigma_{33}^{meas'd}$ as the measured value of σ_{33} (before correction), we may write:

$$\sigma_{33}^{corr'd} = 0 = \left\{2\theta_{\phi=\psi=o} - 2\theta_o^{corr'd} - \frac{(\sigma_{11}-\sigma_{33}) + (\sigma_{22}-\sigma_{33})}{K_2}\right\} \times \frac{1}{\frac{1}{K_1} + \frac{3}{K_2}}$$

$$\sigma_{33}^{meas'd} = \left\{2\theta_{\phi=\psi=o} - 2\theta_o^{meas'd} - \frac{(\sigma_{11}-\sigma_{33}) + (\sigma_{22}-\sigma_{33})}{K_2}\right\} \times \frac{1}{\frac{1}{K_1} + \frac{3}{K_2}}$$

From this, we may deduce that

$$2\theta_o^{corr'd} = 2\theta_o^{meas'd} + \sigma_{33}^{meas'd}\left(\frac{1}{K_1} + \frac{3}{K_2}\right) \quad (6)$$

With the aid of Eqn. (6) and using $K_1 = -308$ MPa/degree and $K_2 = 1449$ MPa/degree, the values of $2\theta_o^{corr'd}$ (Table 1) were readily calculated for each of two specimens studied.

For measurements made to a depth (z) of 140 µm, Fig. 4 shows the effect of this correction on the complete stress tensor. Where the stresses are biaxial, the value of σ_{33} is found to be well enough zero.

Table 1

	$\sigma_{33}^{meas'd}$ (MPa)	$2\theta_o^{meas'd}$ (°)	$2\theta_o^{corr'd}$ (°)
Annealed and turned specimen	-20	156.32	156.34
Quenched, tempered and turned specimen	-150	156.32	156.47

Stress Tensor (Values in MPa)

Measurements at a depth of 140 μm

$2\theta_o$=156.32° (value measured on annealed material)

Annealed-turned sample			Quenched-tempered-turned sample		
-58	-43	-5	-170	8	3
-43	-30	-10	8	-173	5
-5	-10	-21	3	5	-153

Corrected value of $2\theta_o$ so that $\sigma_{33}=0$

$2\theta_o$=156.34°			$2\theta_o$=156.47°		
-37	-43	-5	-25	8	3
-43	-9	-10	8	-29	5
-5	-10	0	3	5	0

Fig. 4 Complete stress tensor before and after the $2\theta_o$-correction

To recapitulate, when analysing a triaxial stress state by the method of X-ray diffraction, the determination of the angle $2\theta_o$ for a stress-free material may be effected as follows:

(i) Obtain a value of $2\theta_o$ by the classical method from a stress-free powder or the specimen itself. Alternatively, "assume" a plausible value of $2\theta_o$.

(ii) Using this value, and treating the problem as a case of triaxial stresses, determine the complete stress tensor at such a point, interior to the specimen if necessary, where the measurements do not show an opening out of the curve $2\theta_{\phi\psi} = f(\sin^2\psi)$. In general, the value of σ_{33} will not be zero.

(iii) Now, where the values of σ_{13} and σ_{23} are found to be zero, the value of σ_{33} must equally be zero. Thus, the non-zero value obtained for σ_{33} may be used to correct the measured value of $2\theta_o$ giving the corrected value of the latter which would make σ_{33} zero.

APPLICATION OF THE DÖLLE-COHEN METHOD TO TURNING

Applying the Dölle-Cohen method to turning of a medium carbon steel (0.38% C), we have been able to illustrate the following:

(i) An opening out of the curve $2\theta_{\phi\psi} = f(\sin^2\psi)$ for measurements obtained in the direction of lay (Fig. 5) which is attributed to, whilst at the same time emphasizing the presence of, a shear stress σ_{23} (Fig. 3).

(ii) The presence of a shear stress σ_{12} in the plane of the specimen (X_1X_2 in Fig. 1) which provokes a rotation of the polar stress distribution diagram for the direct stress σ^{ϕ} (Fig. 6)[5] given by the following relation:

$$\sigma\phi = \sigma_{11} \cos^2 \phi + \sigma_{12} \sin 2\phi + \sigma_{22} \sin^2 \phi \qquad (7)$$

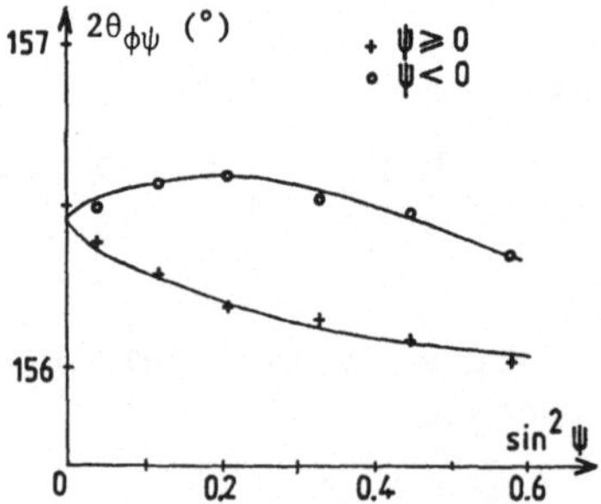

Fig. 5 Opening out of the curve $2\theta_{\phi\psi} = f(\sin^2\psi)$ for measurements made in the direction of lay ($\phi = 90°$)

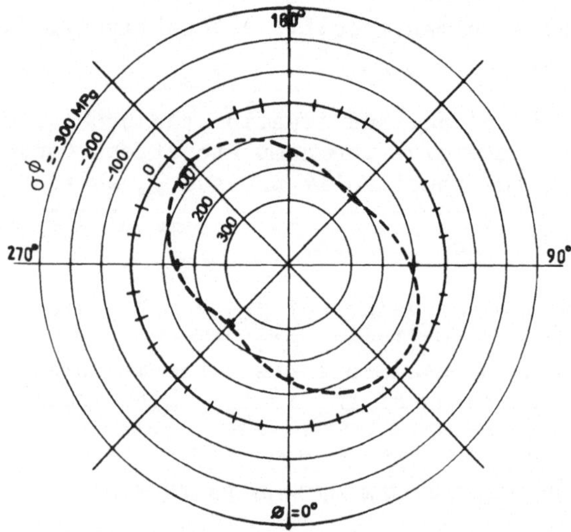

Fig. 6 Rotation of the polar stress distribution diagram for σ^{ϕ} (Eqn. (7)) caused by the shear stress σ_{12}

CONCLUSIONS

For the analysis of a triaxial stress system by X-ray diffraction using the Dölle-Cohen method, prior knowledge of the angle $2\theta_o$ for the stress-free material has normally been considered necessary. By utilizing the fact that for equilibrium requirements the normal stress has to be zero where the shear stress σ_{23} is zero, we have shown that the correct value of $2\theta_o$ may readily be determined together with the complete stress tensor. With this method, we believe that the initial choice of $2\theta_o$ is not very significant.

Application of the Dölle-Cohen method to turning has enabled us to bring to the fore the existence, on the surface of a turned specimen, of the shear stress σ_{12} in addition to σ_{23}.

REFERENCES

1. Dölle, H., J. Appl. Cryst., Vol.12, 1979, pp.489-501
2. Dölle, H. and Cohen, J. B., Met. Trans., Vol.11A, 1980, pp.159-164
3. Maeder, G., Lebrun, J. L. and Sprauel, J. M., N.D.T. Int'l, 1981, pp.235-248
4. Yoshioka, Y., Kawata, M. and Morinaga, M., Adv. in X-ray Anal., Vol.24, 1981, pp.181-185
5. Torbaty, S., Thesis, 1981, Paris VI-ENSAM (Paris)
6. Castex, L., Lebrun, J. L. and Bras, S., Adv. in X-ray Anal., Vol.24, 1981, pp.139-142
7. Sprauel, J. M., Lebrun, J. L., Odent, J. P. and Imbert, G., Mem. Sci. Rev. Met., 1982, pp.73-78

DIRECT DETERMINATION OF STRESS IN A THIN FILM DEPOSITED ON A SINGLE-CRYSTAL SUBSTRATE FROM AN X-RAY TOPOGRAPHIC IMAGE

Wayne S. Berry

IBM General Technology Division
Essex Junction, Vermont 05452

ABSTRACT

X-ray topography is one of several methods used by the semiconductor industry to measure stress of thin films deposited on single-crystal substrates. A procedure to determine stress values directly from the X-ray topographic image by measuring the separation of the $K\alpha_1$-$K\alpha_2$ X-ray peaks is reviewed. Although less sensitive than the single-crystal technique[1], it has been found applicable to monitor stress levels in films such as CVD silicon nitride on silicon wafers. This technique may also be used to quantify the plastic deformation of wafers that is induced by semiconductor processes.

INTRODUCTION

Films deposited on single-crystal wafers, when under sufficient internal stress, can cause the wafer to elastically bow, either in a concave (tensile stress) or a convex (compressive stress) direction. The magnitude of the film's stress can be calculated by stress formula 1 when the radius of curvature (bow) has been determined.

$$\sigma_f = \pm \frac{t_s^2}{6t_f r} K . \qquad (1)$$

Where:
σ_f = film stress in dynes/cm^2,
$\pm$ = + tensile, - compressive,
t_s = substrate thickness,
t_f = film thickness,
r = radius of curvature,
k = 1.805×10^{12} dynes/cm^2 for (100) Si.

The radius of curvature (r) by the described technique is derived from formula 2:

$$r = \frac{L}{1.15 \times 10^{-3}} . \tag{2}$$

Where:
r = radius in cm,
L = distance between $K\alpha_{1-2}$ (cm),
1.15×10^{-3} = angular separation between $K\alpha_{1-2}$ (Mo radiation).

This technique requires that the film(s) be uniformly deposited on one side of a defect-free, single-crystal substrate or wafer. The deposition conditions have to be well controlled to prevent plastic deformation and dislocations in the wafer; i.e., only elastic deformation should occur.

PROCEDURE

A standard single-crystal transmission X-ray topographic camera (Lang Technique) is used with a narrow (0.5-mm) collimating slit on the diverging X-ray beam. Alignment is identical to the standard setup for topography. The wafer is then translated to determine the relative positions of the $K\alpha_1$ and $K\alpha_2$ peaks of the elastically deformed wafer. If necessary, θ may be readjusted to bring the $K\alpha_1$-$K\alpha_2$ separation near the center of the wafer.

The magnitude of the $K\alpha_1$-$K\alpha_2$ separation may be determined from: (1) the topographic image recorded on a film (Fig. 1), (2) a strip chart recording which is calibrated to the translating speed of the wafer (Fig. 1), or (3) visually comparing the position of a graduated scale on the topographic camera stage to the $K\alpha_1$-$K\alpha_2$ peaks as monitored on a rate meter.

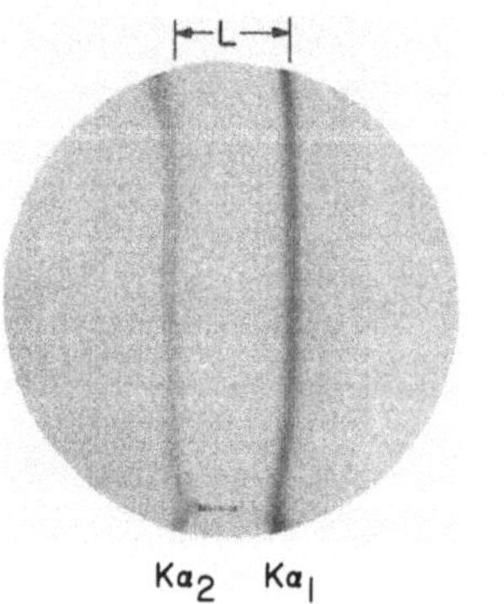

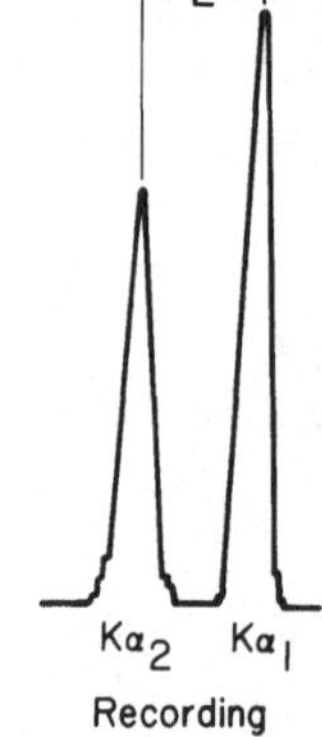

Wafer: 82mm Diameter (100) Si
Topograph: (220) Diffraction Conditions
Film: 0.7μ CVD Silicon Nitride
Radius of Curvature: 1.9×10^3 cm
Stress: 3.5×10^9 Dynes/cm^2, Tensile

Figure 1. Typical Topographic Image and Strip Chart Recording of Stressed Film

DISCUSSION

Determining Direction of Curvature

Figures 2 and 3 show the relative positions of the $K\alpha_1$-$K\alpha_2$ diffraction conditions with respect to direction of curvature and wafer-film orientation. Close examination shows that $K\alpha_1$-$K\alpha_2$ positions are reversed for the opposite direction of curvature. Once a convention is established (wafer-film to X-ray source and diffracting angle), convex and concave directions can be quickly determined.

Accuracy

Several samples of CVD silicon nitride deposited on silicon wafers were used to compare the direct technique to the single-crystal technique for reproducibility and precision. The results are plotted in Fig. 4 showing a maximum deviation of 5%.

Sensitivity

The minimum sensitivity (Fig. 5) is approximately an order of magnitude lower when compared to the single-crystal technique. This limits the technique to films with higher levels of stress, such as CVD silicon nitride.

While the minimum sensitivity is limited by the wafer diameter, the maximum is related to how well the $K\alpha_1$-$K\alpha_2$ lines can be separated (max. r = $\sim 1 \times 10^2$ cm). However, when r approaches 1×10^2 cm, plastic deformation has usually occurred. Consequently, this technique is also useful for quantifying the plastic deformation of wafers that is induced by semiconductor processing.

SUMMARY

The X-ray topographic technique of observing the separation of the $K\alpha_1$ and $K\alpha_2$ peaks to determine stress has been routinely used

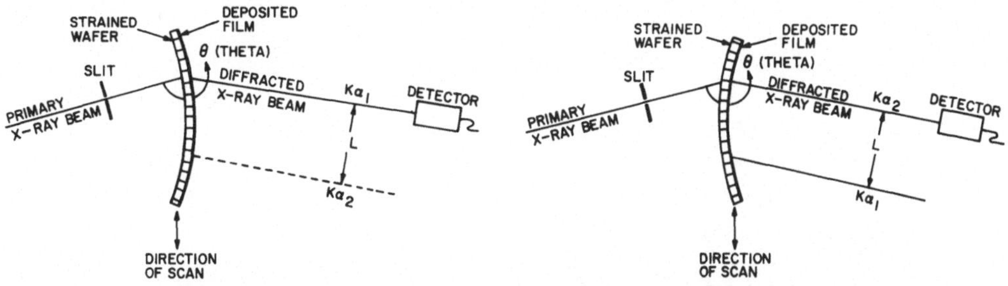

Figure 2. Diffracting Conditions for a Compressive Stressed Film

Figure 3. Diffracting Conditions for a Tensile Stressed Film

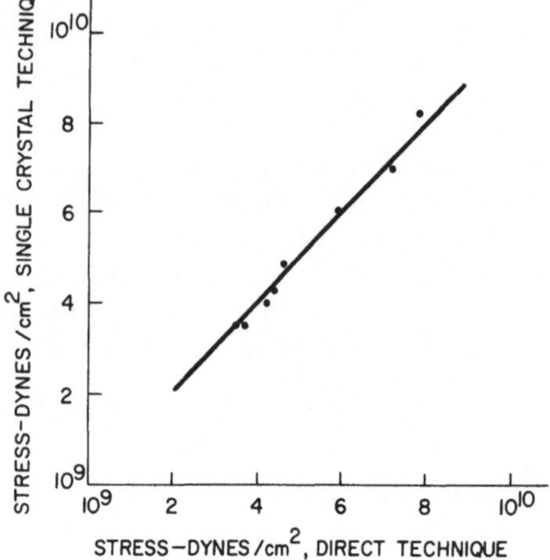

Figure 4. Comparision of Direct Technique to Single-Crystal Technique for Same Samples

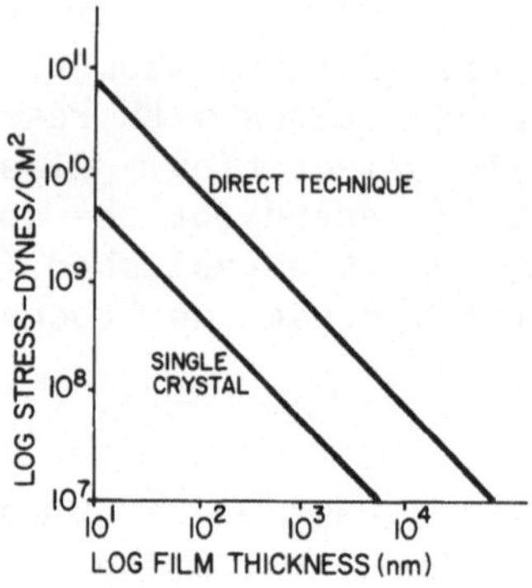

LOG STRESS-DYNES/CM^2

10^{11}
10^{10}
10^9
10^8
10^7

DIRECT TECHNIQUE

SINGLE CRYSTAL

10^1 10^2 10^3 10^4

LOG FILM THICKNESS (nm)

Figure 5. Comparision of Minimum Sensitivities Between Direct Technique and Single-Crystal Technique

in our laboratory. Even though it is less sensitive than the single-crystal technique, it has proven reproducibility with simple data acquisition. The data can be recorded by photographic techniques, strip chart recording, or directly observing the relative wafer positions with respect to the $K\alpha_1$-$K\alpha_2$ peaks.

REFERENCE

1. E. W. Hearn, Stress measurements in thin films deposited on single-crystal substrates through topographic techniques, Advances in X-ray Analysis, 20:273-281 (1977).

THE DETERMINATION OF ELASTIC CONSTANTS USING A COMBINATION OF X-RAY STRESS TECHNIQUES

Charles Goldsmith and George A. Walker

IBM - East Fishkill Facility

Hopewell Junction, NY 12533

ABSTRACT

The powder diffraction x-ray technique commonly used to measure strain in polycrystalline materials requires a knowledge of the elastic constants in order to convert the strain into a stress value. For many materials, these constants are not always known. Another technique to measure strain is the x-ray lattice curvature (substrate bending) method which requires no knowledge of the film elastic constants. The strain is measured in the substrate and requires only the elastic constants of the substrate to convert the measured strain into stress. Using a combination of the powder diffraction technique and a double crystal lattice curvature technique, the elastic constants of $TaSi_2$ and WSi_2 have been determined for various crystallographic directions.

INTRODUCTION

In semiconductor technology, thin films are of great technological importance. These films can vary in thickness from less than a micron to several microns and may be either amorphous or crystalline in nature. Residual stress, one of many important film properties, can be detrimental, leading to loss of film adhesion, film cracking and, at times, generating slip and dislocations in the underlying silicon substrate. Consequently, these adverse effects create a need to measure the residual stresses in new thin film systems.

Two x-ray techniques are employed in our laboratory to measure residual stress. One method is the standard polycrystalline powder diffraction residual stress technique and the other is a seldom used double crystal x-ray lattice curvature (radius of curvature) technique.[1,2] In this paper, we describe how the two techniques can be combined to determine the elastic constants for specific crystallographic directions (planes) and then apply the technique to determine the elastic constants of WSi_2 and $TaSi_2$.

EXPERIMENTAL

The WSi_2 film was prepared by co-evaporating approximately 0.12μm of tungsten and silicon onto a [100] silicon wafer 0.40mm thick, followed by sintering at 1000°C - 30 minutes in N_2. The $TaSi_2$ films were prepared by co-sputtering approximately 0.12 μm of tantalum and silicon onto a [100] silicon wafer 0.40mm thick and sintering at 1000°C - 30 minutes in argon.

The x-ray double crystal lattice curvature technique provides an indirect measure the film stress by measuring the elastic deformation that the film stress induces in the single crystal substrate. For this study, we have assumed that the film is attached to the substrate with perfect adhesion and that there is a uniform stress through the film thickness. Figures 1a and 1b show a schematic of a double crystal diffractometer in transmission-transmission mode. Figure 1a shows the camera with a stress-free crystal in position two. When the crystals are translated, the intensity of the double

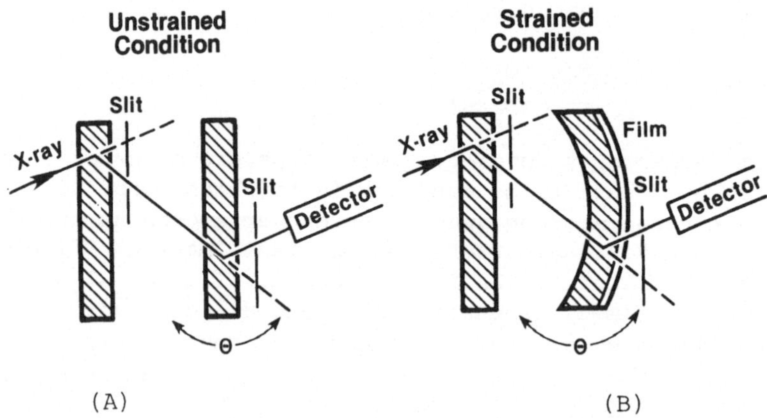

Fig. 1. Schematic of double crystal arrangement.
(A) both crystals are perfect; (B) the crystal in position 2 is strained.

diffracted x-ray beam remains constant since the Bragg angle is aligned at all points in the second crystal. In Figure 1b, the crystal in position two has a curvature due to the stress generated by a thin film. When these crystals are translated, the intensity of the double diffracted x-ray beam decreases rapidly. In this case, the Bragg angle is aligned only in a small area due to the curvature of the lattice. To maintain the reflection from the second crystal, the Bragg angle theta has to be adjusted after a translation. This required adjustment can be used as a measure of the crystal curvature where the radius of curvature (R) = $1/\Delta\theta$ and 1 is the scan length.

The lattice curvature of the substrate is a direct result of the film deposition and is related to the film stress by the following equation:[3]

$$\sigma_F = \pm \left(\frac{t_s^{\,2}}{6\, t_f\, R}\right)\left(\frac{E_s}{1-\nu_s}\right) \qquad (1)$$

σ_F = film stress
t_s = substrate thickness
t_f = film thickness
R = radius of curvature
E_s = Young's modulus of the substrate
ν_s = Poisson's ratio of the substrate
$\pm$ = sign of the film stress (+ = tension)

The lattice curvature was measured using molybdenum radiation diffracted from the silicon [220] planes and was converted into stress using a value of $E_s/1-\nu_s$ of 1.805 X 10^5 MPa.[4] The determination of the sign of the stress follows standard convention.

The polycrystalline powder diffraction residual stress technique provides a direct measure of film strain via the measurement of film lattice spacings. This technique has been in use for a number of years and a detailed description can be found in many papers.[5-7] Basically, the technique requires that the difference in lattice spacings be determined from a set of planes oriented at two or more angles to the specimen surface. A schematic of this arrangement showing the relationship between plane normal and surface normal is shown in Figure 2. The change in lattice spacing with tilt angle is the result of a residual strain and can be related to stress as follows:

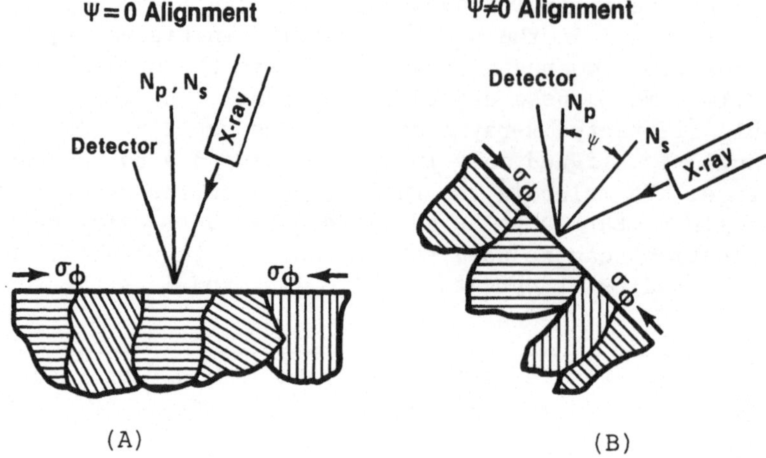

Fig. 2. Schematic of polycrystalline powder diffraction residual stress method. N_p = plane normal; N_s = surface normal.

(A) normal specimen alignment $\psi = 0$; (B) inclined specimen alignment $\psi \neq 0$.

$$\sigma_\phi = -\left(\frac{E}{1+\nu}\right)_{hkl} \left(\frac{\text{Cot}\ \Theta}{2}\right) \left(\frac{\pi}{180}\right) \left(\frac{\Delta 2\Theta}{\sin^2\Psi}\right) \tag{2}$$

σ_ϕ = film stress
E = Young's modulus of the film
ν = Poisson's ratio for the film
Θ = Bragg angle
Ψ = tilt angle

THEORY

The powder diffraction residual stress technique (Equation 2) shows that the measured strain can be converted into stress only if the elastic constants for the specific crystallographic direction we are measuring are known for the film material. At times, bulk elastic constants are used, but this practice can lead to a considerable error in the stress value. Elastic constants for specific directions are known for a few common materials, but many times neither bulk nor specific elastic constants are known for uncommon materials.

Since the x-ray lattice curvature technique (Equation 1) does not require a knowledge of the elastic constants of the film to determine the stress because the measurements are performed on the substrate, the two stress techniques can be combined to determine the elastic constants for a specific crystallographic direction. Equation 2 can be written in the following form:

$$\left(\frac{E}{1+\nu}\right)_{hkl} = -\left(\frac{2\sigma_\phi}{\cot\theta}\right)\left(\frac{180}{\pi}\right)\left(\frac{\sin^2\Psi}{\Delta 2\theta}\right) \quad (3)$$

The elastic constants for a specific plane can now be determined by substituting the strain from the powder diffraction residual technique (in the form of $\Delta 2\theta/\sin^2\psi$) and the stress determined by x-ray lattice curvature technique (assuming $\sigma_F = \sigma_\phi$) into equation 3.

RESULTS

X-ray diffractometer traces were obtained of the sintered refractory metal silicides to be sure that the films were composed of a single phase. This is an important step since the x-ray lattice curvature technique measures the combined stress of multiple films or phases and the powder diffraction stress technique measures the strain in each film or phase separately.

Table 1 contains the stress results and the elastic constants determined for the specific planes listed. Also included in the table are the diffractometer configurations[7,8] and radiations used in the polycrystalline powder diffraction residual stress measurements.

The copper film shown in Table 1 was used as a standard to check the technique. The specific elastic constants determined for these planes correlated well with a set of specific elastic constants that we had calculated previously from single crystal elastic constants.

The double crystal measurements show that the stresses in our metal disilicide systems are very high; however, they agree well with published values for these and other similar metal silicide systems.[9,10]

The measurement of strain by the polycrystalline powder method yielded linear plots of $\Delta 2\theta$ vs. $\sin^2\psi$ regardless of the plane used or diffractometer configuration. The plots show no evidence of ψ splitting, indicating no strain gradients in the films. Typical plots are shown in Figures 3 and 4 for WSi_2 and $TaSi_2$, respectively. The specific elastic constants for both WSi_2 and $TaSi_2$ show little difference from plane to plane, indicating that these materials are not

Table 1. Results

Sample	Stress (MPa) Lattice Curvature	XRPD Stress Tech. Slope of 2Θ vs. $\sin^2\Psi$	Diffractometer Configuration	(hkl)	RAD.	Elastic Const. $(E/1+\nu)_{hkl} \times 10^5$ MPa
Copper	160	-0.3547	1	(220)	Cr	1.04
		-0.3202	2	(331)	Cu	1.43
		-0.5524	2	(420)	Cu	1.04
WSi_2	1560	-1.095	1	(213)	Cr	3.59
		-1.694	2	(316)	Cu	3.58
#1 $TaSi_2$	1725	-2.040	1	(220)	Cr	3.29
		-1.465	2	(400)	Fe	3.54
		-2.125	2	(420)	Cu	3.10
#2 $TaSi_2$	1525	-1.890	3	(220)	Cr	3.13
		-2.072	2	(420)	Cu	2.96

(1) Parallel Beam

(2) Fixed Slit

(3) Para Focusing

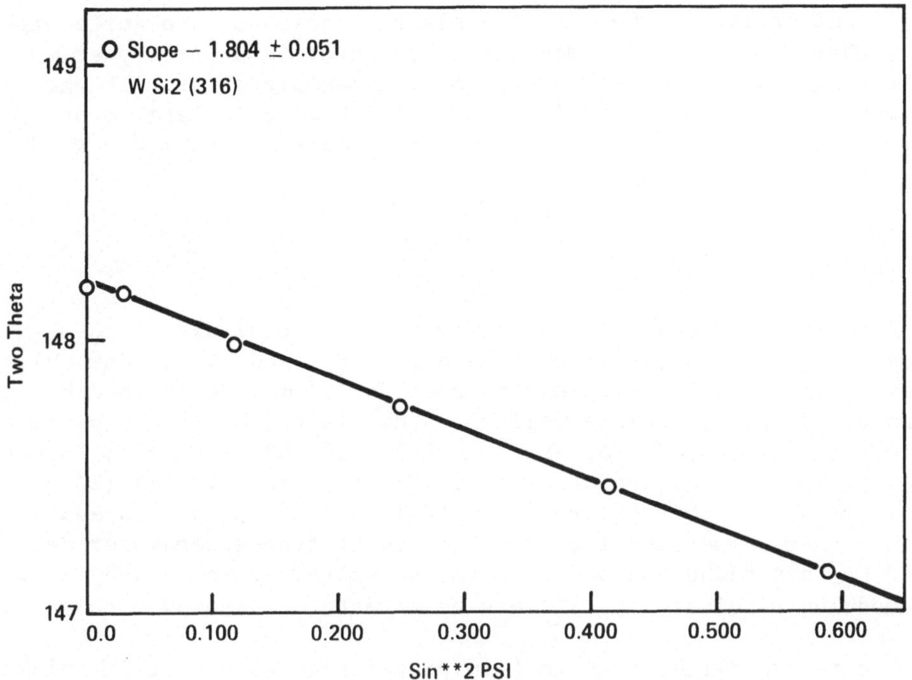

Fig. 3. Typical plot of 2Θ vs. $\mathrm{Sin}^2\Psi$ for WSi_2.

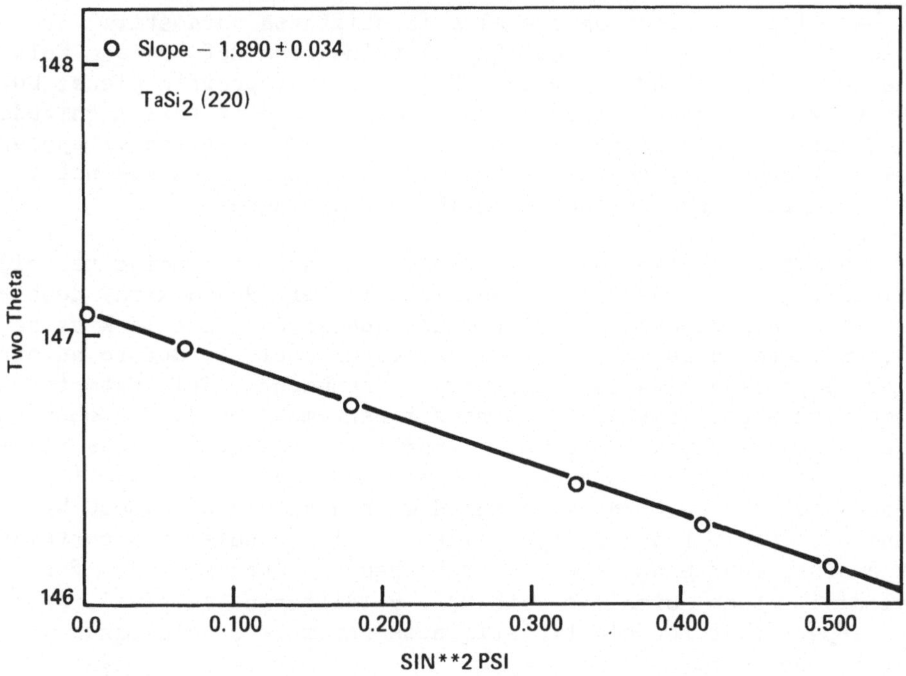

Fig. 4. Typical plot of 2Θ vs. $\mathrm{Sin}^2\Psi$ for $TaSi_2$.

highly anisotropic. The specific elastic constants are quite high, e.g., approximately 3.5 times the value determined for copper, indicating that these refractory disilicides are stiff. If one assumes $\nu = 0.3$, then Young's modulus for WSi_2 calculate to be approximately 4.7×10^5 MPa and for $TaSi_2$ approximately 4.0×10^5 MPa.

DISCUSSION

The stresses measured on our refractory metal disilicides agree well with the published values for WSi_2 and $TaSi_2$ (Ref. 9). However, the elastic stiffness parameters $(E/1+\nu)$ determined by our technique do not correlate well with the elastic stiffness parameters $(E/1-\nu)$ (1.2×10^5 MPa for WSi_2 and 1.1×10^5 MPa for $TaSi_2$) determined by Ref. 9. We cannot compare $(E/1+\nu)$ directly with $(E/1-\nu)$, but if we assume that Poisson's ratio is a constant of 0.3 and substitute that value back into the elastic stiffness parameter determined by this technique and into the parameter determined by Ref. 9, the modulus of elasticity differs by a factor of approximately 5.5.

There are differences in film preparation such as film thickness and sintering ambient. The films in this paper were 1000A thick and sintered in either nitrogen or argon, while Ref. 9 used films 2500A thick and sintered in hydrogen. However, these differences do not give corresponding differences in measured stress and probably have little effect on the elastic stiffness parameter. It should also be noted that the $(E/1-\nu)$ value from Ref. 9 is a bulk value and we determine the value $(E/1+\nu)$ for a specific plane, but anisotropy could hardly account for a difference of this magnitude. Consequently, the differences in measured elastic stiffness appear to be dependent only on the method of measurement. We are not yet able to explain why the two techniques do not agree.

Since the stresses measured by both substrate bending techniques (optically levered laser beam technique in Ref. 9 and x-ray double crystal in this paper) agree, it would appear that the lack of correlation in the elastic stiffness parameter could be due to an error in the polycrystalline residual stress technique. This prompted a request for a polycrystalline stress measurement by J. B. Cohen and C. Noyen[11] as a check on our polycrystalline residual stress technique. The strain measurement by Cohen and Noyen on $TaSi_2$ #1 yielded a $\Delta d/d$ of 0.0028 as compared with a strain of 0.0030 by our technique. Since the polycrystalline strain measurements correlate with another source and the substrate bending stress values for the metal disilicides correlate with Ref. 9 (different samples), it would appear that the elastic stiffness parameters determined in this paper are correct.

CONCLUSIONS

The following conclusions were reached in this study:

1. The combination of x-ray stress techniques can be used to determine elastic constants for specific crystallographic directions (planes).

2. These refractory metal disilicides are much stiffer than previously reported.

3. Based on the limited number of planes examined, these silicides are not highly anisotropic.

ACKNOWLEDGEMENTS

The authors would like to thank D. Campbell for providing the $TaSi_2$ samples. We would also like to thank E. Hearn, T. Nunes, and P. DeHaven for many valuable discussions. We especially thank J. B. Cohen and C. Noyen for measurements and helpful discussions.

REFERENCES

1. E. W. Hearn, Stress measurements in thin films deposited on single crystal substrates through x-ray topography techniques, Adv. in X-Ray Anal. 20:273 (1977).
2. A. Bohg, Phys. Stat. Soc. (A) 46:445 (1978).
3. G. G. Stoney, "Proc. Roy. Soc. A" (London), 82:172 (1969).
4. W. A. Brantley, J. Appl. Phys., 44:534 (1968).
5. H. H. Lester and R. H. Aborn, "Army Ordinance," 6:120-127, 200-207, 283-287, 364-369 (1925).
6. E. P. Marcherauch, "Expl. Mech." 6:140 (1966).
7. J. B. Cohen, H. Dolle and M. R. James, "NBS Special Pub." 567-453 (1980).
8. R. H. Chrenko, X-ray residual stress measurements using parallel beam optics, Adv. in X-Ray Anal. 20:393 (1977).
9. T. F. Retajczyk and A. K. Simha, "Thin Solid Films," 70:241-247 (1980).
10. J. Angilello, F. d'Heurle, S. Peterson, and A. Segmuller, J. Vac. Sci. Technol., 17 (1), 1980.
11. Private communication - Northwestern University.

ONE-DIMENSIONAL, CURVED, POSITION-SENSITIVE DETECTOR FOR X-RAY DIFFRACTOMETRY

B. Sleaford[1], V. Perez-Mendez[1] and C.N.J. Wagner[2]

[1]Lawrence Berkeley Laboratory, University of California
Berkeley, California 94720
[2]Materials Science and Engineering Department
University of California, Los Angeles, California 90024

ABSTRACT

A curved, one-dimensional position-sensitive detector has been designed and constructed for the measurement of the scattering patterns from non-crystalline materials. The chamber is a one-dimensional, pressurized, gas-filled detector with delay line readout for position encoding. It covers an angular range of 45° in 2θ, and its quantum efficiency is 80% and 50% for 17.5 and 60 KeV x-rays, respectively, when using a Xe-20% CO_2 gas mixture at 7 atm.

INTRODUCTION

Position-sensitive proportional detectors have been used for detection of x-rays in small angle scattering[1] and residual stress measurements[2]. In these applications, the parallax present in a straight linear detector is not of grave concern because of the limited 2θ-range required in these measurements. However, when position-sensitive detectors are used in powder diffractometry, a large angular range would be highly desirable, say 45° in 2θ, which can only be accomplished by a curved one-dimensional detector which is parallax-free[3].

In the design of our curved detector, a second condition was introduced, i.e., the quantum efficiency should be as high as possible for x-ray energies up to 60 KeV. This required a relatively large x-ray path in the detector coupled with a high pressure of the Xe-CO_2 gas mixture. This higher pressure also contributes to a good position accuracy, since it limits the range in the gas of the emitted K and Auger electrons from the photon interaction with the xenon.

The combination of large angular range and high counting efficiency of the scattered x-rays will permit us to apply this detector in experiments on the large-angle scattering from non-crystalline materials such as liquids and glasses using mono-energetic x-rays of high energy E to cover a large range in $K = (4\pi/\lambda) \sin\theta = (4\pi e/hc)E \sin\theta$ necessary for the evaluation of the radial distribution function. It is also highly suitable for retained austenite measurements with MoKα radiation (17.5 KeV) which permits the registration of several diffraction peaks for both austenite and ferrite, thus reducing the influence of preferred orientation on the evaluation of the amount of retained austenite in the sample. In this paper, we describe the design and the construction of a curved, position-sensitive detector for large-angle scattering experiments using monoenergetic x-rays in the range from 8 to 60 KeV.

SINGLE-WIRE, LARGE-ANGLE, CURVED CHAMBER

The chamber is a one-dimensional, pressurized, gas-filled detector with delay line readout for position encoding. The x-ray sensitive region of the chamber with a depth of 25 mm (1 in) and a height of 12.5 mm (0.5 in), the beryllium window 0.5 mm (0.020 in) in thickness, the anode wire (gold-plated tungsten) 38 μm (0.0015 in) in diameter, and the delay line are curved with a radius of 360 mm (14.2 in), covering an arc of 45° in 2θ. The anode wire is suspended in this circular arc by the interaction of a current flowing through it and a magnetic field provided by two permanent magnets placed above and below the wire running parallel to it over the full length of the curved chamber.

The permanent magnets are made of rubber-bonded Barium Ferrite[4] which can readily be machined and bent to fit on an arc of the circle. With the iron return yoke as shown in the cross sectional view of the sensitive region of the chamber in Figure 1, the magnetic field is ≈700 gauss at the center of the gap. A current of 130 mA flowing through the anode wire keeps it well centered. The cathode planes consist of a plated circuit delay line, the iron yoke on which the magnet slabs are mounted and the beryllium pressure window in front. The magnetic slabs have two 0.020" wires embedded in their front surfaces which are held at the same potential as the anode wire itself. These 4 high-voltage wires serve to stabilize the position of the anode wire when voltages up to 8 kV are applied, necessary for the chamber to work when filled up to 7 bar pressures. Without these field shaping wires, the anode wire oscillates badly due to the electric forces between the anode and the grounded iron return yoke of the magnet when the chamber is operated above 3 kV. Making the anode-to-delay line and anode-to-Be window asymmetric provides a net electric attractive force between the anode and the delay line and thus helps to inhibit oscillations.

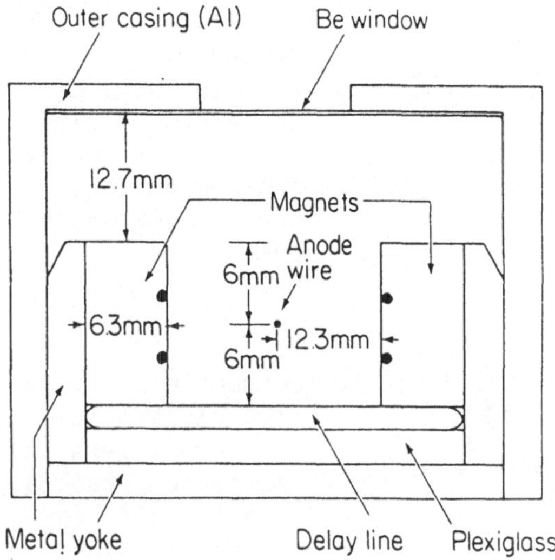

Fig. 1. Cross-sectional view of sensitive region of chamber showing ferrite magnets, stabilizing wires, iron return yoke, delay line, anode wire, and beryllium window.

When the appropriate voltage on the anode is selected for the particular gas composition and pressure, the signal amplitudes on the anode wire produced by 6 KeV x-rays are in the 10-60 mV range with rise times of ≈ 7 ns[1]. The corresponding pulse amplitudes on the delay line are ≈ 20% of the anode signal. The lower amplitudes are in the proportional regime and suitable for energy discrimination. More accurate timing is obtained at the higher anode voltages which produce larger amplitude saturated signals in the 40-60 mV range.

The delay line is a plated circuit line described by LeComte et al.[5] The combination of the specific capacity to ground, compensating pad capacity and plated windings/cm produces an overall delay of 8 ns/cm with total delay of 210 ns. This relatively short total delay implies that, with suitable digitizing electronics, event rates in excess of 10^6/sec can be accepted.

When filled with a Xe-80% CO_2-20% gas mixture at 7 atm, the detector has quantum efficiencies of 50% and 90% for 60 KeV and 17.4 KeV x-rays, respectively, and a position resolution of 0.5 mm or 0.08° in 2θ, which is sufficient for the measurements of broad powder pattern peaks or the diffuse scattering from non-crystalline materials, since it corresponds to a value of $\Delta K/K = \Delta\lambda/\lambda + 7X10^{-4} \cot\theta \leq 1\%$ at $2\theta \geq 8°$.

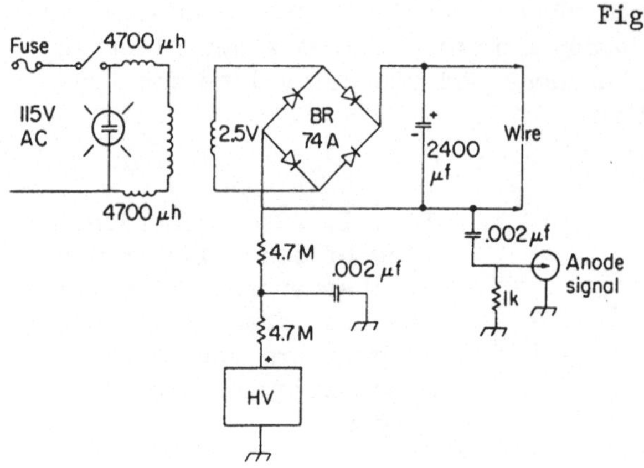

Fig. 2. Schematic of anode wire electronics showing isolation transformer with filters on the primary, rectifier and filters on the secondary. External H.V. is connected to wire with filter network and connections for viewing anode pulses.

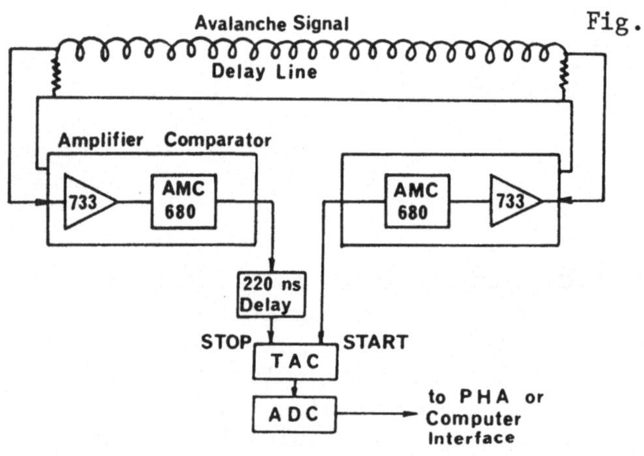

Fig. 3. Schematic of delay line electronics. I.C. amplifiers and timing comparators on both ends lead to external TAC followed by ADC into PHA or computer.

The chamber electronics are shown in Figures 2 and 3. Figure 2 shows the circuit diagram for the anode wire current and high voltage filter. Figure 3 shows the readout electronics for the delay line. Position is determined by digitizing the difference in arrival times of the signal from an interaction event to both ends of the delay line. This time is selected by the fast AMC comparators which produced shaped pulses. Time jitters due to signal amplitude variations are minimal since the shape of the signal on both ends of the lines are almost identical and compensate each other. The shaped comparator output timing signals serve as the start and stop signals to a TAC (Time to Amplitude Converter) which is then digitized by a conventional ADC (10 bit accuracy). Measurements done with low energy γ rays using the ^{55}Fe 5.9 KeV line showed that position accuracies of <0.3 mm were obtained.

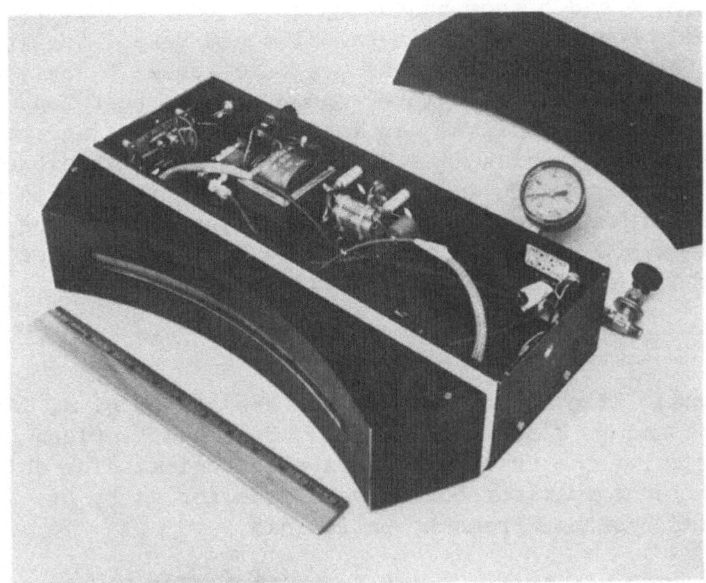

Fig. 4. Assembled chamber showing curved pressure vessel in front with Be window. Electronics and hardware connections are in back sections of metal container.

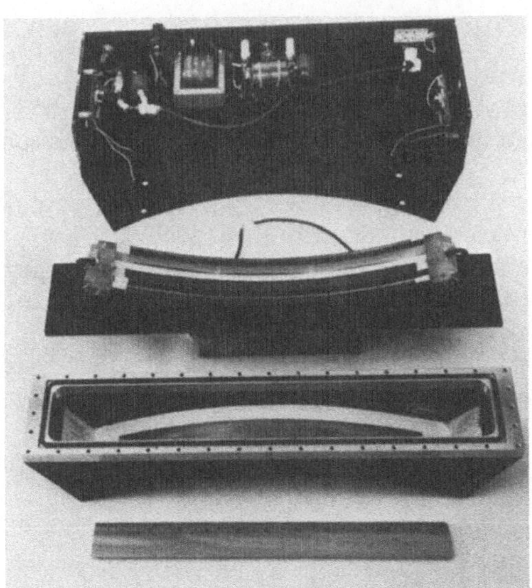

Fig. 5. Disassembled view of the chamber showing rear box with amplifier-comparator cards on the left and right panels, transformer and filter networks on the back panel. Center figure shows iron yoke with ferrite magnets, delay line and anode wire. Bottom figure shows pressure vessel with Be window.

Figures 4 and 5 show photographs of the complete chamber. The assembly consists of two metal aluminum boxes, the front one with the beryllium window is the pressure vessel. The rear one holds the delay line amplifiers, anode wire current source, electrical filters, and the gas fittings. By placing the low signal level electronics in a shielded container in close proximity to the pressure vessel, we can minimize electrical noise problems. The output signals that go to the digitizing electronics are at NIM levels (-850 mV) on 50 ohm shielded cables and hence are not too susceptible to extraneous electrical noise.

ACKNOWLEDGEMENT

We would like to acknowledge the assistance of K. Lee, P. Wiedenbeck and M. Elola at LBL and J. Beck and E. Olsen at UCLA in constructing the chamber and its electronics. The delay line and amplifier comparator boards were made for us by Dr. R. Sparks, Nicolet XRF Systems, Fremont, California.

This work has been supported at LBL by the Director, Office of Energy Research, Office of High Energy and Nuclear Physics, Division of High Energy Physics of the U.S. Department of Energy under Contract No. DE-AC03-76SF00098 and at UCLA by the National Science Foundation, Grant DMR80-07939.

REFERENCES

1. A.R. Forouhi, B. Sleaford, V. Perez-Mendez, D. de Fontaine and J. Fodor, IEEE Trans. Nucl. Science NS-29, 275 (1982).
2. M. James and J.B. Cohen, J. Testing and Evaluation 6, 91 (1978).
3. D. Ortendahl, V. Perez-Mendez, J. Stoker and W. Beyerman, Nuclear Instr. Methods 156, 53 (1978).
4. Plastiform Permanent Magnets, Dielectric Materials and System Division, Minnesota Mining and Manufacturing Corp., Minneapolis, Minn.
5. P. LeComte, V. Perez-Mendez and G. Stoker, Nucl. Inst. and Methods 153, 543-547 (1978).

A PHI-PSI-DIFFRACTOMETER FOR RESIDUAL STRESS MEASUREMENTS

C.N.J. Wagner[1], M.S. Boldrick[1*], and V. Perez-Mendez[2]

[1]Materials Science and Engineering Department
University of California, Los Angeles, California 90024

[2]Lawrence Berkeley Laboratory
University of California, Berkeley, California 94720

ABSTRACT

A ϕ-ψ diffractometer has been designed and constructed to evaluate residual stresses in polycrystalline samples by x-ray diffraction. It permits rotations of the x-ray diffraction apparatus, consisting of an x-ray tube and a position-sensitive proportional counter, about two axes ϕ and ψ. The ϕ-rotation from 0° to 360° is carried out about the normal to the surface of the stationary sample, whereas the ψ-motion consists of a rotation from -45° to +45° about an axis lying in the sample surface and the diffraction plane, but perpendicular to the diffraction vector. This ϕ-ψ diffractometer permits the application of the ϕ- and ψ-differential and integral methods for the evaluation of the strain tensor and its gradient averaged over the depth of x-ray penetration into the sample. Assuming that isotropic elasticity theory is applicable, the stress tensor can then be evaluated from the measured strain tensor.

INTRODUCTION

Residual stresses are stresses which remain in a material when no force is applied. They can be introduced into metals and other materials by any mechanical, chemical or thermal process. The build-up of such residual stresses can be either detrimental or beneficial during industrial applications. For example, initiation and propagation of cracks during fatigue or in stress

*Presently at Boldrick Systems, Manhattan Beach, California 90266

corrosion can be impeded by compressive stresses, but are greatly accelerated by tensile stresses.

The knowledge of the sign and magnitude of residual stresses is thus very important. Methods have been developed to determine residual stresses in metals,[1] based on acoustic and magnetic response of materials to stresses,[2] but these have proved to be rather sensitive to the variations in microstructure. The x-ray method for residual stress measurements, on the other hand, has been successfully employed for many years[3-6]. It is based on the determination of changes in interplanar spacing d by standard x-ray diffraction techniques, i.e., the d-spacing serves as an internal strain gage. In polycrystalline samples, the conventional powder (or Debye-Scherrer) method can be applied when the grains are randomly oriented and their size is less than 0.05 mm (0.002") so that the Debye-Scherrer rings consist of uniformly distributed diffraction spots. Under these conditions it is possible to apply isotropic elasticity theory to convert the measured strains into stresses.

In order to determine the complete strain tensor in the surface region, the specimen must be rotated about two axes. One rotation is characterized by the angle ψ between the normal to the specimen surface $\vec{P}_3$ and the diffraction vector $\vec{s} = (\vec{S} - \vec{S}_o)/\lambda$ where $\vec{S}_o$ and $\vec{S}$ are the directions of the incoming and diffracted x-ray beams, respectively, as shown in Fig. 1. This tilt can be accomplished by a rotation about an axis lying in the specimen surface and perpendicular to the diffraction plane $(\vec{S}_o, \vec{S})$, i.e., the ω-diffractometer, or by a rotation about an axis lying in the specimen surface and diffraction plane, but perpendicular to the diffraction vector $\vec{s}$, i.e., the ψ-diffractometer[7]. The second rotation is characterized by the angle ϕ and is carried out about an axis parallel to the specimen normal $\vec{P}_3$ (Fig. 1).

Only small test specimens can be mounted directly on a two-circle goniostat of conventional diffractometers, which implements either the ω or ψ geometry. For bulk samples, special diffractometers must be employed[8]. Rather than moving the sample, the x-ray tube and detector are rotated about the two axes ϕ and ψ. This has been accomplished in the apparatus described in this paper. Previous portable systems have implemented only the ψ motion, by using either an ω-diffractometer[8] or a ψ-diffractometer,[9] or both[10]. Before describing the ϕ-ψ diffractometer in detail, it is necessary to summarize the basic theory which is required for the determination of the strain tensor ε_{ij}.

X-RAY THEORY FOR RESIDUAL STRESS MEASUREMENTS

The residual (or applied) stress can be determined from a change in interplanar spacing $d_{\phi\psi}$ of lattice planes whose normal

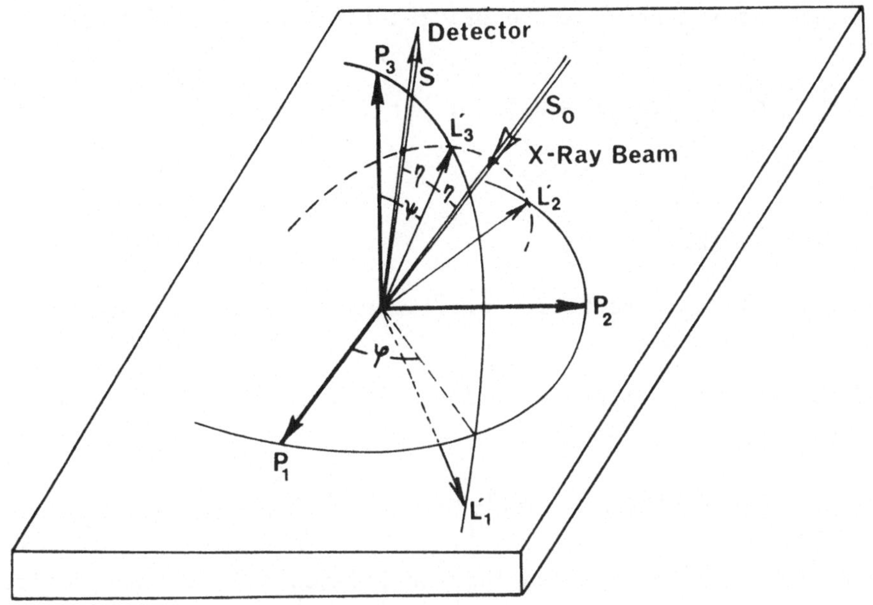

Fig. 1. Diffraction geometry for ψ-diffractometer

$L_{3'}$ forms the angle ψ with the normal P_3 to the specimen surface, and its projection L_{12} on the specimen surface P_1P_2 forms the angle ϕ with the P_1 axis as shown in Fig. 1. The strain $\varepsilon_{\phi\psi}$ is then given by the relation:

$$\varepsilon_{\phi\psi} = (d_{\phi\psi} - d_o)/d_o \tag{1}$$

where d_o is the interplanar spacing of the unstrained material. Since x-rays penetrate the surface of the specimen, we measure the strain $\langle\varepsilon_{\phi\psi}\rangle$ averaged over the penetration depth t, i.e.[11],

$$\begin{aligned}\langle\varepsilon_{\phi\psi}\rangle \equiv \langle\varepsilon_{3'3'}\rangle &= \langle\varepsilon_{11}\rangle \cos^2\phi\sin^2\psi + \langle\varepsilon_{12}\rangle\sin2\phi\sin^2\psi \\ &+ \langle\varepsilon_{22}\rangle \sin^2\phi\sin^2\psi + \langle\varepsilon_{13}\rangle\cos\phi\sin2\psi \\ &+ \langle\varepsilon_{23}\rangle \sin\phi\sin2\psi + \langle\varepsilon_{33}\rangle\cos^2\psi\end{aligned} \tag{2}$$

where $\langle\varepsilon_{ij}\rangle$ is given by:

$$\langle\varepsilon_{ij}\rangle = \int_0^t \varepsilon_{ij} \exp(-z/\tau)dz \Big/ \int_0^t \exp(-z/\tau)dz \tag{3}$$

The absorption factor τ can be written for the ψ-diffractometer as[12]

$$\tau = \tau_o \cos\psi \simeq \tau_o (1 - 0.59 \sin^2\psi) \text{ for } \psi \leq \pm 45 \tag{4}$$

where

$$\tau_o = \sin\theta/(2\mu) \tag{5}$$

and μ is the linear absorption coefficient of x-rays of wavelength λ.

In an isotropic elastic medium, the stress σ_{ij} is related to the strain ε_{ij} by:

$$\sigma_{ij} = [E/(1+\nu)]\{\varepsilon_{ij} - [\nu/(1-2\nu)](\varepsilon_{11} + \varepsilon_{22} + \varepsilon_{33})\delta_{ij}\} \tag{6}$$

where E and ν are Young's modulus and Poisson's ratio, respectively, and $\delta_{ij} = 1$ or 0 if $i = j$ or $i \neq j$.

If we expand the strain $\varepsilon_{ij}(z)$ in a Mac-Laurin-Taylor series as a function of the depth z, then Eqn. (3) yields[13]

$$\langle\varepsilon_{ij}\rangle = \varepsilon^o_{ij} + \sum_{n=1}^{N} \varepsilon^{(n)}_{ij}\tau^n \tag{7}$$

where $\varepsilon^o_{ij} = \varepsilon_{ij}(z = o)$ and $\varepsilon^n_{ij} = d^n\varepsilon_{ij}(o)/dz^n$. The linear strain gradient is measured to a depth of the order of the absorption factor τ.

Since $\langle\varepsilon_{\phi\psi}\rangle$ depends on the angles ϕ and ψ [Eqn. (2)], there are two basic choices in experimental methods which was pointed out by Lode and Peiter[14]. In the conventional method, $\langle\varepsilon_{\phi\psi}\rangle$ is measured as a function of ψ for fixed values of ϕ, which is called the ψ-method. It is also possible to determine $\langle\varepsilon_{\phi\psi}\rangle$ as a function of ϕ at fixed angles ψ which is called the ϕ-method.

It follows from Eqns. (2), (4) and (7), that $\langle\varepsilon_{\phi\psi}\rangle$ is a function of powers of trigonometric functions in ϕ and ψ. It becomes possible to express $\langle\varepsilon_{\phi\psi}\rangle$ as a function of $\sin^n\psi$, or $\sin n\psi$ and $\cos n\psi$ which yield the differential and integral methods, respectively. The differential ψ-method, which includes the well-known $\sin^2\psi$-method as a special case, was developed by Dölle, Hauk, and Cohen[11], whereas the integral method was introduced by Lode and Peiter[12].

We will first introduce an extension of the ψ-differential method[15] to evaluate the strains ε^o_{ij} and their linear gradients ε'_{ij} in the surface of the specimen. Because of its simplicity and ease of computer application we will also describe briefly the ϕ-integral method[13].

The ψ-Differential Method

By combining Eqns. (2) and (7), limiting ourselves to linear terms in τ only, and replacing the absorption factor τ by its approximation in Eqn. (4), we obtain $<\varepsilon_{\phi\psi}>$ as a function of $\cos^m n\psi$ $\sin^p q\psi$ where m, n, p, and q are integers including zero. These trigonometric functions can be converted to powers of $\sin\psi$ only, and the following result[13] is obtained:

$$<\varepsilon_{\phi\psi}> = \alpha + \beta \sin\psi + \gamma \sin^2\psi + \delta \sin^3\psi + \eta \sin^4\psi \tag{8}$$

where

$$\alpha = \varepsilon^o_{33} + \tau_o\varepsilon'_{33} \tag{9}$$

$$\beta = 2[(\varepsilon^o_{13} + \tau_o\varepsilon'_{13}) \cos\phi + (\varepsilon^o_{23} + \tau_o\varepsilon'_{23}) \sin\phi] \tag{10}$$

$$\gamma = (\varepsilon^o_{11} + \tau_o\varepsilon'_{11}) \cos^2\phi + (\varepsilon^o_{22} + \tau_o\varepsilon'_{22}) \sin^2\phi$$

$$+ (\varepsilon^o_{12} + \tau_o\varepsilon'_{12}) \sin2\phi - (\varepsilon^o_{33} + 1.59\, \tau_o\varepsilon'_{33}) \tag{11}$$

$$\delta = -2[(0.59\varepsilon^o_{13} + \tau_o\varepsilon'_{13}) \cos\phi + (0.59\varepsilon^o_{23} + \tau_o\varepsilon'_{23}) \sin\phi] \tag{12}$$

$$\eta = -0.59\tau_o(\varepsilon'_{11} \cos^2\phi + \varepsilon'_{22} \sin^2\phi + \varepsilon'_{12} \sin2\phi - \varepsilon'_{33}) \tag{13}$$

If we measure $<\varepsilon_{\phi\psi}>$ for positive and negative values of ψ, say $-45° \leq \psi \leq +45°$, we can form the following expressions:

$$a^\psi_+ = [<\varepsilon_{\phi\psi}>_{\psi>o} + <\varepsilon_{\phi\psi}>_{\psi<o}]/2 = \alpha + \gamma \sin^2\psi + \eta \sin^4\psi \tag{14}$$

$$a^\psi_- = [<\varepsilon_{\phi\psi}>_{\psi>o} - <\varepsilon_{\phi\psi}>_{\psi<o}]/2 = \beta \sin|\psi| + \delta \sin^3|\psi| \tag{15}$$

The coefficients α, β and γ can be evaluated from the least square quadratic fit of the values of a^ψ_+ when plotted as a function of $\sin^2\psi$. The coefficients δ and η can likewise be obtained from the plot of a^ψ_- vs. $\sin|\psi|$.

The ϕ-Integral Method

Since the absorption factor τ [Eqn. (4)] depends only on the tilt angle ψ, it becomes advantageous to develop $<\varepsilon_{\phi\psi}>$ in Eqn. (2) as a function of ϕ[13].

$$\langle\varepsilon_{\phi\psi}\rangle = A_o^\psi/2 + A_1^\psi \cos\phi + A_2^\psi \cos 2\phi + B_1^\psi \sin\phi + B_2^\psi \sin 2\phi \quad (16)$$

where

$$A_o^\psi = [(\varepsilon_{11}^o + \tau\varepsilon_{11}') + (\varepsilon_{22}^o + \tau\varepsilon_{22}')] \sin^2\psi + 2(\varepsilon_{33}^o + \tau\varepsilon_{33}')\cos^2\psi \quad (17)$$

$$A_1^\psi = (\varepsilon_{13}^o + \tau\varepsilon_{13}') \sin 2\psi \quad (18)$$

$$A_2^\psi = (1/2)\ [\varepsilon_{11}^o + \tau\varepsilon_{11}') - (\varepsilon_{22}^o + \tau\varepsilon_{22}')] \sin^2\psi \quad (19)$$

$$B_1^\psi = (\varepsilon_{23}^o + \tau\varepsilon_{23}') \sin 2\psi \quad (20)$$

$$B_2^\psi = (\varepsilon_{12}^o + \tau\varepsilon_{12}') \sin^2\psi \quad (21)$$

It is readily seen that Eqn. (16) represents a Fourier series whose coefficients can be determined by the relations

$$A_n^\psi = (1/\pi)\int_0^{2\pi} \langle\varepsilon_{\phi\psi}\rangle \cos n\phi \, d\phi \quad (22)$$

$$B_n^\psi = (1/\pi)\int_0^{2\pi} \langle\varepsilon_{\phi\psi}\rangle \sin n\phi \, d\phi \quad (23)$$

when the strains $\langle\varepsilon_{\phi\psi}\rangle$ are measured over the angular range of ϕ from 0 to 360°. The components ε_{ij}^o and ε_{ij}' can be evaluated for at least three tilt angles ψ.

DESIGN OF THE ϕ-ψ DIFFRACTOMETER

To implement the ϕ and ψ-methods of residual stress measurements, a new portable diffractometer was designed which incorporates a movable x, y, z and rotary support for the ϕ and ψ motions. The x- and y translations of ± 150mm and the height adjustment z of ± 300mm permit accurate positioning of the diffractometer with respect to the sample, and the base rotation facilitates the alignment of the normal P_3 parallel to the ϕ-rotation axis. All four motions are motor-driven for remote control [Figure 2].

For convenience in applying the ϕ-method, the ψ motion was chosen to be a rotation about an axis L_2, lying in the surface and the diffraction plane (see Figure 1), and perpendicular to the diffraction vector, i.e., applying the geometry of the ψ-diffractometer[7]. The ψ-motion from -45° to 45° consists of a centerless rotary drive supported by a yoke whose center can be rotated by the angle ϕ from -180° to 180° about an axis perpendicular to the specimen surface and intersecting the ψ-axis as shown in Figure 2.

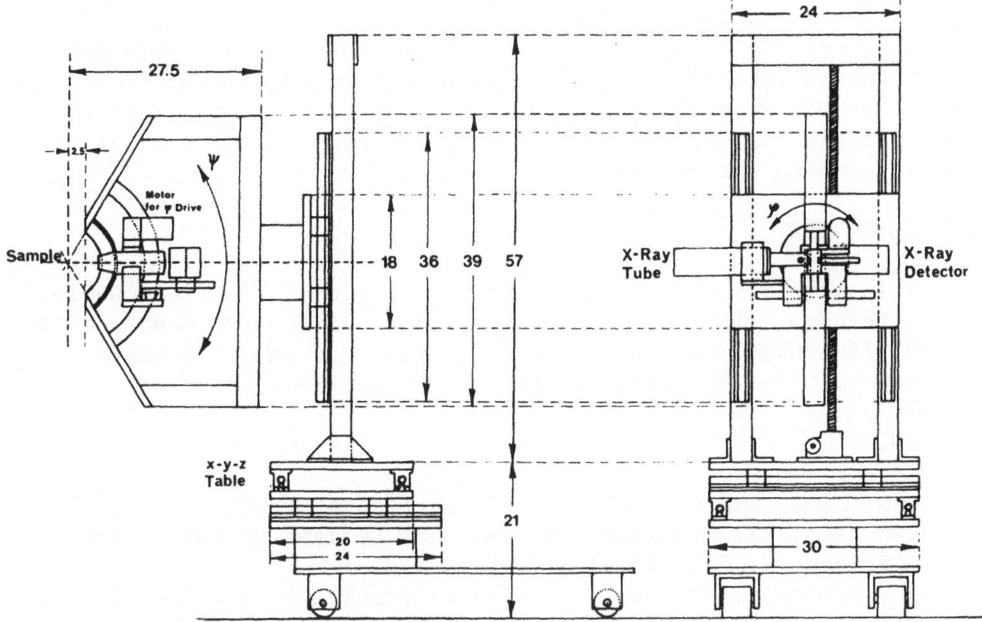

Fig. 2. Sketch of the ϕ-ψ diffractometer. Dimensions are given in inches.

In order to employ computer control, the ψ and ϕ axes are driven by stepping motors.

The x-ray tube and the linear position-sensitive detector are mounted on curved dove-tails, arranged parallel to the diffraction plane which is perpendicular to the plane of the yoke. In the prototype instrument, a conventional Machlett or GE x-ray tube is used powered by a Siemens Kristalloflex II generator.

LINEAR POSITION-SENSITIVE DETECTOR

The position-sensitive detector is a sealed, pressurized, one-dimensional, linear proportional counter filled with a mixture of Ar-10% CH_4 (P-10) or Xe-3% CO_2 at a pressure of 4 atm (60 psi). The x-ray-sensitive part of the chamber has an active length of 130 mm (5.12 in), a height of 12.5 mm (0.5 in) and a depth of 12.5 mm (0.5 in), and is covered with a Be window of thickness 0.5 mm (0.03 in). A 150 mm (6 in) delay line is used for position-encoding[15]. When filling the chamber with P-10 gas at 4 atm, a position-resolution of 0.2 mm could be obtained. At a distance of 360 mm (14.2 in) between the detector and the sample, an angular range of 20° in 2θ can be covered with a position resolution of 0.03°.

The signal from the position-sensitive proportional detector (PSPD) is stored on a multi-channel analyzer (MCA). The dead-time loss of intensity is less than 1%, reaching 3% at 40000 c/s. The position-encoding along the wire is linear within the resolution of the detector. The data acquisition was automated by a microcomputer using the 8080 assembly language. It consists of controlling the ϕ and ψ stepping motors, and the storage and transfer of the data from the MCA to the computer.

ACKNOWLEDGEMENT

This work was supported by the Association of American Railroads, Chicago, Ill., and by the Department of Energy Contract No. DE-AC03-76SF0098. We would like to acknowledge the assistance of J. Beck and E. Olsen in constructing the ϕ-ψ diffractometer.

REFERENCES

1. M.R. James and O. Buck, Quantitative Nondestructive Measurements of Residual Stresses, CRC Critical Reviews in Solid State Science 9, 61, (1980).
2. M. Shibata and K. Ono, NDT International, 14, 227-234, (1981).
3. M.R. James and J.B. Cohen, Experimental Methods in Materials Science, Treatise on Materials Science and Technology, 19A, 2 (1980).
4. B.D. Cullity, "Elements of X-Ray Diffraction," 2nd Ed., Addison-Wesley Publ. Co., Reading, Mass. (1978).
5. Various Authors, Härterei-Technische Mitteilungen, 32, No. 1 and 2 (1976).
6. Residual Stress Measurements by X-Ray Diffraction - SAE J784a, (Society of Automotive Engineers, Warrendale, Pa. 1971).
7. E. Macherauch and U. Wolfstieg, Adv. X-Ray Analysis 20, 369-377 (1977).
8. M.R. James and J.B. Cohen, J. Testing and Evaluation 6, 91-97 (1978).
9. U. Wolfstieg, private communication.
10. R.H. Chrenko, Adv. X-Ray Analysis 20, 393-402 (1977).
11. J.B. Cohen, H. Dölle, and M.R. James, Proc. of Symposium on Accuracy in Powder Diffraction, National Bureau of Standards Publication 567, 453-477 (1979).
12. W. Lode and A. Peiter, Härterei-Technische Mitteilungen, 32, 235-240 and 308-313 (1977).
13. C.N.J. Wagner and M.S. Boldrick, (to be published).
14. W. Lode and A. Peiter, Metall, 35, 758-762 (1981).
15. A.R. Forouhi, B. Sleaford, V. Perez-Mendez, D. DeFontaine, and J. Fodor, IEEE Trans. Nucl. Science NS-29, 275-279 (1982).

X-RAY FRACTOGRAPHY ON FATIGUE FRACTURED SURFACE

Shotaro Kodama and Hiroshi Misawa

Tokyo Metropolitan University, Tokyo 158, Japan

Yuji Sekita

Mitsubishi Heavy Industries LTD, Sagamihara 229, Japan

INTRODUCTION

The technique analysing the cause and mechanism of fracture from the information obtained by X-ray irradiation on the fractured surface is called "X-ray fractography." As a basic study of X-ray fractography the residual stress on the fatigue fractured surface and some parameters of the fracture mechanics were investigated.

SPECIMENS AND EXPERIMENTAL PROCEDURES

The material used was a Ni-Cr-Mo steel (Japanese Industrial Standard SNCM 439: 0.30%C 1.39%Ni, 0.85%Cr, 0.30%Mo and 0.85%Mn) which was normalized at 870°C, oil quenched from 850°C and tempered at 600°C for one hour. The mechanical properties were: U.T.S. = 963MPa, Yield Stress=878MPa, Elongation=15.8% and Reduction of Area=42.3%.

The test pieces were CT specimens with W=51mm and B=12mm. An electro-hydraulic closed loop testing machine was used and the crack length was measured by a traveling microscope with a sensitivity of 0.01mm.

The fatigue tests were conducted under constant load and also under constant range of stress intensity factor (constant ΔK). For the constant ΔK fatigue tests the load was decreased stepwise to maintain the ΔK value within +2.5% limits. The crack closure was measured by the unloading compliance technique to calculate the effective stress intensity factor K(eff). The test conditions are shown in Tables I and II, together with some experimental results.

Table I Test condition
(constant load)

Pmax kN	Pmin kN	R
10.6	0.8	0.08
12.8	1.0	0.08
12.8	2.9	0.23
12.8	4.9	0.39
15.9	6.1	0.39

Table II Test condition
(constant ΔK)

K MPa$\sqrt{m}$	R	da/dN 10^{-7}m/cycle	σ(r) MPa
50	0.08	53.2	260
40	0.08	30.1	305
30	0.08	14.0	370
30	0.23	19.2	380
30	0.39	19.2	390
20	0.08	5.2	370
50-30	0.08	52.1-19.0	260-385

The residual stress along the crack propagation direction on the fatigue fractured surface was measured by a parallel beam X-ray stress measurement system with Cr target, and the peak of the diffraction profile was determined by parabola fitting with LPA correction. The details of the measuring conditions are shown in Table III, where R is the stress ratio ($\sigma_{min}/\sigma_{max}$) and σ(r) is the mean value of the residual stress on the fractured surface.

RESULTS AND DISCUSSION

The crack propagation rate against stress intensity factor range ΔK is shown in Fig. 1. The solid triangle marks represent the results of the constant stress intensity factor range ΔK tests and the smaller ones are of the constant load tests. The results from the constant load test fall in a narrow band. A slight effect of stress ratio R is seen, i.e., a higher stress ratio R gives a faster crack propagation rate. The constant ΔK tests were carried out at ΔK = 20, 30, 40 and 50 MPa$\sqrt{m}$ with R = $\sigma_{min}/\sigma_{max}$ = 0.08. The crack propagation rates in these cases were constant or the crack lengths increase linearly with the number of load cycles. The

Table III Condition of X-ray stress measurement

Characteristic X-ray	CrKα
Diffraction plane	α-Fe(211)
Slit divergence angle	0.26°
Filter	V foil
Counter	Scintillation counter
Voltage	30kV
Current	9mA
Irradiated area	0.8x6.0mm^2
Preset time	20sec.

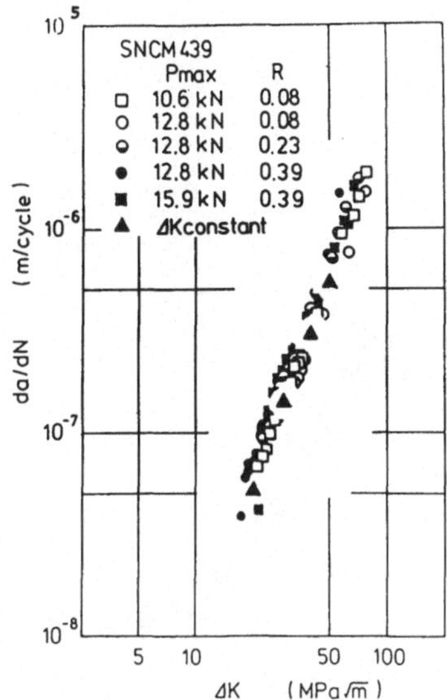

Fig. 1 Crack propagation rate vs. stress intensity factor range

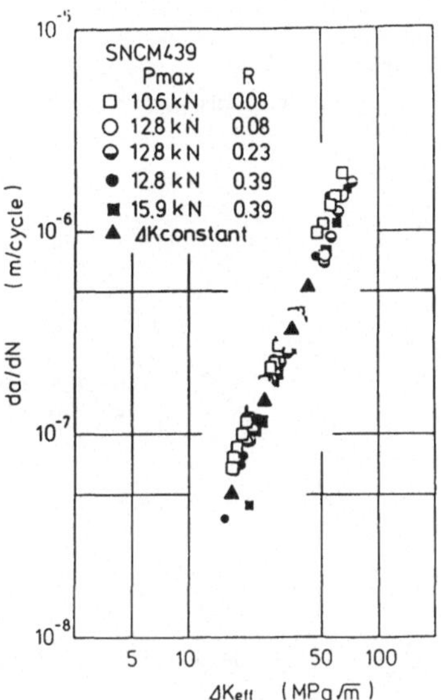

Fig. 2 Crack propagation rate vs. effective stress intensity factor range

crack propagation rate of the constant ΔK test agrees with the results of the constant load tests at the corresponding ΔK value. As shown in Fig. 2 the effective stress intensity factor range ΔKeff which was calculated from the unloading compliance technique is a better parameter for the crack propagation rate. The experimental results fall in a narrower band and the agreement of the results from constant load fatigue tests and the constant stress intensity factor range fatigue tests is better.

Fig. 3 gives the results of the constant stress intensity factor range tests in which the ΔK value was changed from ΔK = 50 MPa√m. The crack propagation rate showed a marked retardation after the change of stress intensity factor; then it approached a constant value. The crack opening stress was increased after the change of the stress intensity factor, which resulted in the decreased effective stress intensity factor.

RESIDUAL STRESS ON FRACTURED SURFACE

Fig. 4 shows the relation between the stress intensity factor range and the residual stress on fractured surface. The residual stress increased with increasing ΔK value up to about ΔK = 30 MPa√m, then decreased monotonously. The effect of stress ratios was seen in the region of the increasing residual stress.

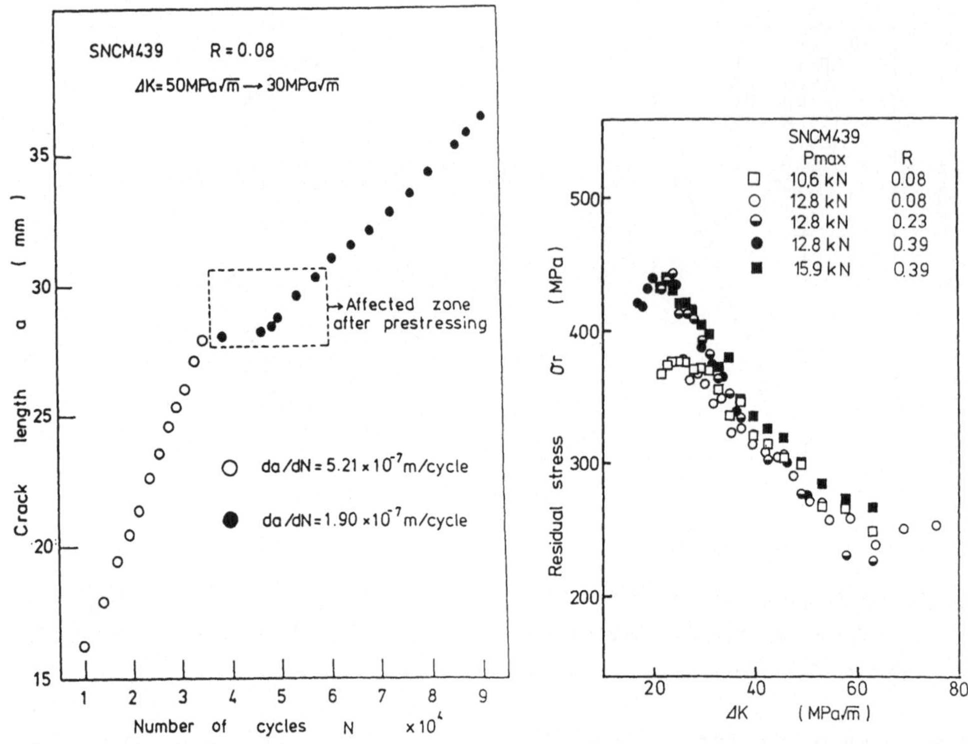

Fig. 3 Crack length vs. number of cycles

Fig. 4 Residual stress vs. stress intensity factor range

Fig. 5 shows the same results as above but plotted against the maximum stress intensity factor. In this case the residual stress also had peaks at about K_{max} = 30 MPa√m but the effect of the stress ratios was very clear. At a given K_{max} value, the residual stress was higher for the higher stress ratios. These results show the possibility that the stress intensity factor at the fracture can be estimated by measuring the residual stress on the fractured surface.

The residual stress on the fractured surface under the constant stress intensity factor range fatigue tests is plotted against the crack length in Fig. 6. The residual stress kept constant values independent of the crack length. The residual stresses in the ΔK = 20 and 30 MPa√m tests were almost the same, and this can be explained from the behavior of the residual stress shown in Fig. 4. For these two constant stress intensity factor range tests, the load control was stopped and the test was continued under constant load. The solid triangle and square marks show the results. Except for the transient period, the residual stress decreased with the crack length or the increasing ΔK values. The residual stresses by ΔK = 40 and 50 MPa√m were slightly unstable but they were constant and independent of crack length.

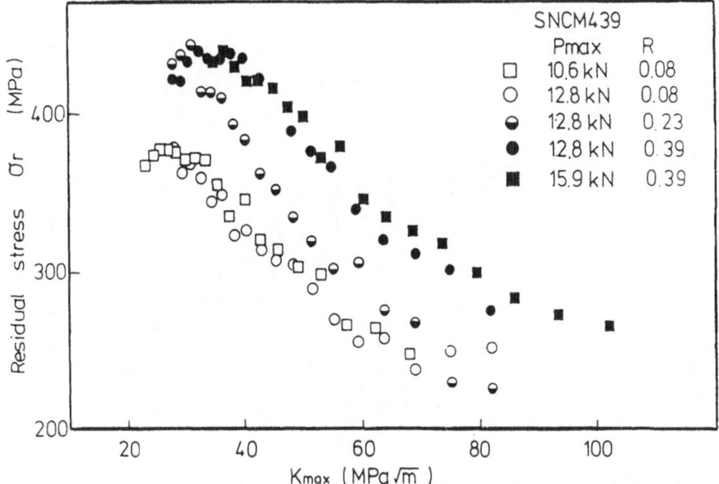

Fig. 5 Residual stress vs. maximum stress intensity factor

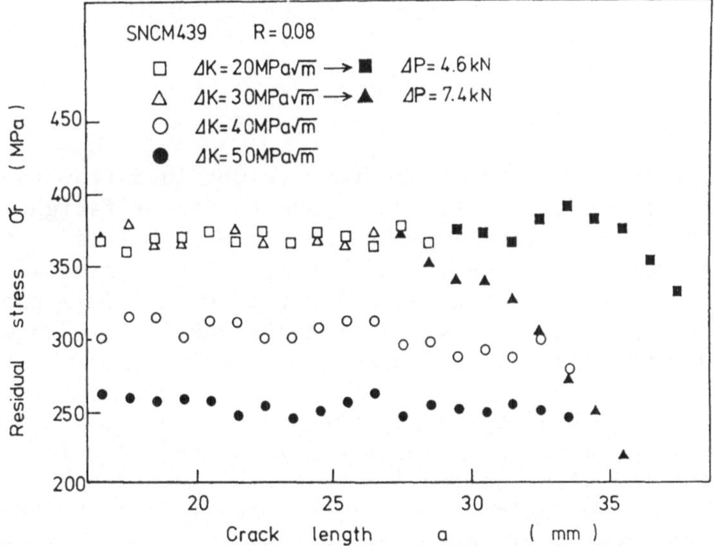

Fig. 6 Residual stress vs. crack length

Fig. 7 shows the effect of the stress ratios on the residual stress under the constant stress intensity factor range fatigue test $\Delta K = 30\ MPa\sqrt{m}$. At each stress ratio, the residual stress was kept constant and the residual stresses were higher for the higher stress ratio. In these experiments the stress control was stopped after the crack length reached about 27 mm. Then the residual stress decreased as expected.

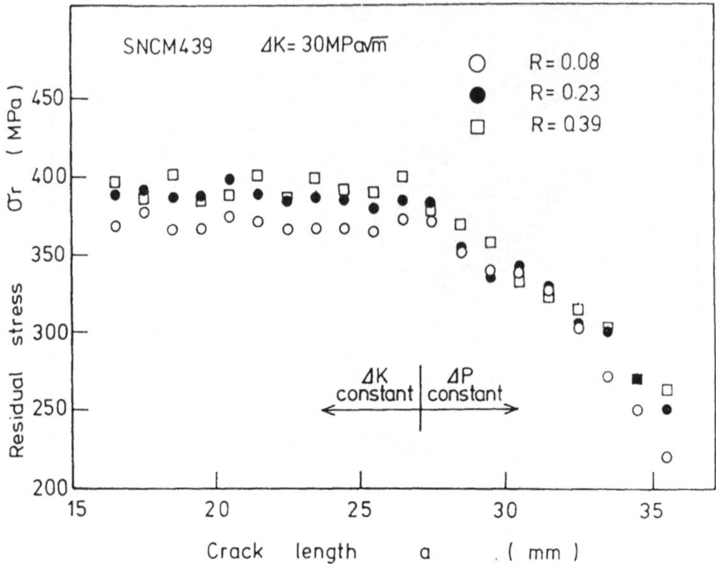

Fig. 7 Residual stress vs. crack length

In Fig. 8 a comparison of the results obtained from the constant load and the constant stress intensity factor fatigue tests is shown. The solid triangle marks are the mean values of the residual stresses shown in Fig. 12. These results agree with the band indicated by the constant load fatigue tests. This can be a proof that the residual stress on fractured surface are determined by the stress intensity factor range.

The results of Fig. 4 are represented by two chain lines which are the results obtained from short crack length and relatively high load compared with the results of Fig. 6 obtained with long crack length and low loads. However, as seen in Fig. 9, the fact that they agree very well indicates that ΔK is a quite useful parameter.

Fig. 10 gives the results of a change of ΔK in a constant stress intensity factor test. In this case ΔK was changed from 50 to 30 $MPa\sqrt{m}$. The residual stress on the fractured surface decreased just after the change of stress intensity factor range but rapidly increased to the steady value which is the same as the one shown in Fig. 6.

Fig. 11 shows the results of the half value breadth measurement of the diffracted beam. In this figure, a rather strong effect of stress ratios is seen. Plotting of the half value breadth against

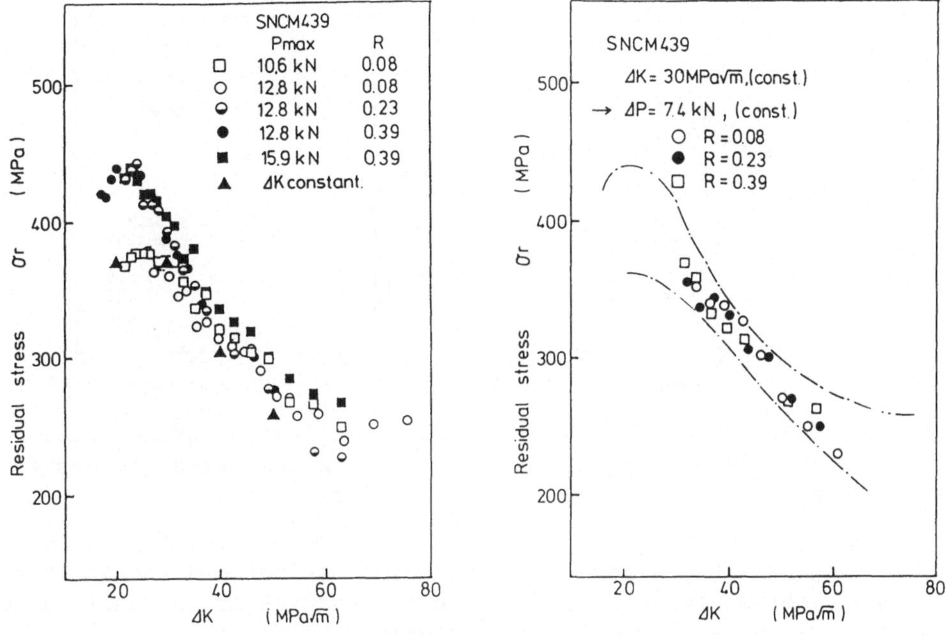

Fig. 8 Residual stress vs. stress intensity factor range

Fig. 9 Residual stress vs. stress intensity factor range

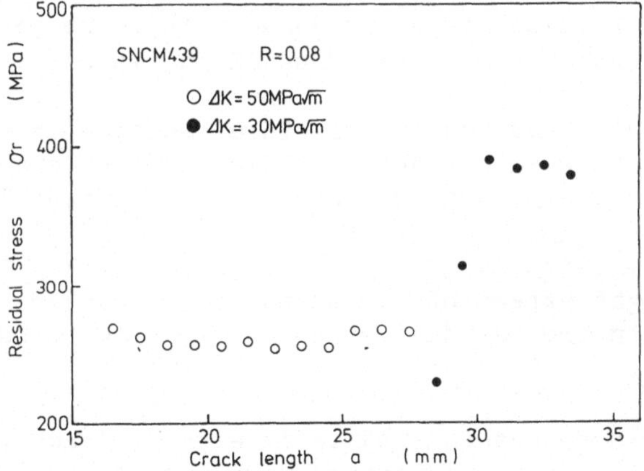

Fig. 10 Residual stress vs. crack length

the maximum stress intensity factor is shown in Fig. 12. The data come in a narrower scatter band compared with Fig. 11. The dependence of the half value breadth on the stress ratio in relation to stress intensity factor range suggests that by combining the residual stress measurements and half value breadth, the stress intensity

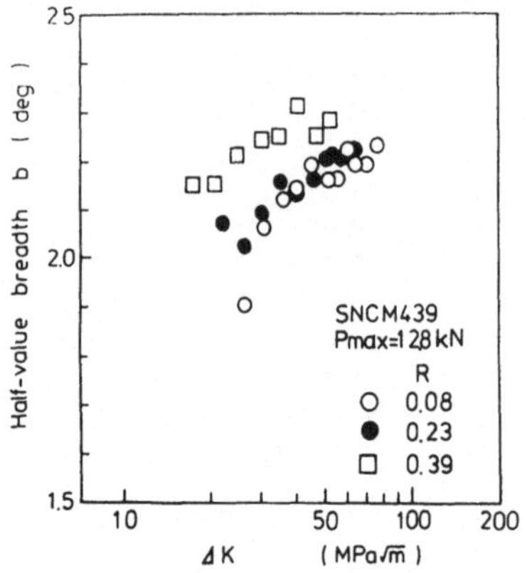

Fig. 11 Half-value breadth vs. stress intensity factor range

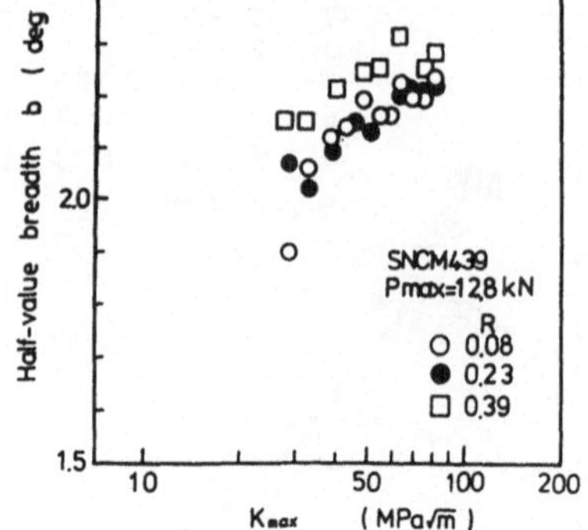

Fig. 12 Half-value breadth vs. maximum stress intensity factor

factor range and the stress ratio could be estimated although there are some problems in accuracy because the changes in the half value breadth were not so large in this material.

CONCLUSIONS

1) The residual stress on the fractured surface $\sigma(r)$ in the constant load fatigue tests increased with ΔK up to $\Delta K = 30$ MPa$\sqrt{m}$, but after that $\sigma(r)$ decreased as ΔK increased.

2) The residual stress on the fractured surface in the constant ΔK fatigue tests was constant regardless of the crack length, and agreed with the values obtained by the constant load fatigue tests.

3) Only a slight effect of the stress ratio was seen inthe relation between the residual stress on the fractured surface and ΔK.

4) In the two-step constant ΔK fatigue test, crack propagation rates were constant and agreed with the respective values obtained by the constant load fatigue tests, except during the transient period after ΔK changing. The residual stress on the fractured surface was also nearly constant and agreed with the results of constant fatigue tests.

X-RAY DIFFRACTION OBSERVATION OF FRACTURE SURFACES OF DUCTILE CAST IRON

Zenjiro Yajima

Faculty of Engineering, Kanazawa Institute of Technology
7-1 Oogigaoka, Nonoichi, Kanazawa 921, Japan

Yukio Hirose

Faculty of Education, Kanazawa University
1-1 Marunouchi, Kanazawa 920, Japan

and

Keisuke Tanaka

Department of Mechanical Engineering and Mechanics
Lehigh University, Bethlehem, Pennsylvania, U.S.A.

INTRODUCTION

X-ray diffraction observation of metal fractures provides fracture analysists with useful information on the mechanisms and mechanical conditions of fracturing. This method is called "X-ray fractography" and has been developed especially in Japan as a new engineering tool for fracture analysis[1,2].

In the present paper, X-ray fractography is applied to fracture surfaces of ductile cast iron (JIS FCD 60) which are widely used as machine parts. The fracture toughness tests were conducted at ambient and low temperatures by using compact tension (CT) specimens with blunt notches and three-point bending (TPB) specimens with fatigue pre-cracks. The line broadening of X-ray diffraction profiles was measured on and beneath fracture surfaces of fracture toughness specimens. The amount of plastic strain and the depth of the plastic zone left on the surface are evaluated from line broadening. The results are discussed in connection with the mechanics and mechanisms of fracture.

EXPERIMENTAL PROCEDURE

Material and Fracture Toughness Tests

The specimens used are as-cast spheroidal graphite cast iron (JIS FCD 60). The chemical composition of the material is as follows (wt. percent): 3.70C, 2.40Si, 0.40Mn, 0.04P, 0.018S. The yield stress and the tensile strength are 392 and 579 MPa, respectively. The matrix of the material mostly consists of ferrite and an extremely small amount of pearlite. The grain size of the ferrite is about 90 μm; the diameter of the graphite is about 60 μm.

The notch in CT specimens was made with an electro-discharge machine carefully so as to minimize the worked layer. The radius of notch tips, ρ, is from 0.12 to 3.0 mm. TPB specimens were pre-cracked by fatigue. The shapes and dimensions of the specimens are shown in Fig. 1. Fracture toughness tests were carried out in accordance with ASTM standards E 399[3] and E 813.[4] TPB specimens were tested at low temperatures down to -150°C; CT specimens were fractured at room temperature. The electrical potential technique is employed to detect the crack extension. The details of the toughness test procedure are described in our previous papers.[5,6]

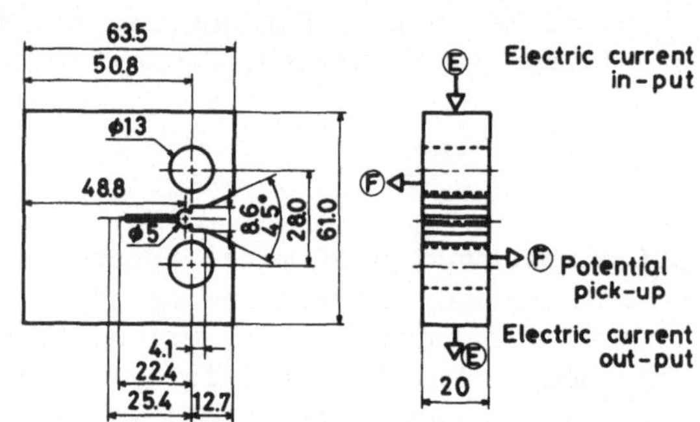

(a) CT specimen

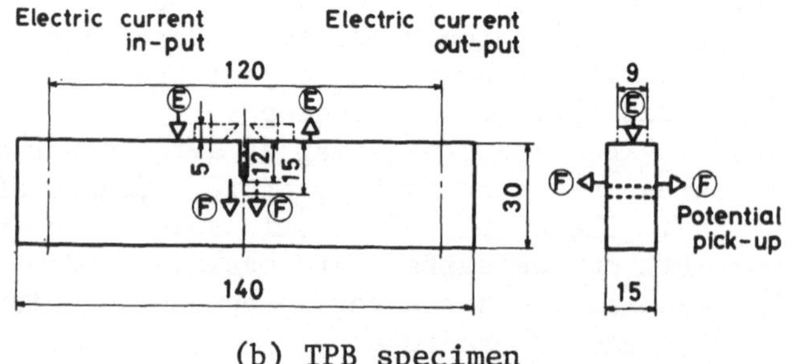

(b) TPB specimen

Fig. 1. Dimensions of test specimens (in mm) and positions of electrical potential pick-up.

In parallel with the above tests, tensile and torsion tests of the same material were carried out to obtain the correlation between X-ray line broadening and plastic strain.[6]

X-Ray Observation

The distribution of X-ray diffraction profiles beneath the fracture surface was measured by using an X-ray diffraction stress analyser. The area irradiated by X-rays was of 1 mm width and 10 mm length at the middle of the specimen thickness, touching the notch tip as indicated in Fig. 2. The half-value breadth measured is the range of diffraction angle at a half peak value of the doublet of K_{α_1} and K_{α_2}. The conditions of X-ray observation are given in Table 1. The distribution of the half-value breadth in the depth direction was measured by removing the surface layer successively by electro-polishing.

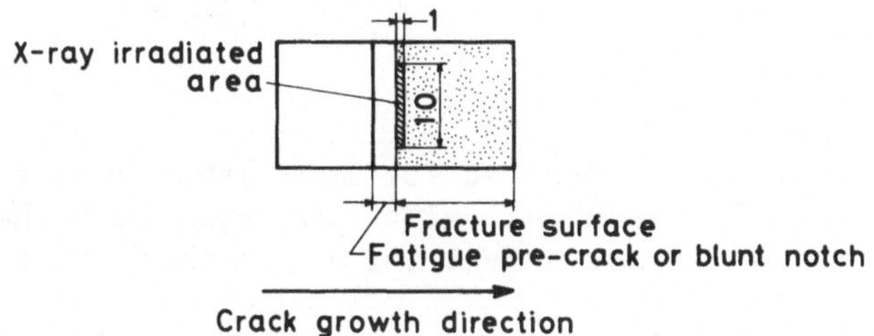

Fig. 2. Schematic illustration of X-ray irradiated area.

Table 1. X-ray diffraction condition.

Characteristic X-ray	Cr-K_α
Diffraction plane	(211)
Filter	V
Tube voltage (kV)	30
Tube current (mA)	16
Scanning speed (deg/min)	4
Soller slit divergence (deg)	0.15

EXPERIMENTAL RESULTS AND DISCUSSIONS

Relation between the Half-Value Breadth and Plastic Strain

The relation between the half-value breadth ratio B/B_o and the von Mises equivalent plastic strain ε_p obtained from tensile and torsional tests is presented in Fig. 3, where B_o is the initial half-value breadth of virgin material (B_o=2.6x10^{-2}rad), and B the value after the deformation mode. It is expressed by

$$B/B_o = 0.77 \log \varepsilon_p + 2.10 \quad (1)$$

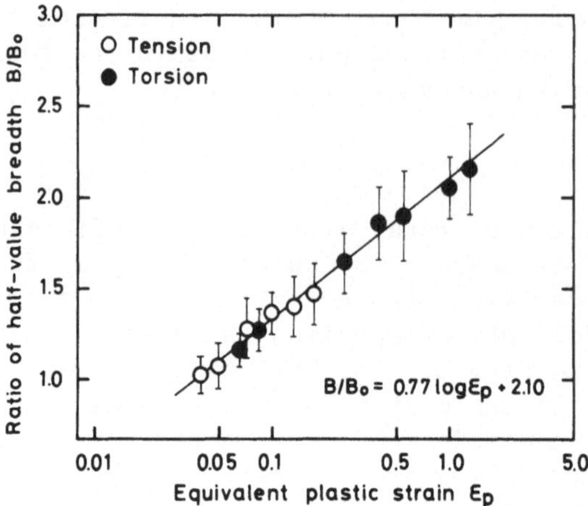

Fig. 3. Relation between half-value breadth and plastic strain.

Similar results were reported for low-carbon steels by Tanaka et al.[7] and Goto[8]. Tanaka et al. also confirmed that the lower testing temperatures did not affect the functional relation.

Plastic Zone Size

The stress intensity factor K_i at crack initiation, detected by the electrical potential method, for TPB specimens decreased with lowering temperature[5]. The K_i value for bluntly notched CT specimens increases as ρ becomes large[6].

Fig. 4 shows the distribution of B/B_0 beneath the fracture surface of TPB specimens fractured at various temperatures. The value B/B_0 is high close to the surface, and approaches to one as the depth increases. Since B/B_0 increases above one due to plastic strain, the size of the plastic zone, ω_y , can be defined as the depth where $B/B_0=1$. The distribution of B/B_0 was also measured for the fracture surface of bluntly notched specimens and the plastic zone size was evaluated.

In Fig. 5 is shown the relation between ω_y and K_i/σ_Y (σ_Y=the yield stress). It is noted that ω_y is proportional to the square of K_i/σ_Y for the cases examined except for that of ρ=1.5 mm. The relation is expressed as

$$\omega_y = \alpha \left(K_i/\sigma_Y \right)^2 \tag{2}$$

where α=0.13. The value of α reported previously for the case of high strength steels is 0.12~0.14[9]. Its value calculated by

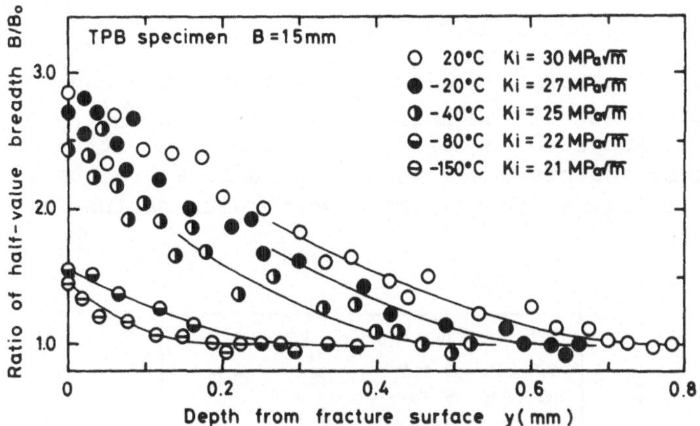

Fig. 4. Half-value breadth distribution near fracture surface.

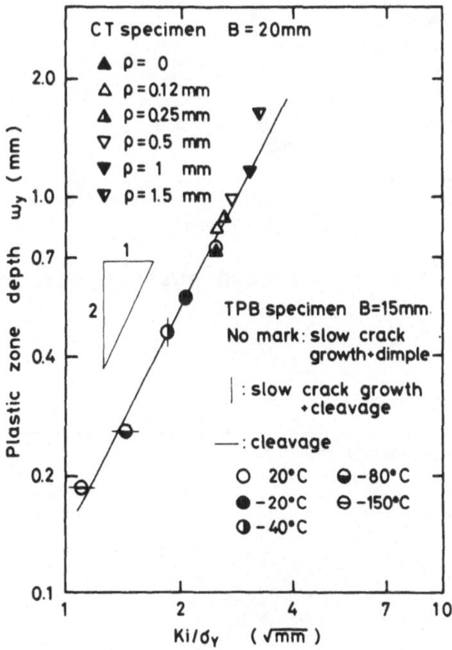

Fig. 5. Relation between plastic zone depth and stress intensity factor divided by yield stress.

Levy et al.[10] with the finite element method is 0.15, which is close to the 0.13 obtained in Fig. 5.

Equation (2) is useful to evaluate the fracture toughness of the fractured material from the measurement of ω_y by X-ray diffraction.

Plastic Strain on Fracture Surface

The plastic strain ε_f on the fracture surface can be estimated from the measurement of the half-value breadth by using Eqn.(1). The value of ε_f is expected to correspond the maximum strain built up in the material at the crack and notch tips before fracture. The maximum strain ε_{max} at the tip of a notch with radius in perfectly

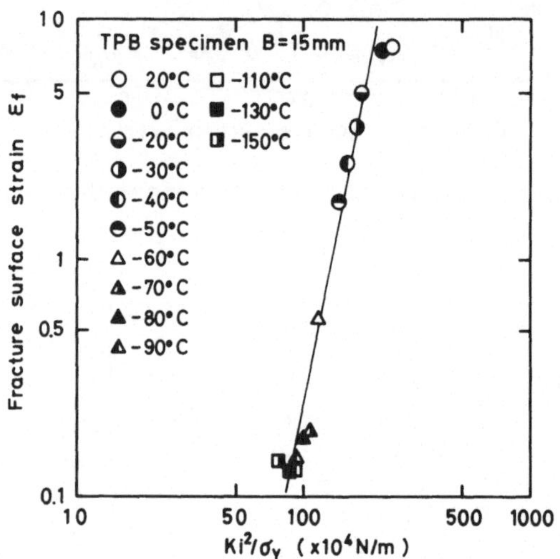

Fig. 6. Relation between fracture surface strain and K_i^2/σ_Y.

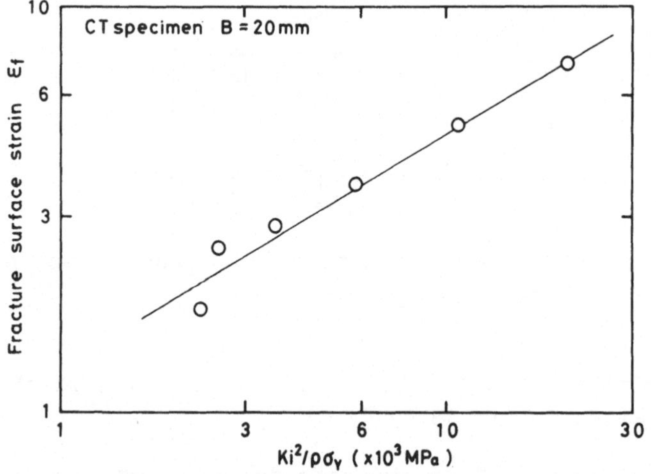

Fig. 7. Relation between fracture surface strain and $K_i^2/\rho\sigma_Y$.

plastic material is given by Rice[11] as

$$\varepsilon_{max} = (3/4)(1/E)(K^2/\rho\sigma_Y) \tag{3}$$

where E is Young's modulus. Therefore, ε_f is expected to be a function of $K_i^2/\rho\sigma_Y$. For the case of a crack, ε_{max} is a function of K_i^2/σ_Y because it is proportional to the crack tip opening displacement (= $K_i^2/E\sigma_Y$).

ε_f is plotted against K_i^2/σ_Y in Fig. 6 for the case of pre-cracked TPB specimens and against $K_i^2/\rho\sigma_Y$ in Fig. 7 for the case of bluntly notched CT specimens. For both cases, a unique relation is obtained, which suggests that ε_f is a unique function of the maximum plastic strain at crack initiation at the tips of notches or cracks.

CONCLUSIONS

The main results obtained in the present study are summarized as follows:

[1] The half-value breadth of X-ray diffraction profiles is related to the equivalent plastic strain ε_p by

$$B/B_o = 0.77 \log \varepsilon_p + 2.10 \tag{1}$$

where B_o and B are the half-value breadth measured with cast iron before and after deformation. This relation is independent of temperature and the mode of deformation, i.e. tension or torsion.

[2] In the distribution of the half-value breadth, the size of the plastic zone ω_y was determined as the depth where the half-value breadth reduced to the initial value. It is related to the fracture toughness K_i through

$$\omega_y = 0.13\ (K_i/\sigma_Y)^2 \tag{2}$$

where σ_Y is the yield stress of the corresponding test temperature.

[3] The plastic strain very close to the fracture surface was evaluated from the measured value of the half-value breadth by using Eqn.(1). The fracture surface strain is a function of the maximum plastic strain at the tip of notches or cracks, thus it is correlated to the square of the fracture toughness divided by the yield stress.

REFERENCES

1. Committee on X-Ray Study on Mechanical Behavior of Materials, "X-Ray Fractography," J. Soci. Mat. Sci. Jap., 31:244 (1982).

2. S. Taira, and K. Tanaka, "Fracture Surface Analysis by X-Ray Diffraction Techniques," J. Iron and Steel Ins. Jap., 65:450 (1979).
3. ASTM Standard, "Standard Test Method for PLANE-STRAIN FRACTURE TOUGHNESS OF METALLIC MATERIALS," Part 10, E 399-81 (1981).
4. ASTM Standard, "Standard Test for J_{IC}, A MEASURE OF FRACTURE TOUGHNESS," Part 10, E 813-81 (1981).
5. Z. Yajima, Y. Hirose, K. Tanaka, and H. Ogawa, "X-Ray Fractographic Study on Fracture Toughness of Ductile Cast Iron at Low Temperatures," To be published in J. Jap. Soc. Strength and Fracture of Materials.
6. Z. Yajima, Y. Hirose, K. Tanaka, and H. Ogawa, "Fracture Toughness of Blunt-Notched CT Specimens of Ductile Cast Iron," To be published in J. Soci. Mat. Sci. Jap.
7. K. Tanaka, K. Fujiyama, and K. Nakamura, "Fracture Toughness and X-Ray Diffraction Observation of Fracture Surface of Structual Low-Carbon Steel," J. Soci. Mat. Sci. Jap., 29:62 (1980).
8. T. Goto, "A Study on the Application of X-Ray Diffraction Technique to Failure Analysis of Metal Components," Proc. 1973 Symp. Mech. Beh. Mat., Kyoto, 265 (1973).
9. Z. Yajima, Y. Hirose, and K. Tanaka, "X-Ray Diffraction Observation of Fractured Surface of Fracture Toughness Specimen of High Strength Steel," J. Jap. Soc. Strength and Fracture of Materials, 16:59 (1981).
10. N. Levy, P.V. Marcal, W.J. Ostergren, and J.R. Rice, "Small Scal Yielding Near a Crack in Plane Strain: A Finite Element Analysis," Int. J. Frac. Mech., 7:143 (1971).
11. J. R. Rice, "Fracture," H. Liebowitz, ed., II, 191, Academic Press, New York (1968).

ANALYTICAL AND EXPERIMENTAL INVESTIGATION OF FLOW AND FRACTURE MECHANISMS INDUCED BY INDENTATION IN SINGLE CRYSTAL MgO

T. Larchuk, T. Kato, R. N. Pangborn and J. C. Conway, Jr.

Department of Engineering Science and Mechanics
The Pennsylvania State University
University Park, PA 16802 USA

INTRODUCTION

The flow and fracture behavior of ceramic and other brittle materials under the influence of contact loading is important to both component fabrication and performance. The ease of machining, severity of residual surface damage and rate of wear during subsequent service are controlled to a large degree by the character and extent of the flow zone and its influence on the fracture mode. This investigation was undertaken to provide experimental verification of the results obtained through elastic/plastic finite element modeling of the stress distribution and deformations introduced by static contact loading. Experimentally, X-ray double-crystal diffractometry (DCD) was applied to obtain a mapping of the distortions produced beneath a Vickers indenter, and hence to evaluate the effect of material and geometric parameters on the flow and fracture mechanisms.

Indentation fracture initiation and propagation phenomena have been studied extensively[1,2] in conjunction with the stress fields described by classical elasticity solutions for point loading.[3,4] As has been pointed out previously,[5] however, this approach presupposes the existence of intrinsic flaws of critical size which are eventually caused to propagate by tensile stresses developed at the elastic/plastic boundary. Neglected are the mechanical anisotropy inherent to crystalline materials, the activation of dislocation sources by favorably oriented shear stresses, and the role of dislocation mobility and interaction in flow and microcrack nucleation, respectively.

Experimentally, etch pit and X-ray topographic analyses have been employed to examine the external surface relief caused by indentation.[6] Transmission electron microscopy[7] and selected-area electron channeling[8] have proven useful in disclosing the extent of plastic deformation beneath an indentation in hard ceramics such as SiC. The X-ray diffraction technique utilized in the present investigation was chosen to provide a more quantitative description of the distortion within the subsurface damage zone.

EXPERIMENTAL PROCEDURE

Specimens with surface dimensions of about 1 x 1.5 cm and thickness averaging 0.4 cm were prepared from clear, fused MgO single crystals by cleavage over {100} planes, followed by chemical polishing to remove residual surface damage. Static indentation to loads ranging from 22 to 220 N was carried out using a Vickers pyramidal indenter mounted in an Instron universal tester. The indenter diagonals were aligned to coincide with the <110> directions on the specimen surface. After unloading, the specimens were sectioned so as to expose a plane normal to the surface of indentation. (001) sections were obtained by cleaving, a fast fracture process thought to introduce very limited deformation.[7] (101) sections were prepared by diamond wheel cutting followed by mechanical and chemical polishing.

The specimen arrangement for conducting X-ray DCD measurements is shown in Fig. 1. $CrK\alpha_1$ radiation was used in the analyses to limit the penetration depth. The specimens were scanned by stepwise translation of the specimen in two directions, and rocking curves were recorded at each position to generate a spatial mapping of lattice distortions in the vicinity of the indentation. A similar procedure was employed by Tsunekawa and Weissmann[9] to examine the microplasticity associated with the fracture of notched silicon crystals. Additional resolution of point-to-point variations in the rocking curve profile widths was obtained by overlapping the areas probed by the incident beam at adjacent positions.

A three-dimensional elastic/plastic finite element code (BOPACE-3D) employing the Huber-Mises yield criterion and the Prandtl-Reuss flow rule was used in the numerical analysis. The model was

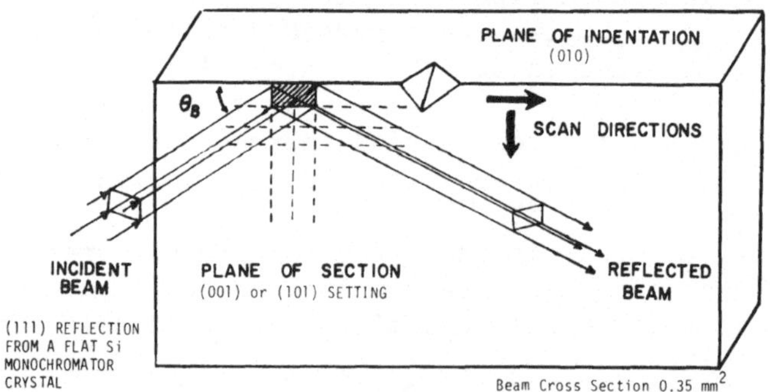

Fig. 1. Schematic illustration of X-ray double crystal diffraction.

constructed to simulate elastic/plastic indentation of a half-space by a rigid, conical indenter with a Vickers included angle. Isoparametric quadrilateral elements incorporating mid-side nodes were utilized in the model with nodal density increasing from far-field to directly beneath the indenter. Loading was applied through displacement of nodes beneath the indenter.

RESULTS AND DISCUSSION

Fig. 2 summarizes the results of the X-ray DCD studies for the two specimen settings. The first of the pair of numerical values associated with each "contour" line is the characteristic rocking curve halfwidth, β, measured for the regions traversed by the line. The second number represents the local lattice distortion, $\Delta d/d$, manifested by the broadening of the rocking curve profiles. This is obtained by first correcting for the intrinsic breadth, β_o, of the undeformed material according to the relation:

$$\overline{\beta} = (\beta^2 - \beta_o^2)^{1/2} ,$$

and then computing the average distortion from an expression derived by differentiation of the Bragg equation with respect to the interplanar spacing, d, and diffraction angle, θ:

$$1/2(\Delta d/d) = \Delta\theta \cot\theta/2 , \text{ where } \Delta\theta \equiv \overline{\beta}.$$

It can be seen that the map for the (001) section exhibits an essentially semicircular pattern of contours as compared to that for the (101) section. This latter setting features a pear-shaped pattern of contours with lobes of higher distortion eminating outward at shallow inclination angles from the indentation surface as well as directly downward in the direction of applied loading. Along the surface of indentation, particularly for the (001) setting, steep decreasing gradients in the induced distortion are manifested by the close spacing of contour lines.

On the grid in Fig. 3a are shown the nondimensionalized stress components of selected elements obtained in the numerical analysis. The modeling involved displacement loading, with nodal displacements prescribed according to experimental measurements. Also input were the material constants and the "indentation yield strength" computed from hardness measurements.[10] The agreement between the program-generated resultant indentation force of 196.6 N and the experimental value of 155.7 N was satisfactory for our purposes. The predicted elastic/plastic boundary is semicircular and located at a radial distance of 0.4 mm from the point of indentation, coinciding with regions of high, tensile, out-of-plane hoop stress, σ_z, on the indentation surface and large, nearly equal in-plane and out-of-plane tensile stresses, σ_x and σ_z, beneath the indenter.

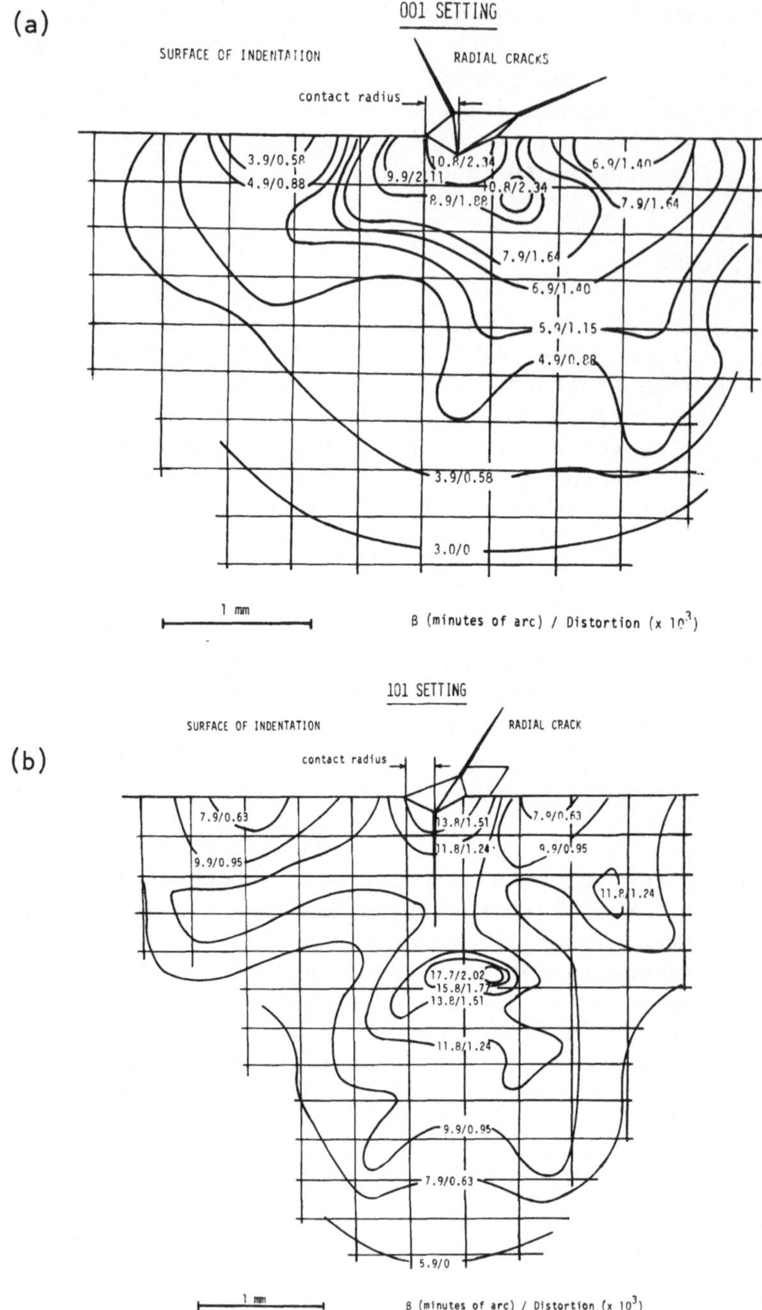

Fig. 2. Contour maps of X-ray rocking curve halfwidths and corresponding lattice distortions for (a) the (001) section, and (b) the (101) section.

[Note the similarity in order of magnitude for distortions derived from X-ray rocking curve data as compared with effective strains calculated through finite element method; see Fig. 3a]

The role of crystallographic slip processes in modifying the actual shape of the flow zone and contributing to crack initiation is best revealed by substituting the calculated stress components for each position into the normal resolved shear stress formulation:

$$\tau_{RES} = n_i b_j \sigma_{ij} ,$$

where n_i and b_j are the direction cosines between the stress direction and the slip plane normal and slip direction, respectively. The six $\langle 101\rangle\{10\bar{1}\}$slip systems shown in Fig. 3b & c have been proposed to account for the plastic deformation of MgO at room temperature.[6,11] Following the usual summation convention for repeated subscripts, the resolved shear stress equations for the six systems and two different specimen settings (see unprimed and rotated, primed coordinate systems of Fig. 3b & c) are given in Table 1. After appropriate substitution of the calculated stress components, ratios between the resolved shear stress and the shear stress required for dislocation multiplication at half-loop sources were calculated for each slip system. This latter stress, referred to as the "activation stress" by Stokes, Johnston and Li[12] was found by them to be significantly less than the macroscopic yield stress obtained from the stress-deflection curve in three-point bending. Less still, was the "motivation stress" for expansion of already existing half-loops also evaluated by these investigators. The inner boundaries in the diagrams of Fig. 4 define the regions within which the resolved-to-activation

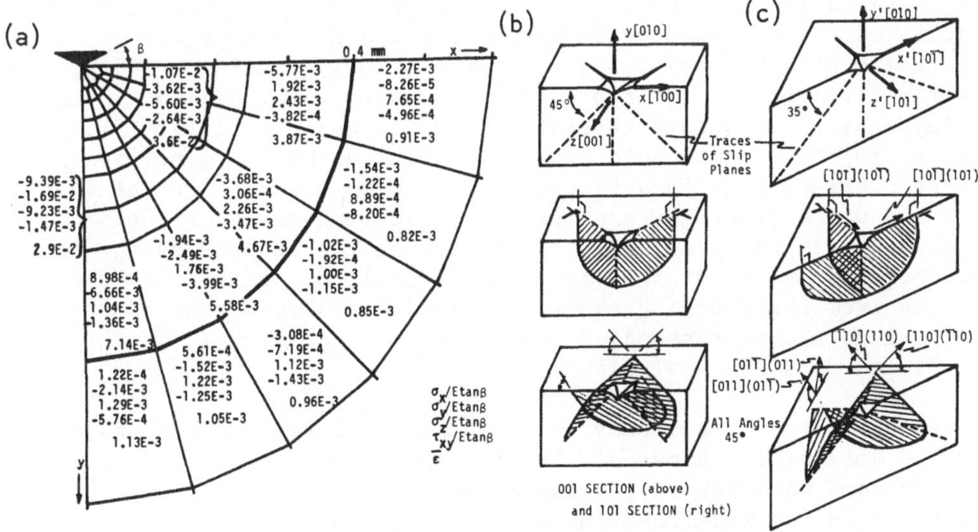

Fig. 3. (a) Finite element grid showing nondimensionalized stress components and effective strains for selected elements; (b) Coordinate directions (top), slip planes normal to surface (middle) and slip planes inclined 45° to surface (bottom) for (001) setting; (c) Similar diagrams, (101) setting.

Table 1. Direction Cosines (a) and Resolved Shear Stress Equations (b) for Six Potential Slip Systems for the (001) and (101) Settings

(a)

	001 SETTING: x[100], y[010], z[001]						101 SETTING: x'[101], y'[010], z'[101]					
SLIP PLANE	$(01\bar{1})$	(011)	$(\bar{1}10)$	(110)	$(10\bar{1})$	(101)	$(01\bar{1})$	(011)	$(\bar{1}10)$	(110)	$(10\bar{1})$	(101)
SLIP DIRECTION	$[011]$	$[01\bar{1}]$	$[110]$	$[\bar{1}10]$	$[101]$	$[10\bar{1}]$	$[011]$	$[01\bar{1}]$	$[110]$	$[\bar{1}10]$	$[101]$	$[10\bar{1}]$
n_1	0	0	$-1/\sqrt{2}$	$1/\sqrt{2}$	$1/\sqrt{2}$	$1/\sqrt{2}$	1/2	-1/2	-1/2	1/2	1	0
n_2	$1/\sqrt{2}$	$1/\sqrt{2}$	$1/\sqrt{2}$	$1/\sqrt{2}$	0	0	$1/\sqrt{2}$	$1/\sqrt{2}$	1/2	$1/\sqrt{2}$	0	0
n_3	$-1/\sqrt{2}$	$1/\sqrt{2}$	0	0	$-1/\sqrt{2}$	$1/\sqrt{2}$	-1/2	1/2	-1/2	1/2	0	1
b_1	0	0	$1/\sqrt{2}$	$-1/\sqrt{2}$	$1/\sqrt{2}$	$1/\sqrt{2}$	-1/2	1/2	1/2	-1/2	0	1
b_2	$1/\sqrt{2}$	$1/\sqrt{2}$	$1/\sqrt{2}$	$1/\sqrt{2}$	0	0	$1/\sqrt{2}$	$1/\sqrt{2}$	$1/\sqrt{2}$	1/2	0	0
b_3	$1/\sqrt{2}$	$-1/\sqrt{2}$	0	0	$1/\sqrt{2}$	$-1/\sqrt{2}$	1/2	-1/2	1/2	-1/2	1	0
	$(110)_{45°}$ SLIP PLANES				$(110)_{90°}$ SLIP PLANES							

(b)

SYSTEMS	001 setting	101 setting
$(01\bar{1})\,[011]$, $(011)\,[01\bar{1}]$	$\tau_{RES} = 1/2(\sigma_{yy} - \sigma_{zz})$	$\tau'_{RES} = 1/4(-\sigma_{xx} + 2\sigma_{yy} - \sigma_{zz} + \cancelto{0}{2\sigma_{xz}})$
$(\bar{1}10)\,[110]$, $(110)\,[\bar{1}10]$	$\tau_{RES} = 1/2(-\sigma_{xx} + \sigma_{yy})$	$\tau'_{RES} = 1/4(-\sigma_{xx} + 2\sigma_{yy} - \sigma_{zz} - \cancelto{0}{2\sigma_{xz}})$
$(10\bar{1})\,[101]$, $(101)\,[10\bar{1}]$	$\tau_{RES} = 1/2(\sigma_{xx} - \sigma_{zz})$	$\tau'_{RES} = \cancelto{0}{\sigma_{xz}}$

stress ratios (shown) exceed unity. The outer boundaries are similarly derived limits enclosing regions where loop expansion would be anticipated. Superimposing diagrams a, b and c corresponding to the three pairs of active slip systems for the (001) setting would give a nearly circular arc for the extreme boundary, very similar to that displayed by the contours generated by X-ray analysis (Fig. 2a). All three pairs of systems could be active simultaneously (i.e., inner regions overlap) only in a narrow area just to the side of the indentation where the steep gradient in distortion was previously noted. The last diagram, Fig. 4d, indicates the boundaries for the four active systems for the (101) setting, all of which share the same resolved shear stress equation. The pear-shaped regions conform well to the corresponding contour pattern of Fig. 2b depicting the results of X-ray analysis.

Although the consistency between the theoretically and experimentally determined shapes of the flow zones beneath the indenter is quite good when the crystallographic conditions for slip are incorporated in the modeling, the X-ray contour maps show evidence of distortion considerably beyond the limiting boundaries predicted for dislocation multiplication and glide. It is therefore proposed that the gross multiplication and interaction of dislocations in the vicinity of the indentation result in numerous pileups and concomitant intensification of long range stresses in the bulk material. The interaction of dislocations moving on intersecting slip planes

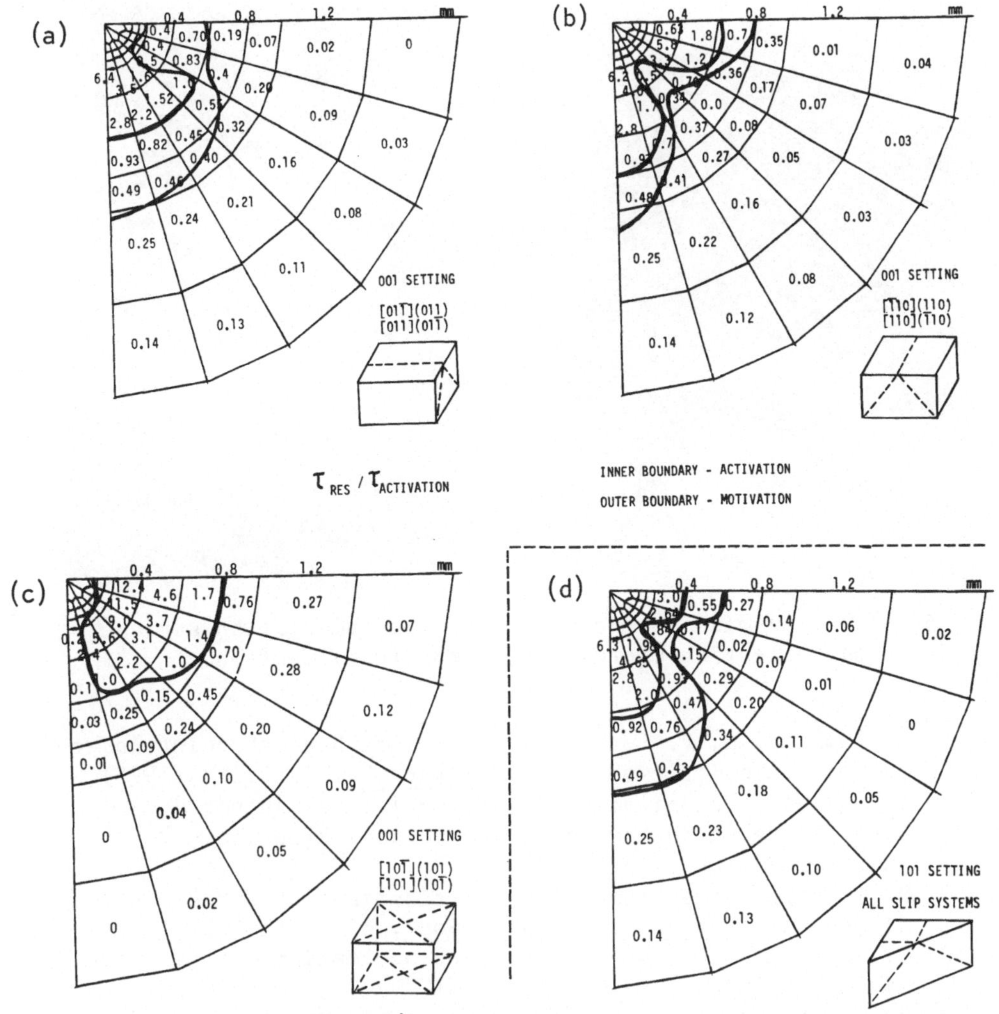

Fig. 4. (a,b,c) Boundaries enclosing regions for which the resolved shear stress exceeds the activation or motivation stresses for 6 active slip systems, (001) setting; (d) Activation and motivation limits for 4 active slip systems, (101) setting. [$\tau_{ACT} \simeq 2.5\tau_{MOT} \simeq 76$ MPa for $\{110\}_{45°}$; $\tau_{ACT} \simeq \tau_{MOT} \simeq 41$ MPa for $\{110\}_{90°}$]

to produce sessile line segments,[6] and condensation of dislocations on the same slip plane when their motion is impeded[11] also contribute to the formation of radial, median and lateral (shear) cracks (see micrographs in Fig. 5). These mechanisms associated with fracture will be discussed in more depth in a subsequent paper.

REFERENCES

1. B. Lawn and R. Wilshaw, Review, Indentation fracture: principles and applications, J. Mater. Sci., 10:1049 (1975).

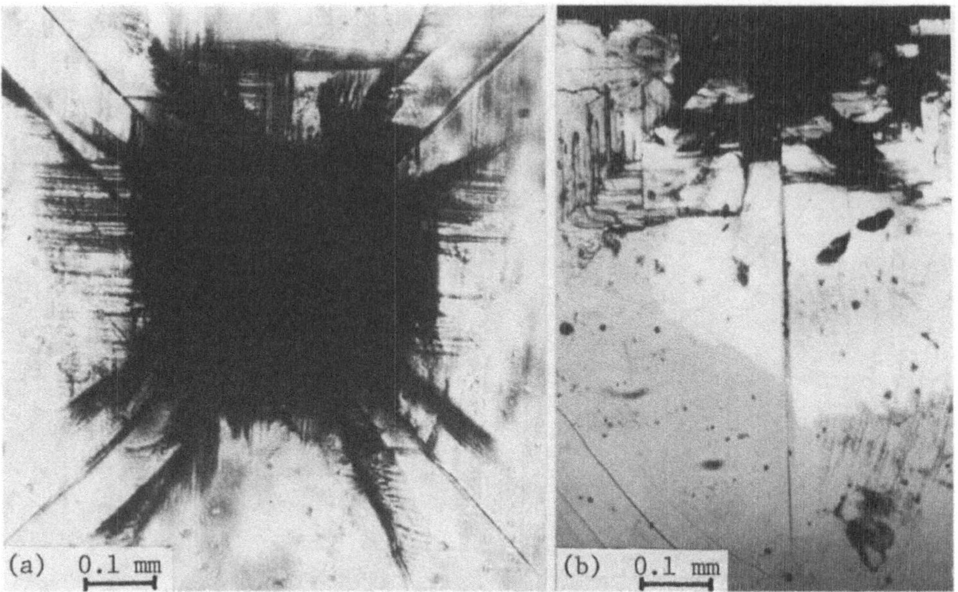

Fig. 5. (a) Transmission micrograph of the indentation surface; (b) Reflection micrograph of the (001) section.

2. B. R. Lawn and M. V. Swain, Microfracture beneath point indentations in brittle solids, J. Mater. Sci., 10:113 (1975).
3. J. Boussinesq, "Application des Potentiels a l'Etude de l'Equilibre et du Mouvement des Solides Elastiques," Gauthier-Villars, Paris (1885).
4. J. H. Michell, Some elementary distributions of stress in three dimensions, Proc. London Math. Soc., 32:23 (1900).
5. J. T. Hagan, Micromechanics of crack nucleation during indentations, J. Mater. Sci., 14:2975 (1979).
6. R. W. Armstrong and C. Cm. Wu, Lattice misorientation and displaced volume for microhardness indentations in MgO crystals, J. Amer. Cer. Soc., 61:102 (1978).
7. B. J. Hockey and B. R. Lawn, Electron microscopy of microcracking about indentations in aluminum oxide and silicon carbide, J. Mater. Sci., 10:1275 (1975).
8. J. Lankford and D. L. Davidson, Indentation plasticity and microfracture in silicon carbide, J. Mater. Sci., 14:1669 (1979).
9. Y. Tsunekawa and S. Weissmann, Importance of microplasticity in the fracture of silicon, Met. Trans., 5:1585 (1974).
10. K. L. Johnson, The correlation of indentation experiments, J. Mech. Phys. Solids, 18:115 (1970).
11. R. J. Stokes, T. L. Johnston, and C. H. Li, Crack formation in magnesium oxide single crystals, Phil. Mag., 3:718 (1958).
12. R. J. Stokes, T. L. Johnston, and C. H. Li, Effect of surface condition on the initiation of plastic flow in magnesium oxide, TMS-AIME, 215:437 (1959).

X-RAY DIFFRACTION STUDY OF SHAPE MEMORY IN URANIUM-NIOBIUM ALLOYS

D. A. Carpenter and R. A. Vandermeer†

Oak Ridge Y-12 Plant*
Union Carbide Corporation, Nuclear Division
Oak Ridge, Tennessee 37830

INTRODUCTION

An x-ray diffraction study of the reversible deformation modes associated with the shape memory effect has been carried out on a series of uranium-niobium alloys near the monotectoid composition (6.2 wt. % Nb). Diffraction patterns were measured as a function of strain, in situ, while the specimens were under stress as part of an attempt to explain the "easy-flow", low-strain plateau in the stress-strain curve.[1] The alloys, consisting of highly twinned, metastable α'' (monoclinic) and γ° (tetragonal) phases derived from the high-temperature BCC γ phase, produced broad, overlapping diffraction lines difficult to analyze by conventional techniques. One solution to this problem was to use a segmented step-scan technique so as to apportion the scan time to concentrate on the most difficult regions. This paper discusses data obtained from an α'' alloy and a dual-phase $\alpha'' + \gamma^\circ$ alloy.

EXPERIMENTAL

Flat tensile specimens were machined from the center of a 9.4-mm-thick plate that had been solution heat treated for 7200 seconds at 1075 K in vacuum and then water-quenched. Samples containing both α'' and γ° phases were prepared by aging a machined, water-quenched specimen containing 6.4 wt. % Nb for

---

* Operated for the U.S. Department of Energy by Union Carbide Corporation, Nuclear Division, under Contract W-7405-eng-26.

† Also affiliated with The University of Tennessee, Knoxville.

14,000 seconds at 475 K. All specimens were electropolished to remove machining deformation and oxide films.

The alloy specimens were positioned in a specially designed tensile jig that was mounted on an automated horizontal x-ray diffractometer. The stress axis was parallel to the horizontal plane and perpendicular to the diffraction vector. A nickel powder standard, calibrated against the SRM 640 Si standard, was used as a 2Θ standard. The x-ray data were collected using copper $K\alpha$ radiation and a diffracted-beam graphite monochromator. Patterns were measured in the unstressed condition and at small strain increments while the samples were under stress until fracture. Occasionally, at intermediate strain levels, patterns were measured after the specimens were unloaded and then again while incrementally reloading. A segmented step-scan technique was used with several specimens because of the complexity of the patterns. In order to scan the pattern as quickly as possible without compromising counting statistics, broad, weak peaks and highly overlapped peak groupings were step-scanned using relatively long counting times compared with regions of the pattern containing stronger, sharper peaks. Regions containing no peaks were skipped. Counting times and scan segments were operator selected for each specimen. The peaks were fit with squared Lorentzians on a PDP-11/70 computer using the Marquardt algorithm in an interactive mode.[2] Lattice parameters were calculated using a least-squares routine.[3]

The α'' volume fraction, $\nu_{\alpha''}$, was calculated for the dual-phase alloy using the direct comparison method with calculated intensities as standards and using peaks selected to reduce the effects of preferred orientation. Those peaks were used which gave an average I/R value invariant to strain in the single-phase data.

RESULTS

Monoclinic α'' $(\bar{h}kl)$ peaks tended to undergo intensity changes with strain in the opposite sense to that of the (hkl) peaks. This is illustrated by $(\bar{1}11)$-(111) peak pair in Figure 1. In addition, the monoclinic angle, $(\hat{\gamma})$, decreased significantly with strain (Figure 2). Unloading at 2 percent and 7 percent produced significant springback in $\hat{\gamma}$, while texture reversibility occurred only after unloading at 7 percent.

A $\gamma° \rightarrow \alpha''$ phase transformation was detected in the dual-phase alloy, initially containing approximately 50 volume % α''. The α'' volume fraction, $\nu_{\alpha''}$, (Figure 3) increased with strain up to about 2 percent, then tended to level off. The partial reversibility of the transformation was shown by the decrease in $\nu_{\alpha''}$ each time the sample was unloaded.

There were discrete changes in the a, b, and $\hat{\gamma}$ lattice parameters (Figures 4 and 5), but no change in c as a result of the phase transformation. The tetragonal $\gamma°$ lattice parameters were converted to monoclinic parameters corresponding to the α'' cell using Hatt's relationship.[4] The $\hat{\gamma}$ parameter decreases continuously in each phase and undergoes large springback on unloading.

DISCUSSION

The principal reversible deformation modes in the single-phase α'' alloys and in the dual-phase $\alpha'' + \gamma°$ alloys correspond largely to those observed in other shape-memory alloys.[5]

Reversible deformation corresponds to the amount of strain which can be recovered by unloading and by heat-activated shape recovery. Reversible deformation in the single-phase α'' alloy corresponds, primarily, to the first plateau in the stress-strain curve, to the initial preferred orientation development, and to the changes in the $\hat{\gamma}$ parameter. A recent optical metallography study suggests that the initial development of preferred orientation corresponds to a reorientation of twin variants.[6] One variant, more favorably oriented with respect to the stress vector than the other variants, increases in volume fraction with respect to other less favorably oriented variants in order to accommodate the strain. Mid-test unloading showed that the low strain variant reorientation was permanent and not reversible with the removal of stress. In another study, complete strain recovery (shape memory) occurred when the unloaded material was heated so as to cause reversion to the crystallographically related $\gamma°$ and γ phases.[1,7]

The reversal of the preferred orientations at higher strains, although not well understood at this time, is associated with the onset of appreciable nonreversible plastic deformation.

In the dual-phase $\alpha'' + \gamma°$ alloys, the strain mechanism corresponds to a thermoelastic martensitic phase transformation. The transformation involves shear on the $\{112\}\gamma$ planes in $\langle 111 \rangle_{\gamma}$ directions along with a lattice contraction along the a_m axis and expansion along the b_m axis. The first plateau in the stress-strain curve is due to the ease with which this diffusionless transformation occurs. Shape recovery would be expected to correspond to the reverse of the $\gamma° \rightarrow \alpha''$ phase transformation.

ACKNOWLEDGMENTS

The authors wish to acknowledge the contribution of A. D. Condrey to the experimental part of this work, and to K. W. Kaiser,

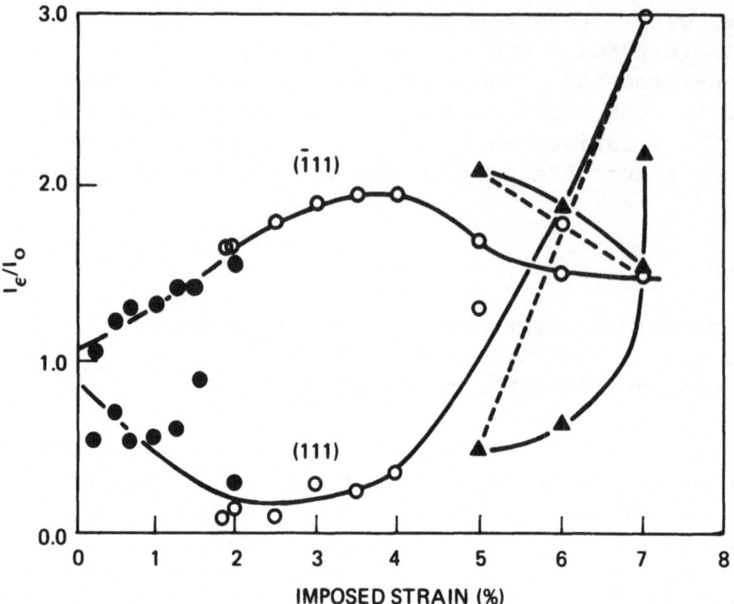

Fig. 1. Intensities of the $\alpha''(\bar{1}11)$ and $\alpha''(111)$ peaks, normalized to zero strain, as a function of strain for a water-quenched U-6.4Nb alloy.

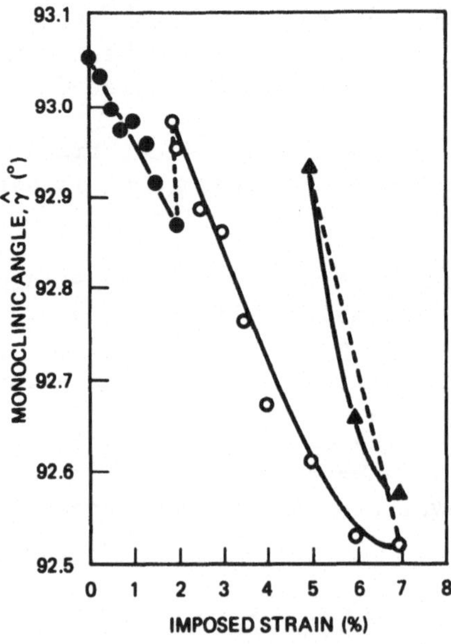

Fig. 2. The monoclinic angle, $\hat{\gamma}$, of the α'' phase as a function of strain for a water-quenched U-6.4Nb alloy.

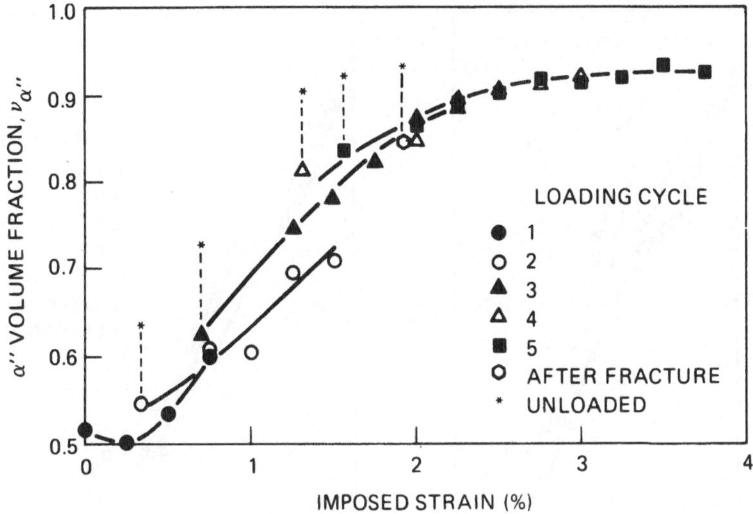

Fig. 3. The α" volume fraction, ν_{α}", as a function of strain for an aged U-6.4Nb alloy containing both γ° and α" phases.

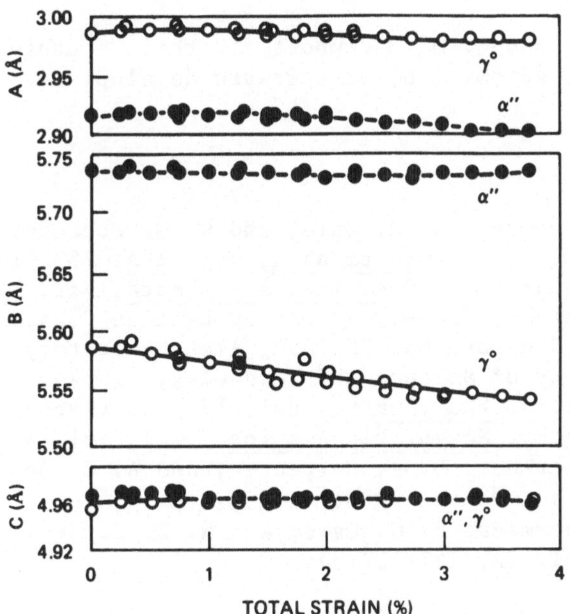

Fig. 4. The <u>a</u>, <u>b</u>, and <u>c</u> lattice parameters of the α" and γ° (based on monoclinic cell related to α") phases as a function of strain for an aged U-6.4Nb alloy.

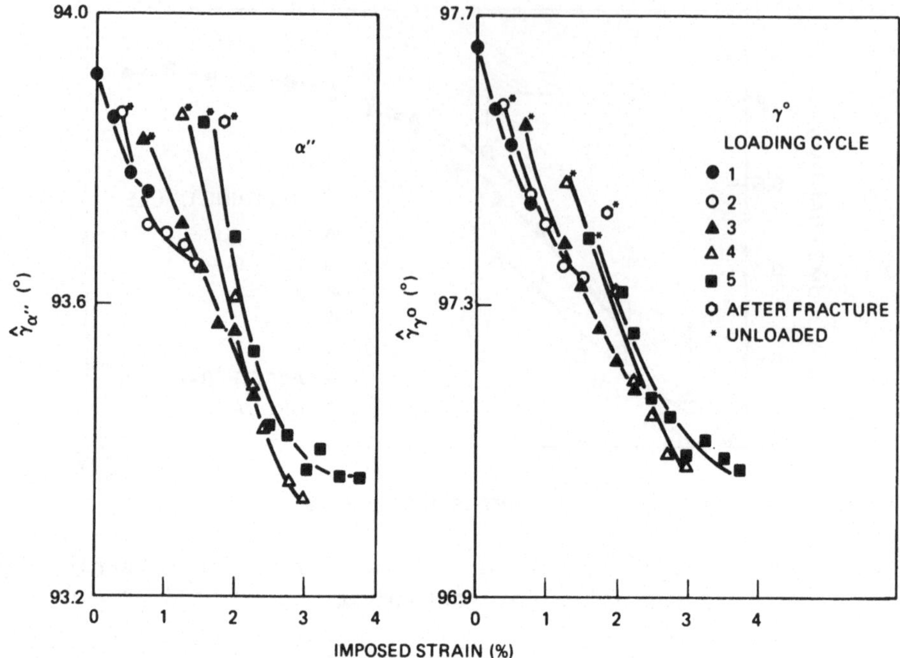

Fig. 5. The monoclinic angle, $\hat{\gamma}$, of the α'' and γ° phases as a function of strain for the aged U-6.4Nb alloy.

J. J. Dunigan, and C. M. Davenport for their contributions to the diffractometer automation and software development.

REFERENCES

1. R. A. Vandermeer, J. C. Ogle, and W. G. Northcutt, Jr., Metallurgical Transactions A, Vol. 12A, 733 (1981).
2. D. W. Marquardt, J. Soc. Ind. Appl. Math, Vol. 11, 431 (1963).
3. D. E. Williams, "LCR-2, A Fortran Lattice Constant Refinement Program," Report No. IS-1052, Ames Laboratory, Iowa State University of Science and Technology (1964).
4. B. A. Hatt, J. Nucl. Matr., Vol. 19, 133 (1966).
5. L. M. Schetky, Scientific American, Vol. 141, No. 5, 74 (1979).
6. R. A. Vandermeer, D. A. Carpenter, and A. G. Dobbins, unpublished data, August 3, 1982.
7. R. A. Vandermeer, J. C. Ogle, and W. B. Snyder, Jr., Scripta Met., Vol. 12, 243 (1978).

X-RAY FLUORESCENCE ANALYSIS USING SYNCHROTRON RADIATION

J.V. Gilfrich, E.F. Skelton, D.J. Nagel and A.W. Webb

Naval Research Laboratory, Washington, DC 20375

S.B. Qadri and J.P. Kirkland

Sachs/Freeman Associates, Bowie, MD 20715

INTRODUCTION

Synchrotron radiation (SR) has several unique properties which cause it to be used for many purposes in science and technology. It is continuous spectrally, collimated spatially, short-pulsed yet continuously-operating temporally, as well as polarized(1). Many applications of SR have already been demonstrated. These range from production of fine scale structures by lithography to numerous studies of materials. Determination of atomic structure by diffraction and other scattering experiments and by absorption spectroscopy (EXAFS), and emission spectroscopic studies of electronic structure have been given most attention(2). Only limited attention has been paid to determination of materials composition by x-ray fluorescence analysis (XRF) employing synchrotron radiation. Sparks et al.(3-5) have performed the most notable experiments, examining mica inclusions for the presence of primordial super heavy elements and irradiating two National Bureau of Standards Standard Reference Materials (SRM 1571, Orchard Leaves and SRM 1632, Coal) to measure the fluorescent x-ray intensity from the trace elements.

The work reported here is the first part of a program to define the capabilities of SR excitation for the analysis of low-concentration "thin" samples of elements spanning much of the periodic table. Energy and wavelength dispersive measurements using only SR continuum have been made to date. Later work will include similar measurements with tunable monochromatic SR and conventional laboratory source measurements on the same samples. Full details of the work to date will be published elsewhere(6). The purpose of this paper is to synopsize the available data and to compare with alternative methods of x-ray analysis. The next

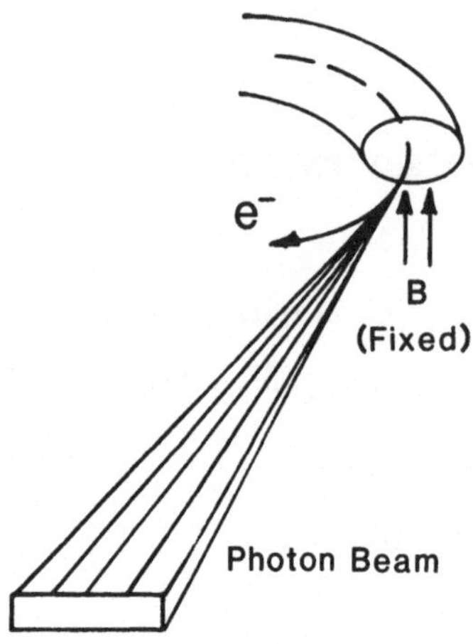

Fig. 1. High energy electrons circulating within an ultra-high vacuum torus in the magnetic field, B, emit synchrotron radiation, which is collimated in the vertical direction.

section is a brief review of the salient features of SR. Then the experimental section is presented and the results given and discussed in succeeding sections.

PRODUCTION OF SYNCHROTRON RADIATION

The generation of SR is illustrated schematically in Figure 1. Electrons with typically several billion electron volts (GeV) of kinetic energy orbit in an ultrahigh vacuum tube between the poles of strong ($\sim 10^4$ Gauss) magnets (not shown). The vertical field deflects (accelerates) the electrons horizontally, causing the emission of SR. Synchrotron radiation is magnetic bremsstrahlung, in contrast to ordinary electronic bremsstrahlung produced when energetic electrons encounter the coulomb field of nuclei. The SR spectrum is characterized by the critical photon energy (E_c), which is the energy above and below which one-half of the total emitted SR power falls. E_c and the energy range of the spectrum are dependent on the electron energy (E) and the orbital radius (R).

$$E_c = \frac{2.218E^3}{R}$$

where E_c is in KeV, E is in GeV and R is in meters

The vertical divergence angle (Ψ) can be calculated from an empirical equation(7) involving the photon wavelength (λ), the critical wavelength ($\lambda_c = 12.398/E_c$), both in angstroms, and the electron energy (E) in GeV.

$$\Psi = \frac{1.54(\lambda/\lambda_c)^{0.38}}{1957E}$$

The small values of Ψ (<1 mrad for E >0.5 GeV) are noteworthy. The horizontal divergence of the SR used in a particular situation depends on the arc length of the orbit being observed (generally several mrad). The pulse length (typically about 10^{-9} sec.) is the time it takes one group ("bunch") of electrons to traverse the viewed arc. The pulse repetition rate is often near 1 MHz. Polarization is complete in the orbital plane (horizontal) and decreases with increasing angles above and below that plane.

The continuous nature of the SR spectrum is illustrated by Figure 2 which shows the spectra computed for two facilities: HASYLAB, in Europe, operated at 3.5 GeV and 200 mA, and the Stanford Synchrotron Radiation Laboratory (SSRL) in Calfornia, operated at 3.0 GeV and 60 mA. The difference in the energy range is due to the different E and R values leading to critical energies of 7.8 keV for HASYLAB and 4.7 keV for SSRL. The difference in intensity is due mainly to the different electron currents.

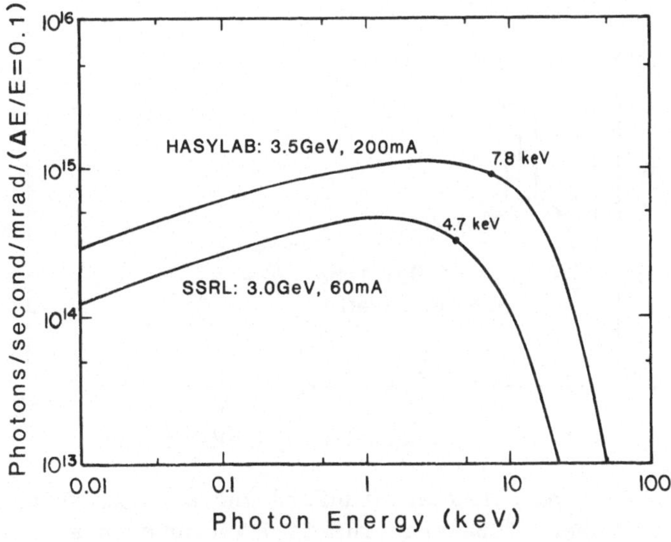

Fig. 2. Spectra at two synchrotron laboratories.

EXPERIMENTAL SETUP

This work was carried out at SSRL on Beam Line II-4 which has no monochromator and delivers one milliradian (horizontal) of continuum over the energy range of about 2 to 60 KeV to the experimental station located 17 meters from the storage ring. The end of the beam pipe is sealed with a beryllium window from which the x-rays exit into an interlocked radiation proof enclosure (hutch). Figure 3 illustrates the spectrum available in the hutch, with the storage ring operated at 3 GeV, corrected for various absorbers present in the beam-line. Note that this spectrum is expressed in units of intensity per kilovolt (rather than per 10% band width, as was Figure 2), in order to better compare with the continuum from a conventional tungsten target sealed x-ray tube, also shown in the Figure. This comparison has been made as carefully as possible; the vertical divergence for SSRL Beam Line II-4 averages about a half-milliradian so the x-ray tube has been normalized to a half-milliradian divergence in the vertical direction (and as the vertical axis of the figure indicates, one milliradian in the horizontal), or more precisely, 0.5×10^{-6} steradians.

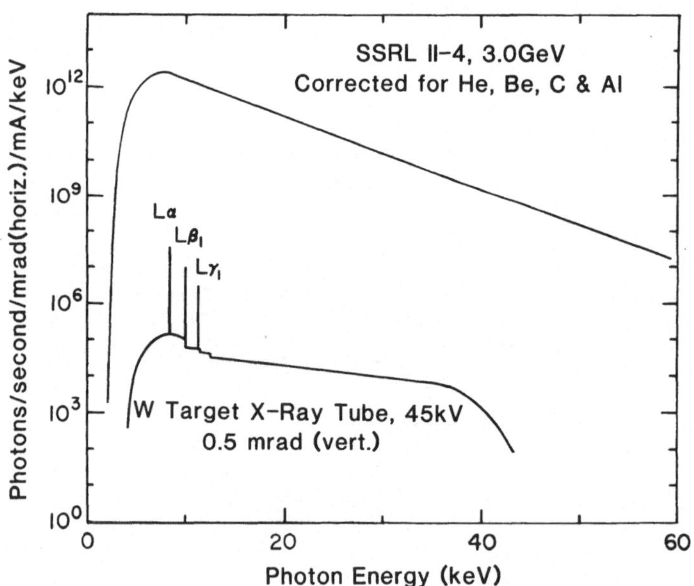

Fig. 3. Spectra from SPEAR at Stanford and a tungsten-target x-ray tube. Tube spectrum is from Reference 8; the intensities of the characteristic lines are represented at their natural line breadth.

The experimental apparatus, set up in the hutch, includes helium-filled plastic pipes with 6 μm mylar windows to minimize absorption and scattering of the x-rays before and after the sample and from the sample to the detector, and a "beam dump" at the rear of the hutch to absorb the transmitted beam. A variable aperture was located adjacent to the beryllium window to control the size of the beam (and hence the intensity). Either a Si(Li) solid state detector or a single-crystal scanning Bragg spectrometer viewed the sample at 90° to the primary beam. The output from the detectors was directed from the hutch to an appropriate set of electronics located outside the protective enclosure.

Samples consisted of elemental and compound thin-films vacuum evaporated onto 6 μm mylar [purchased commercially(9)] and aqueous solutions deposited on millipore filters (prepared at NRL). Table 1 lists the samples and their nominal compositions.

The Si(Li) detector had an energy resolution of 150 eV at 250c/s of MnKα from ^{55}Fe, using an amplifier shaping time of 10 μ s, and was coupled to a Nuclear Data ND-66 Multichannel Analyzer (MCA). The aperture was set to approximately 150 μ m square, limiting the maximum count rate to 2500 c/s. Figure 4 is a photograph of the open hutch with two detectors, one set up to make energy dispersion x-ray diffraction measurements of samples in a high pressure cell (on the left, at a low angle to the primary beam)(10), and the other at 90° to the primary beam for the XRF measurements. For the wavelength dispersion measurements, the crystal spectrometer was located in a helium-filled box in place of the latter Si(Li) detector. This crystal spectrometer used a 125 μ m x 10cm blade collimator, (200)LiF and PET crystals and a 2.3cm

Table 1. Concentrations of Samples (μ g/cm^{2})

ELEMENT	ON MYLAR	ON FILTERS			
V	16	11	0.8	0.07	0.01
Fe	19	8	0.4	0.09	0.02
Cu	18	6	0 5	0.10	
Rb	11		0.7	0.12	0.01
Mo	16		0.9	0.12	
Sn	19	8	0.8	0.28	0.03
Ba	18	4	0.6	0.11	0.02
Tb	16		0.8	0.10	
W	16		0.6	0.06	
Pb	22	9	0.7	0.13	0.05

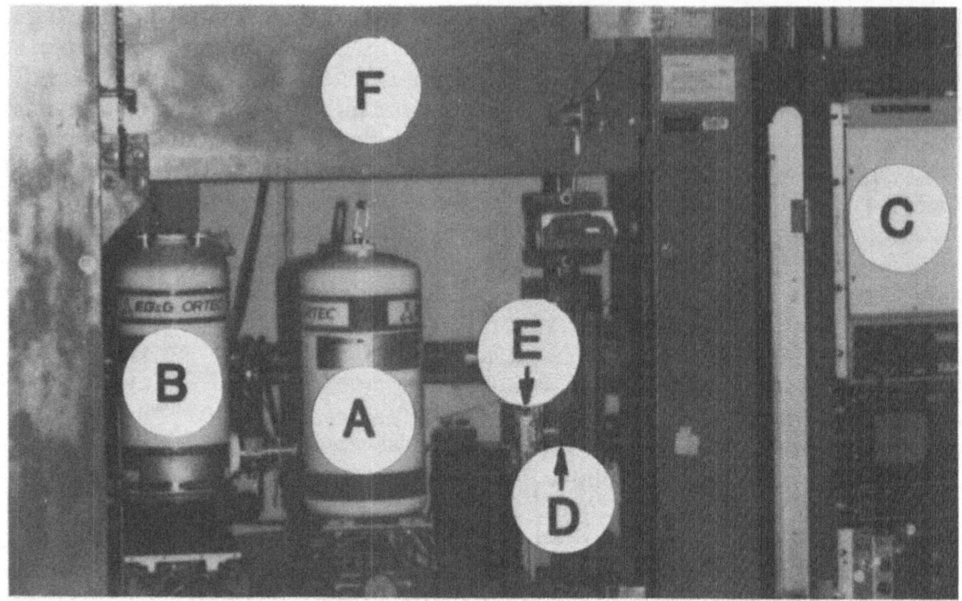

Fig. 4. Photograph of the open hutch: A. XRF Si(Li) detector, B. XRD Si(Li) detector, C. Electronics rack, D. End of the beam line, E. Adjustable aperture, and F. Door of the hutch. The XRF sample is directly on the far side of the detector; the XRD sample is in the diamond cell immediately adjacent to the aperture.

gas-flow proportional counter with one atmosphere of 90% Ar, 10% CH_4. To achieve reasonable counting rates, it was necessary to open the aperture to 0.4 cm^2 for the wavelength dispersion measurements.

RESULTS AND DISCUSSION

The major purpose of these measurements was to determine minimum detection limits (MDL) as a function of atomic number for these "thin" samples. The detector was located at 90° to the primary beam to take advantage of the polarization of the beam to minimize the scattered background. However, the background is measurable because, a.) even the helium atmosphere does scatter slightly, b.) there is some probability of multiple scattering even in these thin samples and c.) polarization is only complete in the <u>exact</u> plane of the electron orbit. Background intensity does increase with increased mass in the beam; thus, the background is higher for the samples on millipore than for those on mylar. Since these "thin" samples suffer essentially no matrix effects, the sensitivity (slope of the linear calibration curve) is the same for both types of samples. From all of this, it follows that the only difference in MDL between the two types of samples is due to the difference in background.

The wavelength dispersive measurements consisted of 100 second counts on the peak of the analytical line and on the background adjacent to the line, with the sample or an appropriate blank at 45° to the primary beam (and 45° to the spectrometer line-of-sight). The data were recorded manually from the digital scaler for off-line reduction. Energy dispersive spectra were collected mostly for 100 seconds live-time (times of 500 and 1000 seconds were used for a few blanks and low-level samples) in the 1024 channel MCA, with the samples in the same geometry. These spectra were transferred to floppy disks, also for off-line data reduction. Figure 5 shows the 100 second, 3σ detection limits(11) as a function of atomic number (Z) with Kα lines being used below Z=50 (Sn) and Lα lines above Z=50. Comparison of the MDL for the two measuring techniques shows them to be remarkably similar for these interference-free samples. The steeper slope of the wavelength dispersive results at short wavelength (Z>30 for Kα and Z>70 for Lα) is due to the low efficiency of the flow-proportional counter for short wavelength radiation. The Si(Li) detector, on the other hand, has relatively uniform high efficiency response over the energy range of interest ($\sim$3.5 to 25 keV).

The 150 μm size beam, used for the energy dispersion measurements, illuminated only 3.4×10^{-4} cm^2 on the sample (at 45°). Thus it is possible to estimate the MDL in absolute mass and to compare the data with some predictions which have been made for XRF with SR excitation(12). These predictions were made for a particular SR spectrum (HASYLAB, shown in Figure 2) exciting similar thin samples on filter paper substrates, energy dispersion measurements being assumed. The higher critical energy for the HASYLAB spectrum contributes to more efficient excitation of the higher atomic number elements, as is borne out by the comparison shown in Figure 6. In general, however, the comparison shows that the prediction was not overly optimistic and that MDL in the 10^{-11} to 10^{-12} gram range are attainable.

Comparisons can also be made with conventional laboratory XRF and with direct excitation x-ray analysis, such as electron probe microanalysis (EPMA) and proton induced x-ray emission (PIXE). Wavelength and energy dispersive XRF using conventional sources can achieve MDL in the 1 to 10 ng/cm^2 range for samples of the type being considered here, under optimum conditions(13). These instruments, however, measure an area on the sample of one to several square centimeters, so that the absolute mass being detected is perhaps 10^{-7} to 10^{-9} grams, significantly poorer than the data reported here. On the other hand, EPMA can achieve MDL of the

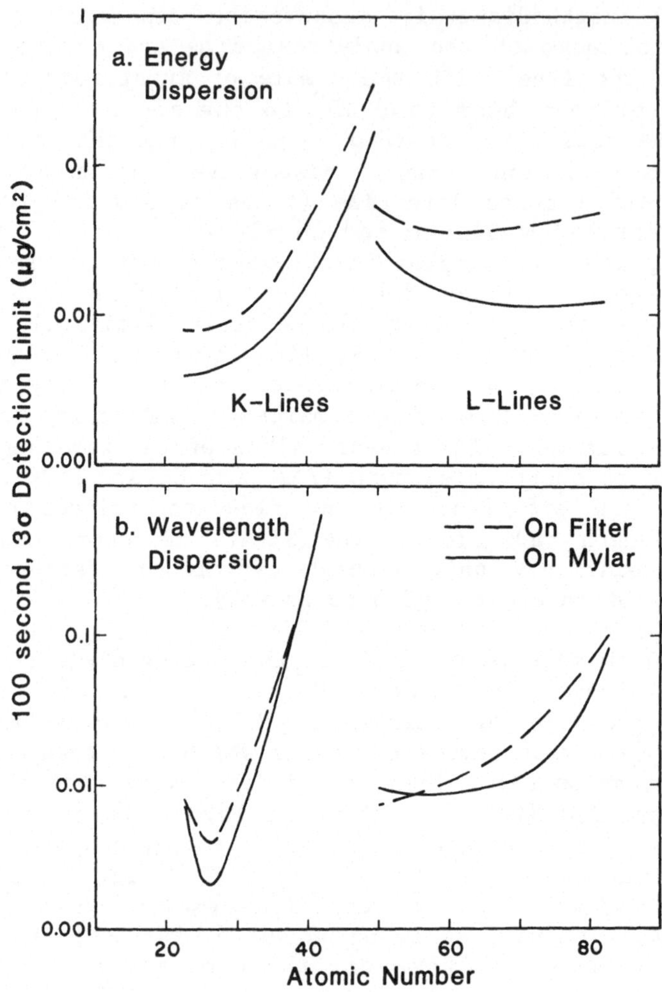

Fig. 5. Detection limits (in $\mu g/cm^2$) for wavelength and energy dispersion measurements.

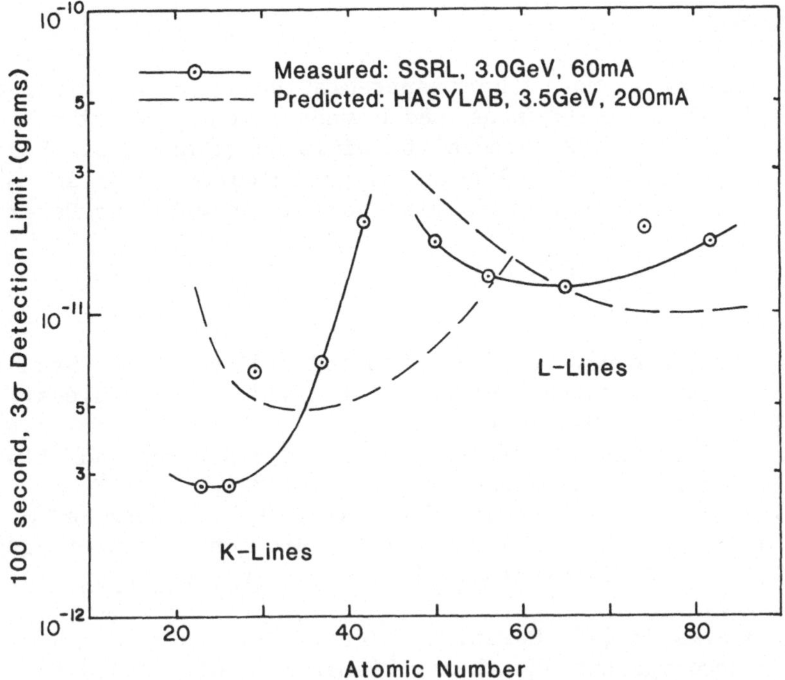

Fig. 6. Measured detection limits (in grams) on millipore filters compared to predictions for HASYLAB.

order of 10^{-14} grams (0.1% in 1 μm^3 at a density of 10 g/cm^3) and, although claims have been made for detection limits of 10^{-12} grams using PIXE(14), an extensive evaluation of proton excitation(15) has shown MDL in the 10 ng/cm^2 range. Considering a proton beam of one square millimeter, this converts to 10^{-10} grams. It must be remembered, as clearly pointed out by Sparks(5), that electron and proton excitation deposit much more energy in the sample than do x-rays, perhaps causing volatilization or other thermal damage leading to questionable analytical information.

SUMMARY

These preliminary experiments demonstrate the capability of continuum SR as a primary source for XRF. The experimental MDL, determined with a very simple experimental configuration fell in the 1 - 100 pg range and agree satisfactorily with predictions.

ACKNOWLEDGMENTS

The authors express their appreciation to SSRL staff, whose enthusiastic cooperation made these measurements possible. SSRL is supported by the NSF through the Division of Materials Research and the NIH through the Biotechnology Resources Program in the Division of Research Resources (in cooperation with the Department of Energy).

REFERENCES

1. H. Winick, Properties of Synchrotron Radiation, in "Synchrotron Radiation Research," Herman Winick and S. Doniach, eds., Plenum Press, New York (1980).
2. Herman Winick and S. Doniach, eds., "Synchrotron Radiation Research," Plenum Press, New York (1980).
3. C.J. Sparks, Jr., S. Raman, H.L. Yakel, R.V. Gentry and M.O. Krause, Search with Synchrotron Radiation for Superheavy Elements in Giant-Halo Inclusions, Phys. Rev. Lett. 38:205 (1977).
4. C.J. Sparks, Jr., S. Raman, E. Ricci, R.V. Gentry and M.O. Krause, Evidence against Superheavy Elements in Giant-Halo Inclusions Re-examined with Synchrotron Radiation, Phys. Rev. Lett. 40:507 (1978).
5. C.J. Sparks, Jr., X-Ray Fluorescence Microprobe for Chemical Analysis in "Synchrotron Radiation Research," Herman Winick and S. Doniach, eds., Plenum Publishing Co., New York (1980).
6. J.V. Gilfrich, E.F. Skelton, S.B. Qadri, J.P. Kirkland and D.J. Nagel, Synchrotron Radiation X-Ray Fluorescence Analysis, submitted to Analytical Chemistry.
7. W.R. Hunter, R.T. Williams, J.C. Rife, J.P. Kirkland, and M.N. Kabler, A Grating/Crystal Monochromator for the Spectral Range 5 eV to 5 keV, Nucl. Instrum. Meth. 195:141 (1982).
8. D.B. Brown, J.V. Gilfrich and M.C. Peckerar, Measurement and Calculation of Absolute Intensities of X-Ray Spectra, Jour. Appl. Phys. 46:4537 (1975).
9. Micromatter Co., Route 1, Box 72B, Eastsound, WA 98245.
10. E.F. Skelton, J. Kirkland and S.B. Qadri, Energy-Dispersive Measurements of Diffracted Synchrotron Radiation as a Function of Pressure: Applications to Phase Transitions in KCl and KI, J. Appl. Cryst. 15:82 (1982).
11. R. Jenkins, Nomenclature, Symbols, Units and Their Usage in Spectrochemical Analysis - IV: X-Ray Emission Spectroscopy, Pure and Appl. Chem. 52:2542 (1980).
12. European Science Foundation, "European Synchrotron Radiation Facility: Supplement I, The Scientific Case," Y. Farge and P.J. Duke, eds., Strasbourg (1979), pp. 78-82.

13. L.S. Birks and J.V. Gilfrich, Evaluation of Commercial Energy Dispersion X-Ray Analyzers for Water Pollution, Appl. Spectrosc. 32:204 (1978).
14. T.B. Johanson, R. Akselsson and S.A.E. Johansson, Proton-Induced X-Ray Emission Spectroscopy in Elemental Trace Analysis, Lund Institute of Technology, LUNP7109, August 1971.
15. J.A. Cooper, Comparison of Particle and Photon Excited X-Ray Fluorescence Applied to Trace Element Measurements on Environmental Samples, Nucl. Instrum. Meth. 106:525 (1973).

ENERGY RESOLUTION MEASUREMENTS OF MERCURIC IODIDE DETECTORS USING A COOLED FET PREAMPLIFIER

Lawrence Ames and William Drummond

Tracor Xray, Inc., Mountain View, CA 94043 and

Jan Iwanczyk and Andrzej Dabrowski

Institute for Physics and Imaging Science
University of Southern California
Marina Del Rey, CA 90291

ABSTRACT

The performance of a room temperature mercuric iodide X-ray detector was investigated as a function of detector bias, amplifier time constant, and detector temperature. A Mn K_α line of 200 eV FWHM was obtained by using low noise electronics developed for Si(Li) detectors, including a cooled input FET. Measurements of the detector's resolution at various X-ray energies result in a Fano factor of 0.20.

INTRODUCTION

There has long been an interest in having a room temperature X-ray detector with good energy resolution. Several different materials have been studied, of which mercuric iodide seems the most promising. Room temperature operation gives mercuric iodide an advantage over currently used Si(Li) detectors in that it doesn't need liquid nitrogen (LN) cooling. Thus HgI_2 holds the promise of a new generation of portable and easily maintained X-ray spectrometers, opening new fields of applications. In addition, HgI_2 spectrometers are potentially cheaper and easier to manufacture.

Several groups have reported on mercuric iodide detectors. The best resolution is reported by the group at the University of Southern California.[1] With both the detector and FET at room temperature, they obtained a Mn K_α line with a full width at half maximum of 295 eV, and Mg K_α FWHM of 245 eV. As the pulser peak was 225 eV wide, it is apparent that the energy resolution of the HgI_2 spectrometer was limited by the electronic noise of the system. The goal of this research project was to study the energy resolution of HgI_2 detectors when the electronic noise of the preamplifier was reduced to the level typical of Si(Li) X-ray spectrometers.

EXPERIMENTAL SET UP

Figure 1 shows the experimental setup, which consists of a standard cryostat slightly modified to accommodate the HgI_2 detectors. A Si(Li) detector would have to be cooled to LN temperatures, then the FET warmed for optimum resolution; for these experiments with HgI_2, both the FET and the detector were heated. Dabrowski's group at USC fabricated the detector used in this experiment. Electrical contacts were applied to a 0.5 mm thick crystal so as to give an active area of about 3 mm^2, and the crystal was mounted on an alumina square. This square was mounted on a grounded copper ring, along with a heater and a thermistor for temperature monitoring and control. Standard Tracor electronics were used: a model 505 pulsed optical feedback preamplifier, a TX 1232 amplifier/pulse processor, and a TN 2010 ADC and computer system.

RESULTS AND DISCUSSION

The performance of the mercuric iodide detector depends on several variables, including detector bias, the time constant of the amplifier, and the detector temperature. Figure 2 plots the effect of bias voltage on the resolution of Mn K_α X-rays and the pulser peak. Note that just cooling the FET permits a Mn K_α FWHM of 251 eV. If the bias is too low, there is incomplete charge collection, resulting

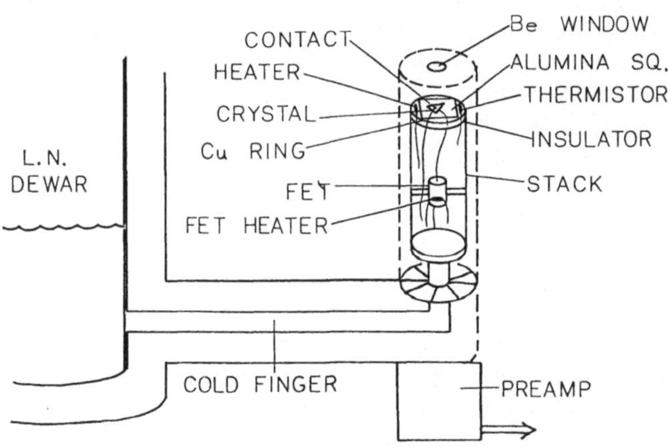

Fig. 1 Experimental Setup

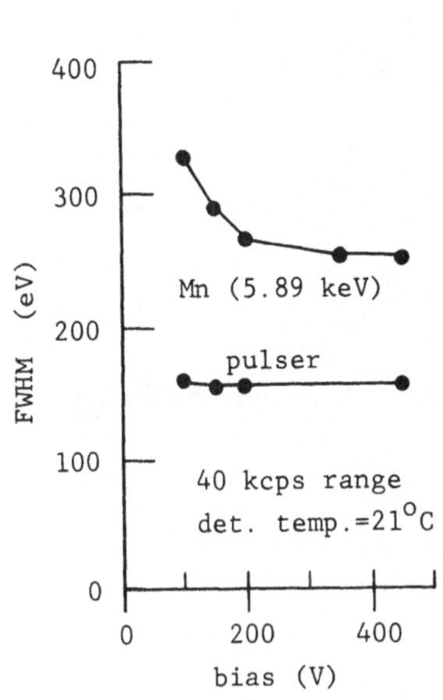

Fig. 2 FWHM vs. bias

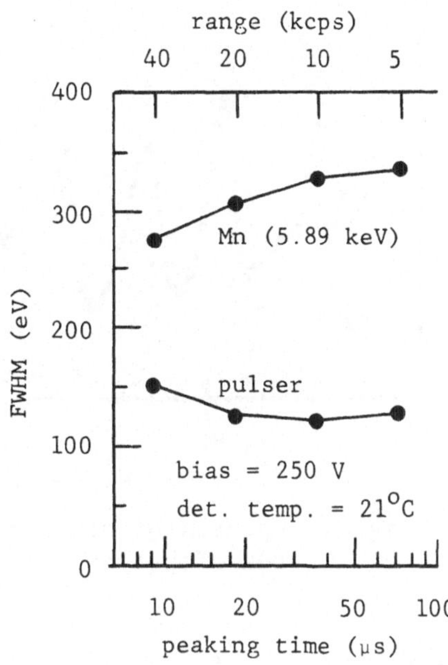

Fig. 3 FWHM vs. time constant

in a low energy tail and poor resolution. High biases can result in detector breakdown, or in higher leakage currents which could increase the electronic noise. This was not a problem, however, as the pulser's width remained constant with bias.

Figure 3 shows that the resolution is also dependent on the peaking time (count rate range) of the amplifier. With Si(Li) detectors, one generally uses the longest time constant the count rate will allow to reduce the serier noise of the system. With HgI_2, a shorter time constant gives better resolution, even though the pulser peak is narrowest with a longer time constant. This allows several interesting deductions: one is that the leakage current (parallel noise) is quite low, for otherwise, the electronic noise would get worse with longer time constants. Another deduction is that this detector has trapping levels: With a short time constant, some fraction of the electrons are trapped and are never counted, and the gain of the system is simply adjusted accordingly. With longer time constants, some of the trapped electrons are released in time to be counted, and the peak develops a high energy tail.

Figure 4 is spectra of old house paint, with the detector at 37°C and at -4°C, showing the resolution is very sensitive to detector temperature. When warm, trapped electrons are released sooner, and again there is a high energy tail, along with an energy shift.

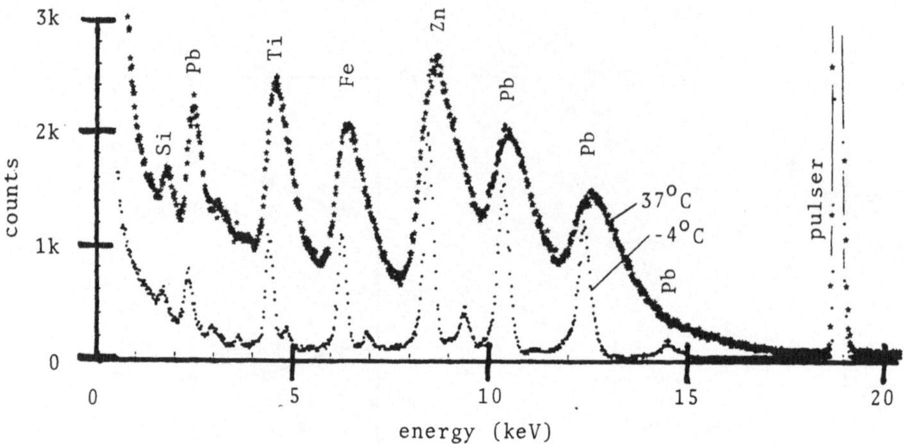

Fig. 4 Spectra of old house paint, illustrating the effect of detector temperature

As figure 5 shows, resolution improves as the detector is cooled: best resolution for this detector was obtained at approximately 0°C. Also, with a cooled detector, the trade off between trapping and electronic noise shifts, giving better data with a longer time constant.

A wide variety of samples were measured with the mercuric iodide detector (Fig. 6). Plotting the square of the resolution vs. the X-ray energy gives a straight line that intercepts the axis

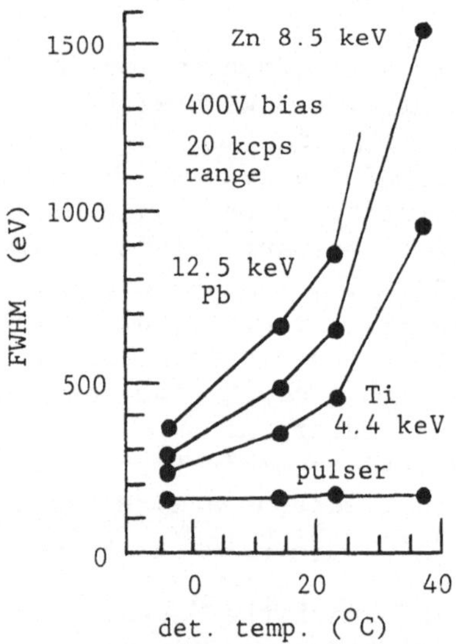

Fig. 5 FWHM vs. detector temperature

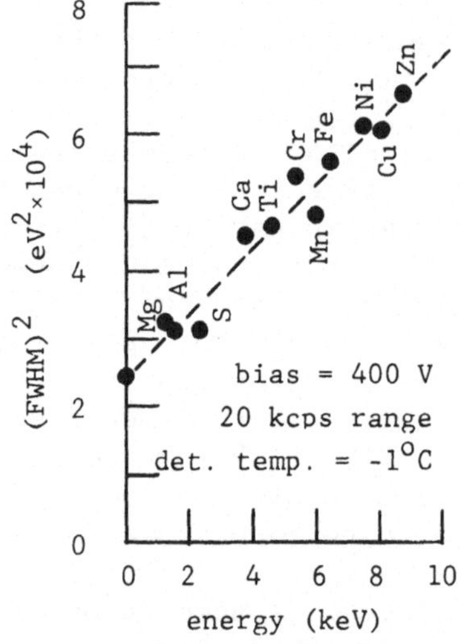

Fig. 6 $(FWHM)^2$ vs. X-ray energy

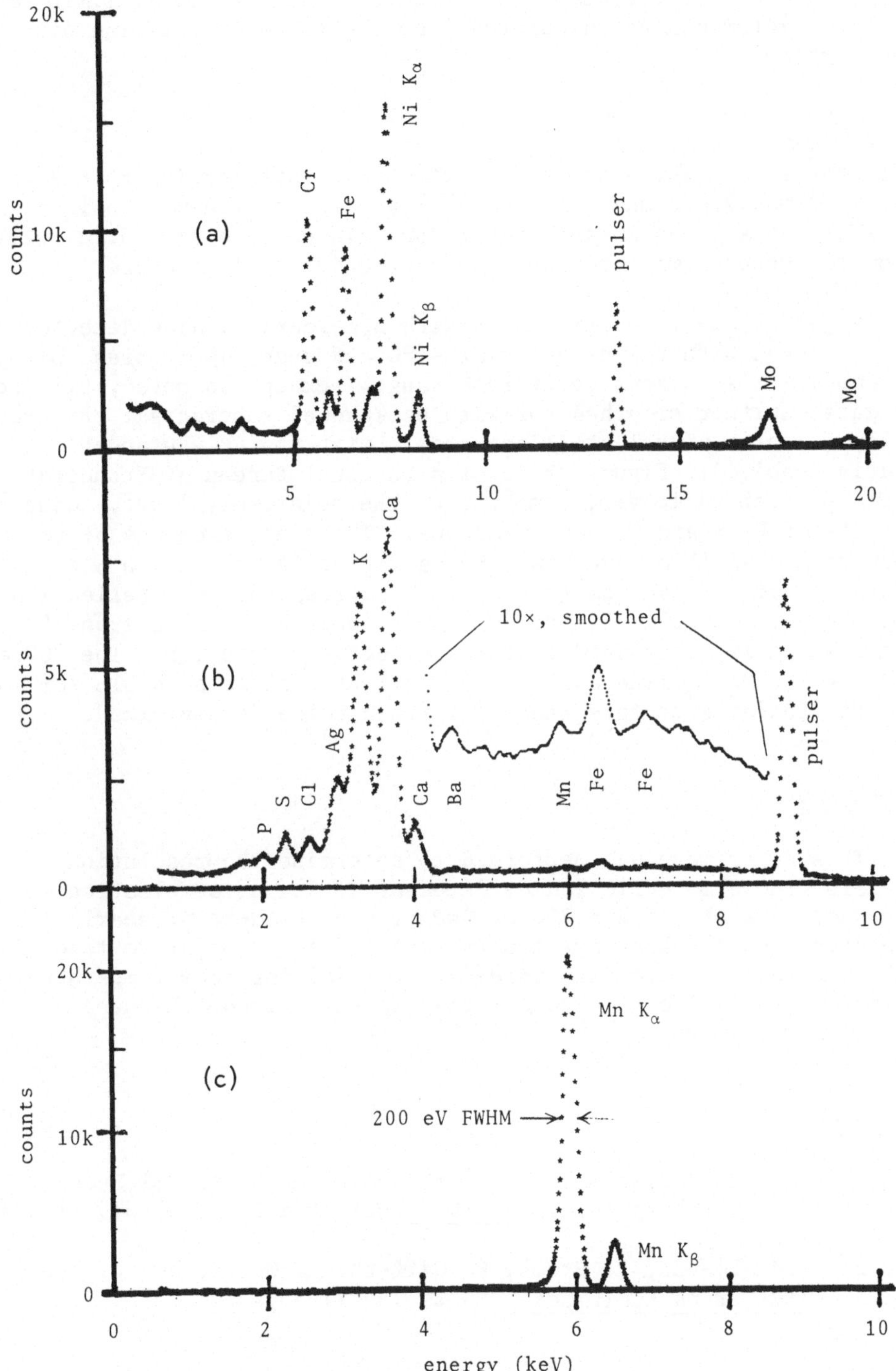

Fig. 7 Spectra taken with HgI_2 detector: (a) stainless steel; (b) N.B.S orchard leaves; (c) iron-55 source

at the square of the system's electronic noise. An upper limit for the Fano factor can be calculated from the slope of the line with the equation:

$$\text{Fano factor} \leq \text{slope}/(8\varepsilon \ln 2),$$

where the slope (which shows the detector's contribution to the resolution) = 4.7 eV, and ε is the energy per electron-hole pair = 4.2 eV.[2] This gives a Fano factor for HgI_2 of less than 0.20, well below the upper limit previously reported in the literature.[1]

Figure 7 shows spectra taken with a mercuric iodide detector. All are taken with the detector at -1°C and under 400 V bias, using the amplifier's 20 kcps count rate range: except as noted, all are raw data, neither smoothed nor with background subtracted. Figure 7a is of a high nickel stainless steel, with the Cr, Fe and Ni clearly resolved. Figure 7b is of a National Bureau of Standards sample of orchard leaves, composed of the relatively light, adjacent elements of P, S and Cl, and K and Ca. The trace elements of Fe, Mn an Ba are visible, even though the latter is present in a concentration of only 44 ppm (the Ag peak corresponds to a reflection from the X-ray tube). Figure 7c is the spectrum from an iron-55 source which is a standard test for detector resolution. The 200 eV FWHM for the Mn K_α line shows the excellent results possible for mercuric iodide detectors when electronic noise is minimized.

CONCLUSION

This experiment was performed to determine the resolution possible from a mercuric iodide detector if low noise electronics were used. We obtained a 200 eV FWHM for Mn K_α X-rays, nearly 100 eV better than the best previously reported.[1] In order to take full advantage of mercuric iodide's potential for room temperature operation, low noise electronics that operate without LN need further development.

REFERENCES

1. J.S. Iwanczyk, A.J. Dabrowski, G.C. Huth, A. Del Duca, and W. Schnepple, IEEE Trans Nucl. Sci., NS-28:579 (1981)

2. J.P. Ponpon, R. Stuck, P. Siffert, B. Meyer, and C. Schwab, IEEE Trans Nucl. Sci., NS-22:182 (1975)

X-RAY POLARIZATION: BRAGG DIFFRACTION AND X-RAY FLUORESCENCE

John D. Zahrt

Chemistry Department
Northern Arizona University
Flagstaff, Arizona 86011

INTRODUCTION

Recent, state of the art, x-ray spectrometers have made use of polarizing the source x-rays by scattering through 90° (1). One then observes the analyte fluorescence in a direction perpendicular to the scattering plane in which the polarized x-rays are generated. The signal/noise ratio at the detector is much improved. Unfortunately there is a concomitant loss of intensity and analysis times increase. This adversely affects the minimum detection limits.

Various approaches have been taken to increase the polarized x-ray intensity at the sample. Scattering off the interior of a B_4C cylinder has been reported as a means of increasing the intensity of the polarized bremsstrahlung (2). Using Bragg diffraction (with $2\theta_B = 90°$) has met with success (3) and using a bent crystal for Bragg diffraction has further increased the power of polarized characteristic x-rays (4). A critical review of these methods has been given (5).

The purpose of this paper is to investigate the theoretical power of a polarized x-ray beam prepared by 90° Bragg scattering from flat and bent crystals. In particular we are concerned with CuK_α radiation diffracting from Cu(113) planes where the reported power due to the bent crystal is about three times that of the flat crystal (4).

THEORY

The theory developed herein makes use of the kinematic theory of x-ray diffraction and Zachariasen's development of Darwin's mosaic crystal theory (6). Both the flat and bent crystals are treated with both a point divergent source and a finite perfectly parallel source of incident unpolarized x-rays.

δ-ζ Relations

The problem considered here is slightly more complicated than the treatment of Zachariasen (flat crystal-parallel radiation) because here the angle of incidence is a function of the position of the scattering event on the diffracting crystal. Consider Figures 1, 2 and 3 where O, A, P and So lie on a circle of radius R and angle SoPA is 90°. The angle of incidence of a ray from the source undergoing 90° scattering at S and ending up at the analyte will be something other than θ_B = 45° and we denote this discrepancy by δ. This discrepancy is in general a complicated function of ζ which in turn gives the scattering point S. A summary of the approximate (first non-zero term in the Maclaurin series expansions) δ-ζ relationships are given in Table 1.

Table 1. Relations between the discrepancy, δ, and the scattering location angle, ζ. See Figures 1, 2 and 3.

	Flat	Bent
Parallel	$\delta \equiv 0$	$\delta \equiv \zeta$
Point divergent	$\delta \cong -\zeta$	$\delta \cong \frac{1}{2}\zeta^2$

Mosaic Crystals

The real crystal, especially the bent one, is in general an imperfect crystal. The imperfect crystal can be treated as a mosaic of a large number of perfect crystallite blocks. These blocks are then assumed to be randomly oriented with respect to the crystal face. For simplicity we assume cylindrical symmetry. The distribution function W which defines the random orientation is a function of one variable Δ which is the angle between the block face and the crystal face. In principle $W(\Delta)$ is experimentally determinable but in the absence of such determinations we have investigated the four distributions given in Table 2. Each distribution function has associated with it a parameter σ which is a measure of its width. We make no claim as to the physical reality of any of these functions (in fact the mosaic crystal is only a workable model whose physical reality may be questioned).

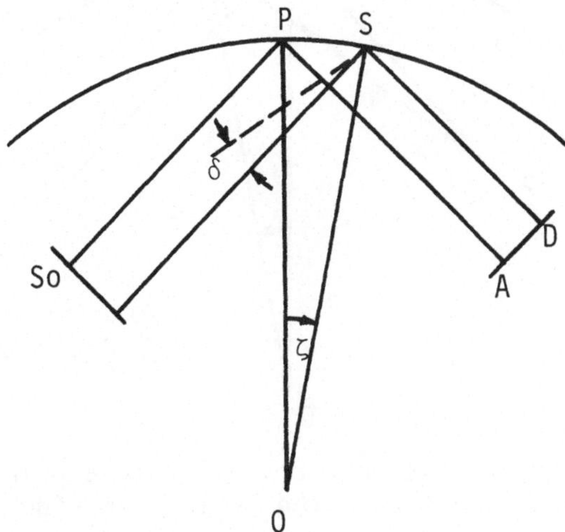

Figure 1. Geometry of a bent crystal with a parallel source of x-rays. OP = 2R. The discrepancy δ is the angle between line SoS (incident ray) and the dashed line (45° to the crystal surface).

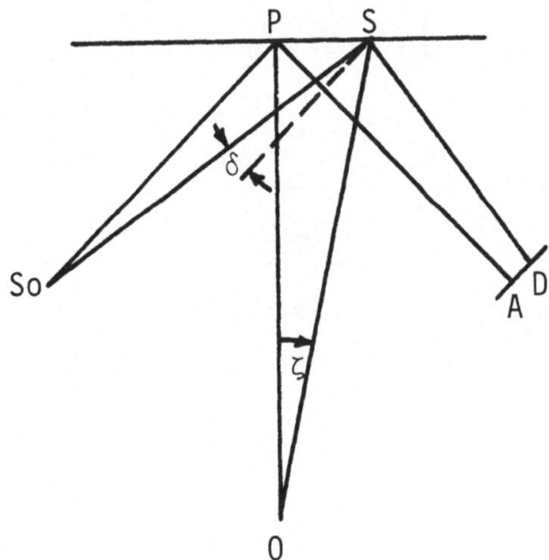

Figure 2. Geometry of a flat crystal with a point divergent source of x-rays. OP = 2R.

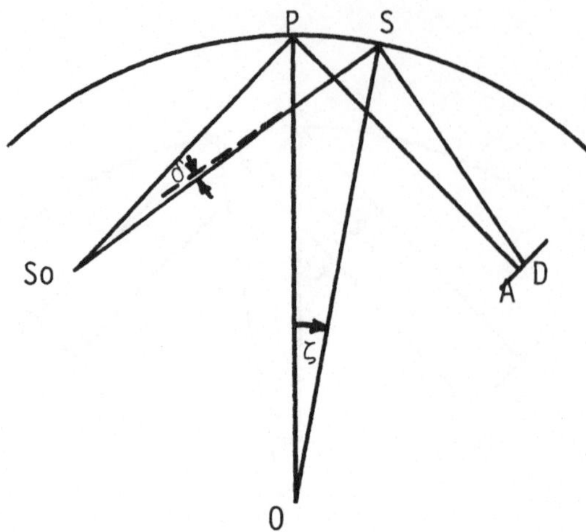

Figure 3. Geometry of a bent crystal with a point divergent source of x-rays. OP = 2R. The angle ζ in Figures 1, 2, and 3 is the same.

Table 2. Mosaic distribution functions and their full widths at half maximum (FWHM).

			FWHM
Polynomial	$W^P(\Delta) = \frac{15}{16\sigma_P^5}(\Delta^2 - \sigma_P^2)^2$	$\lvert\Delta\rvert \leqslant \sigma_P$	$2\sqrt{1-\frac{1}{2}\sqrt{2}}\,\sigma_P$
	$W^P(\Delta) = 0$	$\lvert\Delta\rvert \geqslant \sigma_P$	
Gaussian	$W^G(\Delta) = \frac{1}{\sqrt{\pi}\sigma_G} e^{-\Delta^2/\sigma_G^2}$		$2\sqrt{\ln 2}\,\sigma_G$
Cauchy	$W^C(\Delta) = \frac{\sigma_C}{\pi} \frac{1}{\Delta^2 - \sigma_C^2}$		$2\sigma_C$
Exponential	$W^E(\Delta) = \frac{1}{2\sigma_E} e^{-\lvert\Delta\rvert/\sigma_E}$		$2 \ln 2\,\sigma_E$

Reflecting Power

Let $P_{H}(x)$ be defined as the intensity function of a diffraction line due to planes H of a perfect crystal where $x = \theta_B - \theta$. Now a x-ray beam incident upon the crystal surface will be diffracted only if one or more of the mosaic blocks are oriented such that the angle of incidence with that block is at or very near the Bragg angle. For certain geometries and sources where $\delta \neq 0$, Δ must assume values very near δ for diffraction to occur.

We can thus write

$$\Omega_H(x) = \iint P_H(x+\Delta+\delta)W(\Delta)d\Delta d\zeta \tag{1}$$

where $\Omega_H(x)$ is the reflecting power of a block. If $P_H(x)$ is narrow compared to $W(\Delta)$* then

$$\Omega_H(x) = R_H \int W(x+\delta)d\zeta \tag{2}$$

where

$$R_H = \int P_H(x)dx \tag{3}$$

and where

$$\int P_H(x+\Delta)W(\Delta)d\Delta \cong R_H\, W(x) \tag{4}$$

If $\Omega_H \ll 1-\alpha$ where $\alpha = e^{-\mu_o t_o/2}$ and μ_o is the true absorption coefficient and t_o is the thickness of a mosaic block then the integrated intensity is given by

$$I_H = \int \Omega_H(x)\, dx/(1-\alpha^2). \tag{5}$$

Using Tables 1 and 2 and equation (2) we obtain the results listed in Table 3.

CONCLUSIONS

Obviously the flat crystal with a perfectly parallel source of x-rays is the most efficient scattering system considered. The bent-divergent system is only about 20% as efficient. The other two systems, flat-divergent and bent-parallel are essentially the same and can scatter only about 0.4% the radiation of the flat-parallel case.

The point divergent source has not been investigated in the laboratory and any finite real source of x-rays is a superposition of the parallel and divergent sources. Thus we do not expect the

* Typically the FWHM for $P_H(x) \sim 1$ sec of arc while that for $W(\Delta)$ is ~ 1 min of arc.

integrated intensities computed here to agree with the experimental value of $I_{bent}/I_{flat} \sim 3$ (4). Consideration of real sources, finite in extent is near completion and seems to verify the factor of 3.

Only the case of bent-divergent shows a dependence upon the mosaic distribution function, the broader functions leading to greater integrated intensities (see Table 3).

It seems to be important to pursue the determination of $W(\Delta)$ and how this changes upon bending. Complete experimental control of the FWHM and of the functional form of $W(\Delta)$ is impossible but some modification may be possible. Could increasing the FWHM of the mosaic distribution function of the flat crystal abrogate much of the advantage of the bent crystal?

Table 3. Values of the integrated intensity I_H for negligible secondary extinction.*

	FLAT	BENT
Parallel	$I_H = R_H/(1-\alpha^2)$	$I_H = 0.004R_H/(1-\alpha^2)$
Point Divergent	$I_H = 0.004R_H/(1-\alpha^2)$	$I_H^P = 0.132\ R_H/(1-\alpha^2)$
		$I_H^G = 0.125\ R_H/(1-\alpha^2)$
		$I_H^C = 0.171\ R_H/(1-\alpha^2)$
		$I_H^E = 0.279\ R_H/(1-\alpha^2)$

* Secondary extinction is power loss due to diffraction before the x-ray beam reaches the block under consideration. For CuK_α radiation incident upon Cu(113) planes this implies that the FWHM of $W(\Delta)$ be $\sigma >> 0.5$ min of arc.

REFERENCES

1. Richard Ryon, Adv. in X-Ray Analysis, 20, 575-590 (1977).
2. R. Ryon and J. Zahrt, Adv. in X-Ray Analysis, 22, 453-460 (1979).
3. H. Aiginger, P. Wobrauschek and C. Brauner, Nuc. Instr. and Meth. 120, 541-542 (1974).
4. P. Wobrauschek and H. Aiginger, submitted to X-Ray Spectrometry.
5. R. Ryon, J. Zahrt, P. Wobrauschek and H. Aiginger, Adv. in X-Ray Analysis 25, (1982).
6. William Zachariasen, "Theory of X-Ray Diffraction in Crystals," Dover Publishing Co., New York, 1967.

A NEW TECHNIQUE FOR RADIOISOTOPE-EXCITED X-RAY FLUORESCENCE*

J.J.LaBrecque and W.C.Parker[+]

Instituto Venezolano de Investigaciones Científicas

Apartado 1827, Caracas 1010-A, Venezuela
+Atomic Energy of Canada Ltd.,P.P.Box 6300,Ottawa,K2A 3W3

SUMMARY

In this work we plan to describe a new simple technique for exciting characteristic X-rays for X-ray fluorescence. This technique was originally designed for use with β-emitting radioisotopes (1) but now has been extended to other types of isotopes.The technique involves mixing directly with the sample (about 20mg) a small amount of a radionuclide (about 100μCi). An equal atomic mixture of the following elements was prepared: Ti,Mn,Zn,Br,Ag,Sn,and Ba. This mixture was investigated with the following isotopes: Ni-63, S-35, Pm-147,Na-22,Co-137,I-125,Co57,Cd-109,Zn-65,Fe-55 and Am-241.A composite table of the corrected integrated Kα-lines of the elements in the mixture and Graphs showing the ratio of the Kα line intensity of the element of interest divided by the Kα line intensity of Ti will also be presented.

INTRODUCTION

One method to improve detection limits in radioisotope excited X-ray fluorescence is to optimize the source-sample-detector arrangement. One technique to optimize the source-sample-detector geometry is to minimize the distance between the source and sample as well as the distance between the sample and detector. In other words maximizing the excitation efficiency and detector efficiency. But it is necessary that the arrangement also minimize the transmission of the primary X-rays or radiation directly or scattered from the source and/or structural materials of the system. Thus,the requirement of fluorescencing the sample while avoiding direct or scattered radiation from entering the detector,has made the dis-

*This work was funded in part by CONICIT (Venezuela;S1-1149) which forms part of an international project with the NSF (USA).

tance between the components relatively large, thus a poor source-sample-detector efficiency. In this work, a new technique is described in which the distances between the source-sample and the sample-detector window approach zero.

EXPERIMENTAL

Mixture L, an equal atomic weight mixture, of the following elements: Ti, Mn, Zn, Br, Zr, Ag, Sn and Ba was prepared by mixing the appropriate amounts of their respective oxides. About 20 mg of mixture L is inserted in a 1 cm aperture in a 2x6 cm piece of IBM card between two pieces of "Scotch Magic Tape" but before the second piece of tape is secured in place, 100μCi of the selected radionuclide is added from a 2mCi aqueous solution. The X-ray detection system was a Tracor Northern TN-11 which has been described in detail elsewhere (2), as well as the sofware. Finally, the source-sample is placed directly on the face of the detector window which is protected by a thin polypropylene film from contamination.

RESULTS

The results for the corrected peak intensities for the Kα peaks of the eight elements with the different radionuclides as excitation sources is given in Table I. A comparison of the Kα peak intensities of mixture L to the intensity of the $Ti_{K\alpha}$ peak is shown in figure 1 for three pure beta emitters. It can be seen that the absolute peak intensities increases as the beta maximum energy end

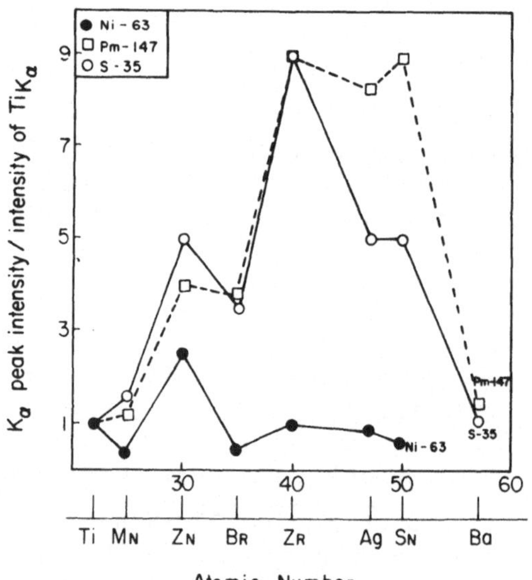

Figure 1

Table I. Corrected Intensities for the Kα - Peaks in Mixture L (∿ 20mg) from 100μCi of Different Isotopes for 1000 Seconds Fluorescent Time*.

Isotope (Z)	Corrected Invensity for the K_α Peak							
	Ti(22)	Mn(25)	Zn(35	Br(35)	Zr(40)	Ag(47)	Sn(50)	Ba(56)
Na - 22	115	34	437	613	2356	1803	1592	967
S - 35	166	229	750	413	1590	1791	1333	101
Fe - 55	4577	-	-	-	-	-	-	-
Co - 57	381	-	790	1159	680	1295	1515	2326
Ni - 63	411	157	-	184	404	357	118	49
Zn - 65	178	332	-	-	-	-	-	-
*Cd -109	86	142	1693	1639	24929	-	-	72
*I -125	78	695	4857	11830	38657	36227	33488	-
Cs -137	242	305	1318	4760	13496	20740	22406	-
Pm -147	183	200	817	786	1813	1533	1580	306
Am -241	240	542	4771	12262	8440	5868	6887	2757

*Only 100 seconds fluorescent time because of overflow.

-When no corrected peak intensity appeared or when the peak intensity was interfered with from the radioisotope source.

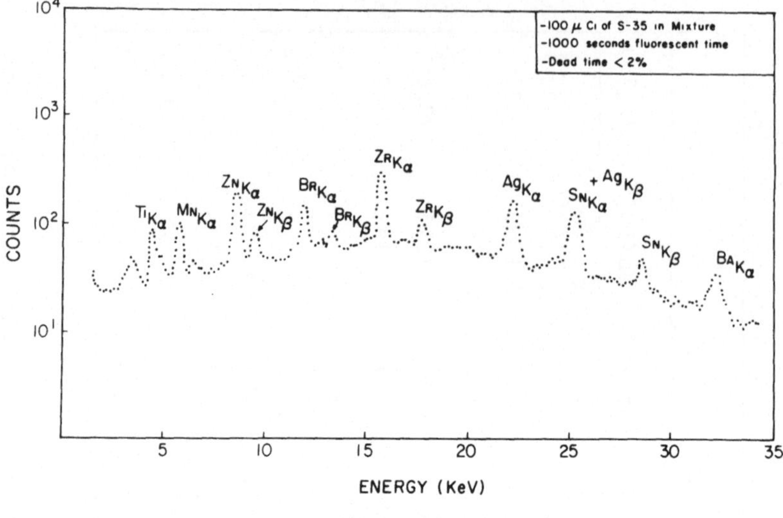

Figure 2

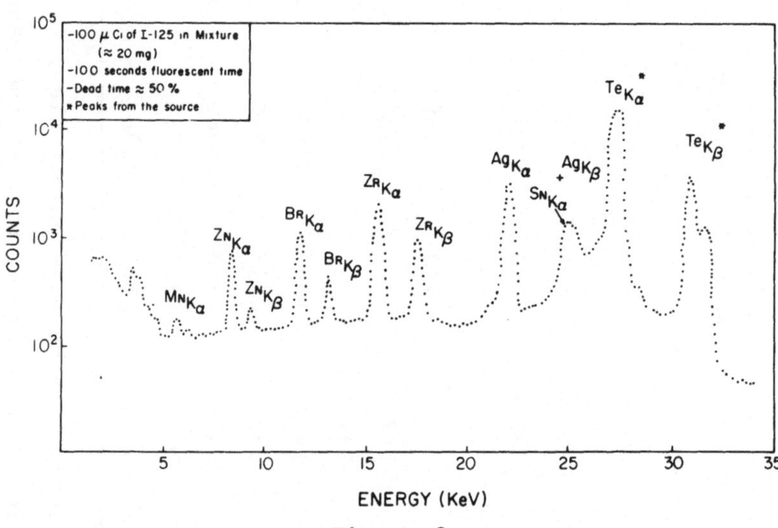

Figure 3

point increases of the pure beta emitters from Table I. Finally, a typical spectrum of a pure beta emitter (S-35) and a pure electron capture isotope (I-125) are displayed in figures 2 and 3. Note that the background is relatively small and flat and a reasonable signal to noise ratio for the pure beta emitter.

REFERENCES

1) J.J.LaBrecque and W.C.Parker, Radiochem. Radioanal. Letters, 49 (1981) 261-270.
2) J.J.LaBrecque, Proc.of the III Intern. Conf. on Computer in Chemical Research, Education and Technology (1976) P.61-84.

BRAGG-BORRMANN X-RAY SPECTROSCOPY FROM A LINE SOURCE*

K. Das Gupta

Radiation Research Laboratory

Texas Tech University, Lubbock, Texas 79409

Earlier, I reported[1] in this conference the results obtained with (i) a three crystal spectrometer (ii) two curved crystal spectrometer of Cauchois type and (iii) spherically bent crystal spectrometer. These instruments were developed to obtain higher resolution that helped us to observe fine structures[2] of $K\alpha_1$, $K\alpha_2$ lines of transition elements. Our work has been followed in other laboratories.

Here I present the essential points of a very simple spectrometer hitherto unreported. The most attractive feature of the spectrometer is its simplicity of adjustment in a minimum time. The essential parts of the instrument are: (i) a fixed line source, (ii) a sliding slit system that can be fixed in a calculated position to cover a particular range of wavelengths and an adjustable pinhole located on the slit, (iii) a fixed monocrystal for dispersion and (iv) a fixed cassette for holding the film. The cassette is an arc of a circle with the center of curvature at the pinhole. For a fixed position of the sliding slit, the analyzing crystal covers a wide range of wavelengths. In a typical case with a line focus tungsten target and for a fixed calculated position of the slit, the entire L-series spectra of tungsten appear on the film covering from 1A to 1.5A.[3] The position of the slit is read against a millimeter scale fixed to the shutter. Without much modification of the slit system currently in use in our laboratory it is possible to study W $K\alpha$ lines and on the long wavelength side, the $K\alpha$ line of sulphur. For much shorter wavelengths corresponding to 1 MeV the slit is to be modified by increasing the thickness that puts

*Work supported by the Robert Welch Foundation.

restriction on the range of wavelength that can be studied in one fixed position of the slit. Nevertheless, the spectrometer can be used for very short and long wavelengths by controlling the thickness of the material of which the slit is made and using thin monocrystals of low atomic number elements. Thus gamma rays from radioactive line source and characteristic x-rays from heavy elements and transuranium elements can be studied both photographically and with a counter. Radioactive nuclei of short life time could be studied with a counter selecting a proper thickness of the analyzing crystal to obtain a maximum value of the signal to noise ratio.

So far we have used the following monocrystals: Si (220), Ge (220), and LiF (200). With all of these analyzing crystals we recorded photographically the tungsten Lα, -β, and -γ lines for a fixed calculated position of the slit. For an accurate measurement of the wavelengths, it is helpful to increase the dispersion by moving the curved film to a larger distance from the analyzing crystal to improve the signal to noise ratio. We have used 50, 80, 100 and 200 cms distances from the crystal with good results. To increase the resolution we had to sacrifice intensity by increasing the thickness of the analyzing crystal for Borrmann transmission.[4] Increasing the thickness also improves the signal to noise ratio. The spectral lines are sharper in anomalous Borrmann transmission (Figure 1(a)) and a mirror image of the reflection spectra is obtained also on the direct beam side as usual in Borrmann transmission spectra, Figure 1(b).

Details of the instrument modification for commercial applications for elemental analysis have been worked out using a movable counter about a vertical axis centering the slit and a movable target for elemental analysis.

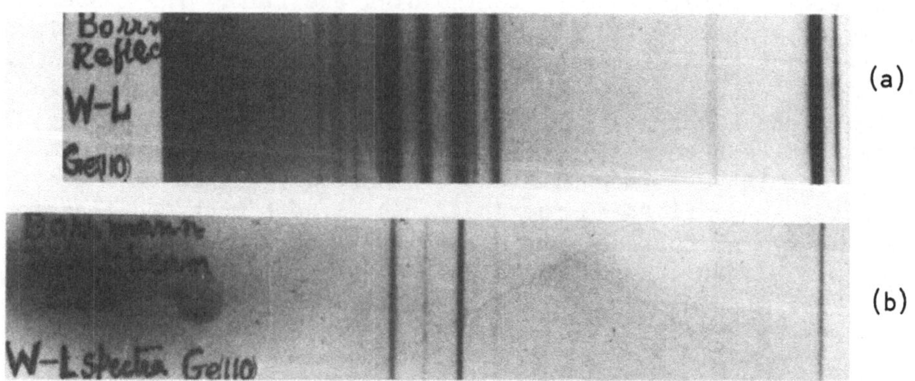

Figure 1. Bragg-Borrmann spectra of tungsten Lα, β, and γ lines: (a) Borrmann reflection spectra, (b) Borrmann direct beam side spectra.

REFERENCES

1. K. Das Gupta, Herbert Welch, P. F. Gott, John F. Priest, Sunny Cheng, and Edmond Chu, Advances in X-Ray Analysis, Vol. 16, Plenum Publishing Corp., pp. 251-259.
2. M. Shah and K. Das Gupta, Physics Letters 29A, 570 (1969).
3. K. Das Gupta, Advances in X-Ray Analysis, Vol. 25, Plenum Publishing Corp., pp. 325-328.
4. G. Borrmann, Phys. Z. 42, 157-162, (1941).

AUTOMATED QUALITATIVE X-RAY FLUORESCENCE ELEMENTAL ANALYSIS

Mary F. Garbauskas and Raymond P. Goehner

General Electric Corporate Research and Development
P.O. Box 8
Schenectady, NY 12301

ABSTRACT

A series of FORTRAN IV programs have been written to aid in qualitative x-ray fluorescence (XRF) analysis. These programs have been implemented on a DEC PDP 11/34 computer with an RSX-11M operating system and access the NIH elemental data base (1). The programs and their application to XRF analysis in our laboratory will be described.

INTRODUCTION

With the advent of energy-dispersive XRF spectrometers, qualitative elemental analysis has become extremely rapid and easy. However, because of the poor resolution of these systems, qualitative identifications are often difficult if line overlaps are present. On the other hand, although wavelength dispersive spectrometers have adequate resolution in most cases, the use of these systems for qualitative work is tedious, requiring a large amount of operator time in the acquisition and analysis of strip chart traces.

The use of computers for qualitative phase identification in powder diffraction has become extremely popular as evidenced by the numerous programs available to search the JCPDS file. This same type of pattern identification is possible for elemental identification in qualitative XRF analysis. Indeed, in the elemental case, the search/match procedure is much simpler than in the diffraction case, since the number of possibilities is much smaller. The major stumbling block to this type of analysis has been the lack of an easily obtainable and computer readable data base. Small data bases and associated programs exist (2) but have limited usefulness since many observable elemental lines are not accounted for. A more nearly complete data base and set of programs has been described (3) for use in an interactive mode on an SEL 32 computer. Recently, a complete elemental data base has been assembled at NIH (1) and can be obtained free of charge on various types of magnetic media. This data base has been loaded onto a DEC PDP 11/34 computer operating under RSX-11M and a series of FORTRAN IV programs has been written to access the data base and interface to stepscan and graphics programs already in use in our laboratory. A description of these programs and their use in XRF analysis is the subject of this paper.

The Data Base

The NIH elemental data base consists of x-ray line and absorption edge information from under 100 eV to over 120 keV. In addition, a relative intensity is assigned to each line within a given family. Over 3000 entries are present and represent a compilation of information from various sources (4-7). It was obtained for our uses from NIH as three ASCII files ordered by element, by wavelength, and by line designation. In the interest of both execution speed and disk space, the original data base was packed into direct access binary files with certain constraints. Since the programs were to be used with a commercial XRF spectrometer, lines with wavelengths longer than 25 angstroms and with relative intensities less than 0.06 were excluded. This allowed the storage of a maximum of 75 lines and 15 absorption edges per element.

The Programs

SPECPLOT

The interactive graphics program SPECPLOT has been described in detail elsewhere (8,9). The version that is implemented on the PDP 11/34 is used for both XRF and x-ray diffraction data reduction. The program will provide graphics output on a Tektronix 4010 or Tektronix 4027 terminal and/or a Tektronix 4662 color plotter. The scale of the y-axis can be either linear or logarithmic. Markers for the positions of the elemental lines can be displayed on the spectrum. An example of the graphics output is presented in Figure 1. The program can be directed to subtract background and perform a second derivative peak search. The result of the peak search is a list written to an ASCII file that contains peak position in 2-theta, wavelength, and energy and intensity above zero in counts per second.

XRFQED

The program XRFQED is the qualitative elemental determination routine that accesses the NIH data base and the SPECPLOT peak search

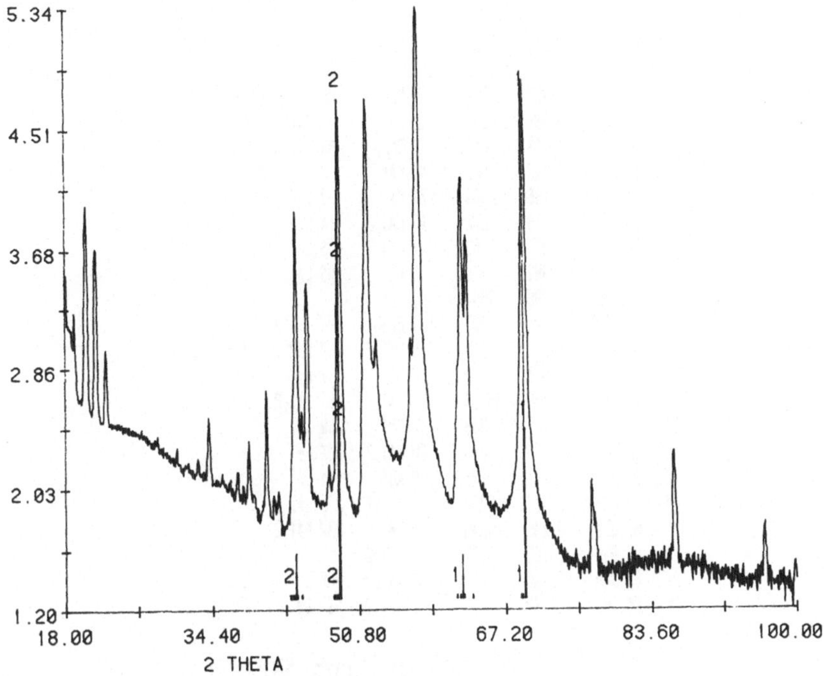

Fig. 1. SPECPLOT display. Semi-log scale with elemental markers for Cr and Ni.

result file to perform the elemental search/match. The input parameters for the search include the error window in 2-theta, the maximum order to be considered, and the excitation potential of the x-ray tube. The program then determines for each element in the data base the most intense line capable of being excited that should occur within the range defined by the observed pattern. If the most intense line exists in the observed pattern within the specified error window, the element is designated a match. If it does not, the element is rejected. In this way, trace phases where only a single line may be observable are included while false matches are minimized. The program allows up to 12 elements per observed line in its final output. It is left to the analyst to select the correct assignment for a particular line from the list provided based upon other lines present, sample form, sample origin, etc. An example of this final output is presented in Figure 2.

ELF and ALPIR

Two utilities programs, ELF and ALPIR, have been written to retrieve information from the NIH data base in an interactive

```
 15  4-15-82 XRF      1234 NBS STEEL  SRM 1152          50/50 SS30 1FN6
START ANGLE=  18.000 STEP= 0.050 TIME=  4.0 CRYSTAL=200S 2D=  4.027
ISM= 0  BKGR=99 0  ND2= 9  SENS= 2.0  NUM= 1  PF=3
ERROR(DEG)= 0.100

 N   2THETA   LAMBDA   HEIGHT     ELEMENTAL LINES MATCHED

 1   18.627   0.652    1313.   NB KB2      MO SKB10   2PM KA2
 2   19.006   0.665    1678.   NB KB1     2ND KA1
 3   20.299   0.710    9617.   ZR SKBN     MO KA1     2ND KA2
 4   21.367   0.747    4867.   NB KA1
 5   22.516   0.786     909.   ZR KA1      PB LG4
 6   33.953   1.176     262.   AS KA1,2    TA LG5      PB LA1
 7   38.470   1.327     163.   TA LB1      OS Ln
 8   40.447   1.392     449.   CU KB1      OS LA1
 9   43.739   1.500    8641.   NI KB1
10   44.401   1.522     305.   NI SKBN     CU SKA^4    TA LA1
11   44.999   1.541    2746.   CU KA1,2   2SR KB2     2TH SLB2^6
12   47.461   1.621     100.   CO KB1      DY LB2      2BI SLG1'
13   48.641   1.658   38933.   NI KA1,2   2Y  KA1,2   2TH LB6
14   51.734   1.757   51988.   FE KB1     2SR KA1,2
15   52.758   1.789    1095.   CO KA1,2   2BK Ls
16   56.653   1.911    1125.   MN KB1      DY LA1     2BI LB2
17   57.492   1.937  216992.   FE KA1,2   2TH LA2
18   62.370   2.085   14849.   CR KB1     2BR KA1,2
19   62.949   2.103    5871.   MN KA1,2
20   69.329   2.290   69654.   CR KA1,2   2HF LG2     2BI LA1
21   86.103   2.749     168.   TI KA1,2    I  LB2     2HF LB1
```

Fig. 2. Output from XRFQED.

fashion. ELF retrieves elemental line and absorption edge information by element symbol and displays it in units of either energy, wavelength, or 2-theta. If 2-theta is selected, the program requests the crystal or d-spacing and the order desired. The output is formatted to fill a 24 line CRT terminal without scrolling.

ALPIR displays all the possible lines in a given range of an input position. The position can be in units of either energy, wavelength, or 2-theta and with a user specified minimum relative intensity. This utility is useful when trying to resolve an unmatched line in an observed pattern.

Examples of the output from ELF and ALPIR are presented in Figures 3 and 4.

ANALYSIS STRATEGIES

Stepscan data is collected under user specified conditions using a Rigaku S/MAX spectrometer controlled by the PDP 11/34 computer. Data reduction can proceed in either an interactive or a batch mode.

```
>ELF
 ENTER ELEMENT SYMBOL <CTRL Z> WHEN DONE>CU
 INFORMATION: 1: LINES
              2: ABSORPTION EDGES
>1
 DISPLAY MODE: 1: WAVELENGTH
               2: ENERGY
               3: TWO-THETA
>1
 ELEMENT: CU    AT.NO.:  29   WAVELENGTHS
SKBN^4   1.369  1.00 SKA'     1.536   1.00
SKBN'''  1.372  1.00 SKA''    1.538   1.00
SKBN''   1.376  1.00 KA1      1.541 100.00
SKBN'    1.377  1.00 KA1,2    1.542 151.39
SKB'''   1.379  1.00 KA2      1.545  51.40
SKB'''   1.380  1.00 LB3     12.119   4.21
KB2      1.381  0.10 LB4     12.119   2.29
SKB6     1.383  1.00 SLB1''' 12.911   1.00
SKB''    1.385  1.00 SLB1''  12.957   1.00
SKB7     1.387  1.00 SLB1'   12.989   1.00
SKB10    1.391  1.00 LG5     13.010   0.21
KB1      1.392 13.41 LB1     13.051  51.28
KB3      1.392  6.84 SLA4    13.176   1.00
SKB'     1.394  1.00 SLA3''  13.233   1.00
SKB      1.398  1.00 SLA3''' 13.261   1.00
SKB1^4   1.402  1.00 SLA3'   13.277   1.00
SKBN     1.408  1.00 LB6     13.317   0.58
SKBN     1.409  1.00 LA1     13.331 100.00
SKA^4    1.522  1.00 LA2     13.331  11.47
SKA'''   1.528  1.00 SLA^5   13.379   1.00
SKA4     1.533  1.00 SLA^6   13.397   1.00
SKA3'    1.534  1.00 Ln      14.902   2.75
SKA3     1.535  1.00 Ll      15.287   6.38
```

Fig. 3. User interface and output from ELF.

```
>ALPIR
 ENTER POSITION OF LINE >52.0
 UNITS FOR ENTERED POSITION: 1 - ANGSTROMS
                             2 - KEV
                             3 - DEGREES (2-THETA)
>3
 ENTER RANGE (+/-) TO BE DISPLAYED>0.5
 ENTER CRYSTAL: 1 - LiF 200
                2 - LiF 220
                3 - PET
                4 - TAP
                5 - Ge
                6 - ADP
                7 - Other
>1
 ENTER ORDER [ D:1 MAX:9 ]>
 ENTER MIN. INTENSITY CUTOFF [ D:1.]>
   52.000 +/-0.500 DEGREES 2-THETA
Fe SKB''  51.469  1.00
Gd SLB14  51.586  1.00
Fe KB1    51.734 13.48
Fe KB3    51.734  6.84
Er Ln     51.734  1.53
Fe SKB'   51.817  1.00
Tb SLB1'  52.119  1.00
Tb LB1    52.370 55.13
Er SLA^X  52.401  1.00
Co SKA3'  52.465  1.00
Er SLA'   52.481  1.00
Co SKA4   52.484  1.00
```

Fig. 4. User interface and output from ALPIR.

In interactive mode, data for a particular sample or sets of samples is collected and stored on disks with a separate raw data file for each sample and each set of conditions (crystal, scan range, etc.). The operator can then examine the data and perform a second derivative peak search using SPECPLOT. The result of the peak search is written to an ASCII file. XRFQED can then be run with this file as input to obtain a list of these peaks with possible elemental identifications.

In order to better utilize operator time, the stepscan program can be instructed to operate in batch mode. In this mode, data is collected as before, but when one set of data is complete, the stepscan program logs the names of the data set in a control file and starts a task named SEEK as it continues taking data. SEEK obtains the name of the raw data file from the control file and performs the second derivative peak search and elemental search/match using input parameters from its own control file. The results are identical to those obtained from SPECPLOT and XRFQED. When SEEK finishes one set of data, it checks to see if the stepscan for another data set has been completed. If it has, SEEK will operate on these files. If it has not, SEEK will exit and the stepscan program will invoke it again. In this way, a series of analyses can proceed unattended.

In addition, the intensities determined in the peak search can be used as inputs to programs such as NRLXRF (10) and XRF11 (11) to provide semi-quantitative analysis from the stepscan data when data from standards obtained under similar conditions is available.

SUMMARY

A series of FORTRAN IV programs have been written to access the newly released NIH elemental data base and aid in qualitative XRF analysis. The programs include graphics, peak search and elemental identifications and data retrieval. Utilities have been implemented in both interactive and batch modes under an RSX-11M operating system on a PDP 11/34. The programs which contain instrument control routines are protected under a data agreement with Rigaku/USA, Inc. and must be obtained from them. The other programs are available from the authors.

REFERENCES

1. Data base available from Charles Fiori, Room 3W13 - Building 13, National Institute of Health, Bethesda, MD 20205.
2. T.C. Huang, W. Parrish, G.L. Ayers, Advances in X-ray Analysis 24, 407 (1981).
3. G. Platbrood, M. Serbruyns, J.M. Quitin, X-ray Spectrometry 11, 83 (1982).
4. B.L. Doyle, W.F. Chambers, T.M. Christensen, J.M. Hall and G.H. Pepper, Sine Theta Settings for X-ray Spectrometers, Atomic Data and Nuclear Data Tables, Vol. 24, No. 5, 1979.
5. E.W. White, G.V. Gibbs, G.G. Johnson, Jr. and G.R. Zechman, X-ray Wavelengths and Crystal Interchange Settings for Wavelength Geared Curved Crystal Spectrometers, Report of the Pennsylvania State Univ., 1964.
6. J.A. Bearden, X-ray Wavelengths and X-ray Atomic Energy Levels, Rev. Mod. Phys., Vol. 39, No. 78, 1967.
7. J.A. Bearden and A.F Burr, Reevaluation of X-ray Atomic Energy Levels, Rev. Mod. Phys., Vol. 31, No. 1, 1967.
8. R.P. Goehner, Advances in X-ray Analysis 23, 305 (1980).
9. R.P. Goehner, General Electric Technical Information Series, #82CRD043, 1982.
10. J.W. Criss, L.S. Birks, J.V. Gilfrich, Anal. Chem. 50, 33 (1978).
11. J.W. Criss, Advances in X-ray Analysis 23, 93 (1980).

A NEW METHOD FOR QUANTITATIVE X-RAY FLUORESCENCE ANALYSIS OF MIXTURES OF OXIDES OR OTHER COMPOUNDS BY EMPIRICAL PARAMETER METHODS

Michael Mantler[†]

IBM Research Laboratory
San Jose, California 95193

Two principal mathematical methods are used for quantitative XRFA: fundamental parameter calculations and the evaluation of empirical parameter equations. A comprehensive computer program based upon fundamental parameter equations was introduced in 1976 by D. Laguitton and M. Mantler (LAMA-I)[1] and improved by T. C. Huang in 1979 (LAMA-II).[2] The present paper describes the features of the theoretical background of a computer program using a new type of empirical (alpha*-) parameter equations. It is essentially designed for convenient analysis of compounds including those containing chemical elements, that cannot be directly measured by conventional X-ray spectrometers, such as oxides, nitrides, and others. The program also communicates automatically with LAMA in order to establish theoretical tables of alpha*-coefficients as well as conventional alpha-coefficients.

Among the unique features of this method is that in the case of compound analysis the alpha*-coefficients are calculated as "element-element coefficients" rather than as "compound-compound coefficients." The advantage of this is, that any standards may be used to establish the alpha*-coefficients without the requirement of a stoichiometric match of elements in the components of the standards and of the analyzed specimen. This also means that more parameters are available to approximate the relationship between concentrations and count rates with greater accuracy and over a wider useful range of concentrations.

In addition, the composition of the specimens can be entered and calculated in terms of weight fractions of its components as well as in the conventional way in terms of weight fractions (or number of atoms) of the individual elements.

In order to establish the alpha- (or alpha*-) coefficient table from standards, pure element standards may be used, but are not necessarily required. Any number of standard specimens may be used for count rate calibration of the X-ray spectrometer. Such standards may be pure elements or consist of any elements that are included in the actual alpha*-(or alpha-) coefficient table.

†Permanent address: Technical University Vienna, Vienna, Austria.

THEORY

The relationship between counts n_i from element i in a compound, pure element counts from element i,N_i, and element weight fractions c_i, is described by

$$-1 = -\frac{c_i}{n_i}N_i + \sum_{j\neq i}^{\ell} \alpha_{ij}c_j + \sum_{j\neq i}^{\ell}\sum_{k\geq j}^{\ell} \beta_{ijk}c_jc_k . \quad (1)$$

This corresponds to the model of Claisse and Quintin,[3] if all coefficients are considered; to the model of Lachance and Traill,[4] if all β_{ijk} are zero; and to the model of Rasberry-Heinrich,[5] if the applicable coefficient selection rules are applied. The pure element count rates can be calculated like additional alpha-coefficients, if they are unknown and only composite standards are used.

In the case of compound-analysis, the known stoichiometric composition of the standard is used to calculate the total weight fraction c_i, of each of the chemical elements as the sum of the partial weight fractions, x_{ip}, in all components, p, of the specimen (c^*_p is the weight fraction of component p in the specimen):

$$c_i = \sum_{p=1}^{g} c^*_p x_{ip} . \quad (2)$$

This relationship is used to substitute for the c's in Eq. (1):

$$-1 = \sum_{p=1}^{g}\Big(\sum_{j\neq i}^{\ell} \alpha_{ij}x_{jp} - x_{ip}n_i/N_i\Big)c^*_p \quad (3)$$

$$+ \sum_{p=1}^{g}\ \sum_{q=1}^{g}\Big(\sum_{j\neq i}^{\ell}\sum_{k\geq j}^{\ell} \beta_{ijk}x_{jp}x_{kq}\Big)c^*_pc^*_q$$

$$-1 = \sum_{p=1}^{g} \alpha^*_{ip}c^*_p + \sum_{p=1}^{g}\ \sum_{q=1}^{g} \beta^*_{ipq}c^*_pc^*_q \quad (i = 1,\ldots,\ell) . \quad (4)$$

The alpha*- and beta*-coefficients are calculated as defined by Eqs. (3),(4) and contain the actual count rates. For the Lachance-Traill model, all beta*-coefficients vanish. The number of such equations is equal to the number of elements, ℓ. If the number of unknown weight fractions of the components, g, is less than ℓ, up to (ℓ-g) equations may be skipped. Up to (ℓ-g) elements may therefore be included in the analyzed specimen and standards, for which no count rates have to be obtained. The pertinent alpha(*)- and beta(*)-coefficients are then never used and remain undetermined. If, however, more than the minimum number, g, of count rates have been measured, a least square fit algorithm is applied to make use of all available information.

For count rate calibration of the X-ray spectrometer, pure element count rates are calculated from measured count rates of all available standards using Eq. (1), and the average value for each element is used as reference.

The alpha(*)-coefficient equations are exact for the composition of a single standard that contains all analyzed elements and is used as the only count rate calibration standard for all measured lines. In this case, analysis of specimens of similar composition as the standard will be highly accurate. This is equivalent to the "delta-coefficients-methods"[6,7] with the added possibility of using any number of standards.

In the case of element analysis, where count rates have been measured for all elements, the equations for the alpha*-method become identical to those for conventional alpha-coefficients.

EXAMPLES

A specimen is assumed to consist of a mixture of three oxides, $KAlSi_3O_8$, Fe_2O_3,CaO. Each component contains at least one element that is not contained in any other component. The component-component alpha-coefficient-table for this case (if K represents the first component) and the element-element alpha*-coefficient-table are:

$$\begin{pmatrix} \alpha_{K\leftarrow KAlSi_3O_8} & \alpha_{K\leftarrow Fe_2O_3} & \cdots \\ \alpha_{Fe\leftarrow KAlSi_3O_2} & \alpha_{Fe\leftarrow Fe_2O_3} & \cdots \\ \vdots & \vdots & \end{pmatrix}, \begin{pmatrix} \overset{*}{\alpha}_{K\leftarrow Al} & \overset{*}{\alpha}_{K\leftarrow Si} & \overset{*}{\alpha}_{K\leftarrow 0} & \cdots \\ \overset{*}{\alpha}_{Si\leftarrow Al} & \overset{*}{\alpha}_{Si\leftarrow Si} & \overset{*}{\alpha}_{Si\leftarrow 0} & \cdots \\ \vdots & \vdots & \vdots & \end{pmatrix}.$$

LAMA was used to calculate the theoretical intensities from standards with compositions shown in Table 1, and the values for the two alpha-coefficient tables were established from these data (Lachance-Traill model). The specimen indicated by + in Table 1 was then used as "count rate calibration standard" and the compositions of all standards were recalculated from the alpha(*)-coefficients and the theoretical count-rates.

Results obtained with this method are shown in Table 1. Despite the wide range of concentrations the results from alpha* are in very close agreement with the original data, and in all cases are significantly better than than those from the conventional type of calculation.

In the second example, the specimen was assumed to consist of $KAlSi_3O_8$, Al_2O_3, and SiO_2. It is important to notice that in this case only one component contains a unique element. The same calculations as in the first example were applied. Again, the results from the alpha*-coefficients agree very well with the original values.

Evaluation of this system using conventional alpha-coefficients, however, reveals significant limitations. It is impossible in this case to obtain correct results because the count rates of Al and Si cannot be assigned to only one component. The alpha-coefficient equations are not adequate in this situation: In contradiction to Eq. (1), a zero concentration of the representative element does not necessarily correspond to zero counts for this element. The alpha*-method, however, may be used in such cases with no theoretical restrictions.

Table 1. Comparison of Results from the Alpha- and Alpha*-Method (Data Are Weight Fractions of the Components)

Example 1	Theor.	Alpha*	Alpha	Example 2	Theor.	Alpha*	Alpha
$KAlSi_3O_8$	0.667	0.667	0.667	$KAlSi_3O_8$	0.667	0.667	0.667
Fe_2O_3 +	0.167	0.167	0.167	Al_2O_3 +	0.167	0.167	0.167
CaO	0.167	0.167	0.167	SiO_2	0.167	0.167	0.167
$KAlSi_3O_8$	0.167	0.167	0.149	$KAlSi_3O_8$	0.167	0.167	0.166
Fe_2O_3	0.667	0.673	0.672	Al_2O_3	0.667	0.665	0.746
CaO	0.167	0.166	0.165	SiO_2	0.167	0.166	0.188
$KAlSi_3O_8$	0.167	0.168	0.161	$KAlSi_3O_8$	0.167	0.167	0.160
Fe_2O_3	0.167	0.167	0.154	Al_2O_3	0.167	0.167	0.150
CaO	0.667	0.668	0.691	SiO_2	0.667	0.667	0.947
$KAlSi_3O_8$	0.400	0.401	0.373	$KAlSi_3O_8$	0.400	0.401	0.402
Fe_2O_3	0.500	0.503	0.512	Al_2O_3	0.500	0.498	0.477
CaO	0.100	0.100	0.098	SiO_2	0.100	0.099	0.178
$KAlSi_3O_8$	0.100	0.101	0.093	$KAlSi_3O_8$	0.100	0.101	0.097
Fe_2O_3	0.400	0.400	0.377	Al_2O_3	0.400	0.399	0.406
CaO	0.500	0.500	0.513	SiO_2	0.500	0.499	0.570
$KAlSi_3O_8$	0.500	0.501	0.498	$KAlSi_3O_8$	0.500	0.500	0.491
Fe_2O_3	0.100	0.100	0.095	Al_2O_3	0.100	0.101	0.122
CaO	0.400	0.400	0.406	SiO_2	0.400	0.401	0.456

REFERENCES

1. D. Laguitton and M. Mantler, Adv. X-ray Anal. 20, 515 (1977).
2. T. C. Huang, X-ray Spectrom. 10, 28 (1981).
3. F. Claisse and M. Quintin, Can. J. Spectroscopy 12, 129 (1967).
4. R. J. Traill and G. R. Lachance, Can. J. Spectroscopy 11, 43 (1966).
5. S. D. Rasberry and K. F. J. Heinrich, Anal.Chem. 46, 81 (1974).
6. W. K. deJough, X-ray Spectrom. 2, 151 (1973).
7. J. Kramer, H. Ebel, and F. Tschismarov, X-ray Spectrom. 6, 30 (1972).

FPT: AN INTEGRATED FUNDAMENTAL PARAMETERS PROGRAM FOR BROADBAND EDXRF ANALYSIS WITHOUT A SET OF SIMILAR STANDARDS

D. A. Gedcke, L. G. Byars, and N. C. Jacobus
EG&G ORTEC
Oak Ridge, Tennessee 37830 U.S.A.

THE NEED FOR STANDARDLESS ANALYSIS

The x-ray fluorescence (XRF) method is well known for its capability to perform fast and accurate quantitative analysis for all elements with atomic numbers greater than ten. Energy dispersive x-ray fluorescence (EDXRF) adds to this capability the benefit of quick qualitative analysis, due to its simultaneous sensitivity to all the elements. The method has the potential for rapid and complete chemical analysis of any sample which arrives on the analytical chemist's doorstep. Although the method has been a productive tool for fast and accurate repetitive analysis of similar samples, its applicability to unique unknowns has been rather limited. The limitation arises from the usual need to calibrate the instrument's response with a set of 6 to 12 standards, whose compositions must be similar to the unknown sample. Anyone who has struggled to develop and maintain such a suite of accurately certified standards knows that a great deal of effort and expense is involved. This effort is well justified when the analyst expects to analyze the same type of material frequently over an extended time period. However, for a unique sample analysis, the task of developing a suite of similar standards simply makes the analysis impractical. What is needed is a method that requires minimal standards, or uses no standards at all.

THE INGREDIENTS FOR A SOLUTION

Although the fundamental parameters equations to serve this need were developed as long ago as 1952[1-4], it wasn't until 1979 that the general solution using minimal standards became available on the minicomputers incorporated in laboratory x-ray fluorescence analyzers[5]. Substantially increased computing power in minicomputers at a reasonable cost is what made this development possible. Earlier approaches depended on large central computers[6-8] or simplifying approximations which limited the accuracy for the general solution with broadband excitation[9-12].

The fundamental parameters solution normally begins with the fundamental theoretical equations relating the measured x-ray counting rate, I_i, for the ith element to the sample composition[1-4].

$$I_i = I_{pi} + I_{si} \tag{1}$$

where I_{pi} is the primary fluoresced intensity[13],

$$I_{pi} = \frac{S_i \eta(E_i) W_i}{\sin \Psi_1} \int_{E_o = \Phi_{pi}}^{E_{max}} \frac{\tau_{pi}(E_o)\omega_{pi} f_{pi} I_o(E_o) dE_o}{\mu(E_o)\csc\Psi_1 + \mu(E_i)\csc\Psi_2} \tag{2}$$

and I_{si} is the secondary fluoresced intensity.

$$I_{si} = \sum_j \frac{S_i \eta(E_i) W_i W_j}{2 \sin \Psi_1} \int_{E_o = \Phi_{qj}}^{E_{max}} \frac{\tau_{pi}(E_j)\omega_{pi} f_{pi} \tau_{qj}(E_o)\omega_{qj} f_{qj} I_o(E_o) dE_o}{\mu(E_o)\csc\Psi_1 + \mu(E_i)\csc\Psi_2}$$

$$\times \left\{ \frac{\sin\Psi_1}{\mu(E_o)} \ln\left[\frac{\mu(E_o)\csc\Psi_1}{\mu(E_j)} + 1\right] + \frac{\sin\Psi_2}{\mu(E_i)} \ln\left[\frac{\mu(E_i)\csc\Psi_2}{\mu(E_j)} + 1\right] \right\} \tag{3}$$

Note that the matrix mass absorption coefficients depend on the elemental concentrations through equations (4) and (5).

$$\mu(E) = \sum_j \mu_j(E) W_j \tag{4}$$

$$\sum_i W_i = 1 \tag{5}$$

Table 1. Symbol Definitions

Symbol	Definition
E_o	An X-ray energy in the excitation source spectrum.
E_i	Energy of characteristic x-ray line being measured for element i.
E_j	Energy of characteristic line emitted by element j.
E_{max}	Maximum x-ray energy in the excitation spectrum.
f_{pi}	Fraction of characteristic radiation selected for measurement and generated by the pth atomic subshell of the ith element.
f_{qj}	Fraction of characteristic radiation generated by the qth atomic subshell of the jth element which can fluoresce the measured line of the ith element.
$I_o(E_o)dE_o$	Number of x-ray photons per second incident on sample from the excitation source in the energy range from E_o to E_o+dE_o.
S_i	Sensitivity factor scaling predicted counting rate to match measured counting rate for element i.
W_i	Weight fraction of element i in the sample.
W_j	Weight fraction of element j in the sample.
$\eta(E_i)$	The Si(Li) detector efficiency for detecting x-ray photons of energy E_i.
$\mu(E)$	Sample matrix mass absorption coefficient at energy E.
$\mu_j(E)$	Mass absorption coefficient of element j at energy E.
$\tau_{pi}(E)$	Partial photoelectric mass absorption coefficient for ionization of the pth subshell of element i at energy E.
$\tau_{qj}(E_o)$	Partial photoelectric mass absorption coefficient for ionization of the qth subshell of element j at energy E_o.
ϕ_{pi}	Absorption edge energy of the pth subshell of element i.
ϕ_{qj}	Absorption edge energy of the qth subshell of element j.
Ψ_1	Angle between sample surface and direction from which x rays from excitation source impinge.
Ψ_2	Angle between sample surface and direction emitted characteristic x rays travel to detector.
ω_{pi}	Fluorescent yield of the pth subshell of element i.
ω_{qj}	Fluorescent yield of the qth subshell of element j.

Equations (1) through (5) cannot be solved in closed form for the set of weight fractions representing the sample composition. The solution is obtained by using estimated concentrations in equations (1) to (5) to predict the intensities. The differences between predicted and measured intensities are used to refine the concentration estimates, and the process is repeated until the solution converges. The final set of weight fractions represents the composition of the sample.

If the mathematical model accurately represents what is happening in the fluorescence analyzer, standardless analysis becomes practical. Achieving this objective places demands on the accuracy of the following items.

a) The excitation spectrum description, $I_o(E_o)dE_o$.

b) The model for the detector efficiency.

c) Fundamental physical parameters (E_i, E_j, f, μ, τ, Φ, ω).

d) The geometry assumptions[13].

e) Ignoring multiple scattering effects in the sample[13].

f) Sensitivity factors S_i are equal for all elements and lines.

For the development of the FPT (Fundamental Parameters Technique) program, a primary objective was accurate standardless analysis. Consequently, items a) through f) were handled as follows.

Because broadband excitation is the most time efficient method[13] for analyzing the major elements in a sample, accurate modelling of the x-ray tube spectrum was thoroughly researched. The geometry of the EG&G ORTEC TEFA III dual anode x-ray tube was duplicated with a scanning electron microscope. Spectra recorded on pure Mo, Rh, and W samples were used to test theoretical predictions of the characteristic line intensities and the bremsstrahlung continuum as developed in the ZAP program[14]. These equations were adjusted and determined to be sufficiently accurate to predict the energy and voltage dependence of the x-ray tube spectra. Confirming tests of the characteristic line to continuum intensity ratios were made on the TEFA III using a low atomic number sample to scatter the x-ray tube spectra.

The model for computing the detector efficiency has been published previously in the ZAP program[14]. One of the basic benefits of EDXRF is the fact that the detector efficiency is easily predicted, stable, and repeatable over a time span of the order of years. This is not true of WDXRF instruments, which makes standardless fundamental parameters analysis impractical for WDXRF.

The most recent compilations of fundamental parameters[15-18] were incorporated as a data file for each element. Any errors in these parameters propagate into the accuracy of the calculated sample composition, since standardless analysis is sensitive to the fundamental parameters. The geometry assumptions amount to a presumption that the dimensions of the excitation source, the detector and the excited volume in the sample are all very small compared to the distances between the major spectrometer components (x-ray tube, sample, and detector)[13]. These assumptions are least valid when analyzing the K lines of a high atomic number element in a matrix having a very low atomic number (e.g., 1% silver in water). This is also the most sensitive case for multiple scattering effects.

If the model is absolutely accurate, the sensitivity factors S_i should be the same for all lines from all elements. S_i would simply be a scaling factor accounting for the solid angles and tube current employed. Under such conditions, standardless analysis is possible since equation (5) eliminates the need to know the common value of S_i. The standardless analysis option in the FPT program proceeds on that basis. For cases where an unmeasured element or compound must be calculated by difference, it is necessary to use a measured value for S_i. This can be accomplished using any conveniently available pure element or multielement standard. As before, a common value for all S_i's may be assumed. The intensity predicted for the standard with $S_i = 1$ is compared to the measured intensity on the standard to derive S_i.

Sometimes the analytical chemist can obtain one or a few standards with compositions similar to the unknown sample. For each element in the unknown, the sensitivity factor measured on the standard(s) can be used. In this way any systematic errors in the fundamental parameters model tend to cancel. The closer the match between standard and unknown compositions, the more completely the errors will cancel. This occurs because the same systematic errors are incorporated into the predicted intensities for the standard and the unknown. The empirical sensitivity factor absorbs these systematic errors. Using similar standards the accuracy of FPT can approach that obtained by the calibration curve method, while using drammatically fewer standards.

THE IMPORTANCE OF SPECTRAL ANALYSIS

If the net peak intensities fed to the fundamental parameters equations are wrong, then the computed concentrations will be wrong. Therefore, extracting accurate net intensities from the spectrum is half of the task. The spectrum in Figure 1 demonstrates the typical spectral analysis problems: severe K series overlaps for the transition metals, escape peaks, and a variable background. At 2.3 keV the Mo L line falls directly on top of the S K line making it very difficult to separate the intensities from these two elements. In

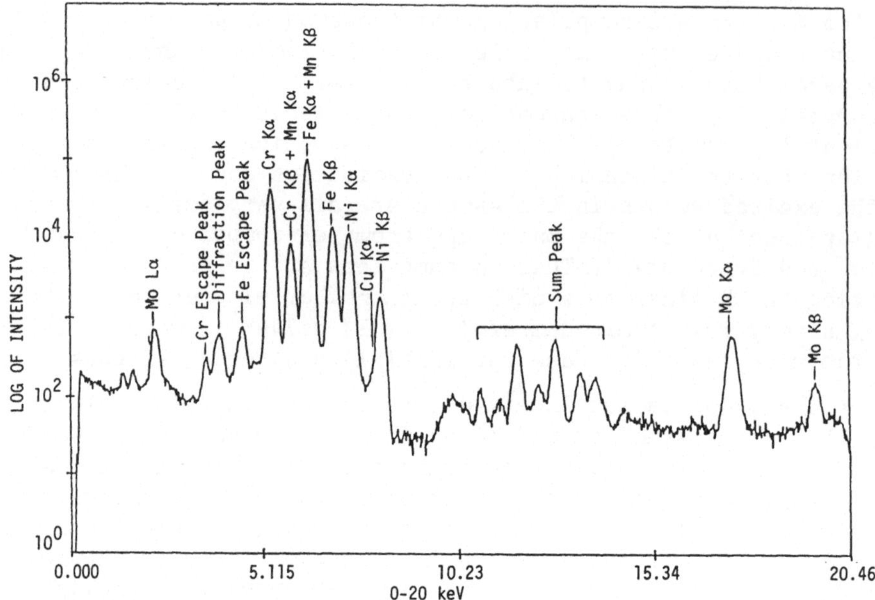

Fig. 1. The x-ray energy spectrum from an NBS 1185 stainless steel standard using 25 kV broadband excitation.

order to solve these problems efficiently for the analyst an automatic spectrum analysis program must be incorporated in the fundamental parameters program.

MAJOR CHARACTERISTICS OF THE FPT PROGRAM

The FPT program was written in FORTRAN IV and runs as object code under RT-11 in a DEC LSI-11/02 computer with one or two floppy disks. It utilizes 64K bytes of memory. The major features summarized in Table 2 distinguish the FPT program from the various programs previously published such as CORSET[9], EXACT[11], QUAN[10], QUANT[12], NRLXRF[6], XRAY 95[19], and XRF-11[5].

Table 2. Major Features of the FPT Program

1)	Quantitative analysis can be performed with: (a) no standards; (b) dissimilar standards; or (c) a similar standard.
2)	Automated spectral deconvolution and analysis is included in the concentration iteration loop.
3)	Automated handling of multiple excitation conditions (including broadband and monochromatic excitation).
4)	Accurate modelling of the current, voltage, and energy dependence of the x-ray tube continuum and characteristic line intensities for the specific tube employed.

The convenience of no standards analysis is available for analyzing unique unknowns. For situations where an unmeasured element or compound must be computed by difference, a conveniently available dissimilar standard can be used. This is typically a pure element metal such as titanium, iron, nickel, or copper. A dissimilar standard is also used when multiple excitation conditions are employed. When a similar standard is available, it can be used to ensure optimum accuracy.

Once the elements to be analyzed and the best lines to use are specified, the spectral deconvolution and concentration computation proceed automatically (see Figure 2). By including the spectral deconvolution within the concentration iteration loop, the intensities of all the lines from each element are calculated at each step of the iteration, including an accurate accounting for the effects of matrix absorption and secondary fluorescence. This leads to a more accurate unraveling of the spectral overlaps illustrated in Figure 1. It also makes it possible to solve severe overlaps such as the Mo L and S K interference in Figure 1. In this case, the Mo K lines are used to measure the Mo concentration, and the Mo L line intensity is predicted from the measured concentration. This L line intensity is stripped from the combined Mo L + S K peak to leave the net S K line intensity. Consequently, the sulfur concentration can be measured in spite of the complete overlap. Results from this example are shown in Table 3. The spectral deconvolution also includes background subtraction and corrections for escape peak overlaps.

Table 3. The FPT Program with Two Excitation Conditions analyzing CARTECH 80C Steel

Element and Line	Percent Concentration: Certified	FPT†	Absolute Error	Excitation Condition
SiKα	0.55	0.50	-0.05	Mo anode, 20 kV, 10 μA, no filter, 100 seconds
SKα	0.329	0.50	0.171	
CrKα	17.24	17.10	-0.14	
MnKα	1.78	1.82	0.04	
FeKα + remainder	(71.44)*	71.47	0.03	
NiKα	8.16	8.22	0.06	
CuKα	0.20	0.20	0.00	
MoKα	0.30	0.18	-0.12	W anode, 45 kV, 3 μA, no filter, 50 seconds
Total	99.999	99.99		
Standard Error on Total			0.27	

*Fe concentration not certified. (71.44) value calculated by difference.
†Ti pure element reference standard used.

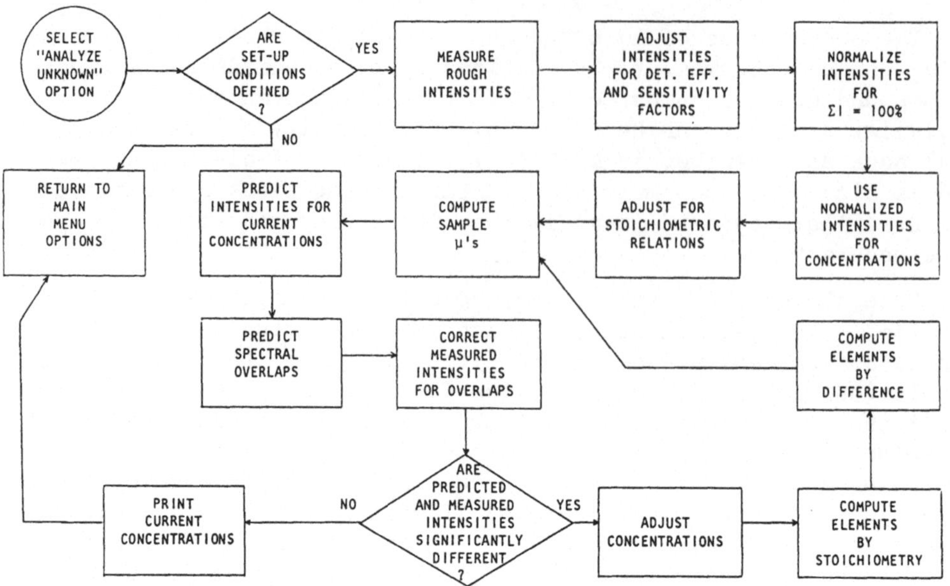

Fig. 2. The FPT program flow chart for analyzing unknown samples.

Normally, broadband excitation is the most efficient means of analyzing a sample for its total composition. Occasionally, a second or third excitation condition is used to optimize determinations on a few trace elements. In this case, the program must consolidate the information from all the excitation conditions used. This is done automatically in the FPT program. To accomplish accurate minimal standards analysis over such a wide range of excitations, the development of an accurate model in FPT for the actual output of the x-ray tube under all operating conditions has been crucial.

USEFUL OPTIONS IN FPT

The FPT program requires no software skills from the operator. All actions are initiated by selecting an option from a menu displayed on the instrument's video terminal. Initially, the user selects one of three main option categories: a) the analysis of an unknown sample; b) the analysis of a standard to obtain sensitivity factors; or c) selection of the basic set-up conditions. At the end of each option the system returns to the main menu display. Table 4 summarizes options which give the analyst flexibility in handling diverse sample analysis problems. Options 1), 2) and 3) are solutions for the problem of unmeasured elements below atomic number 11. They are also useful when the analyst knows a portion of the sample composition. Option 4) makes repetitive unknowns analysis efficient and item 5) covers a variety of operating conditions. Options 6) and 7) are useful for studying the x-ray physics involved or for predicting interelement effects to be considered with a calibration curve approach to quantitative analysis.

Table 4. Options in the FPT Program

Option #	Description
1	The concentration of an unmeasured element or defined compound can be calculated by difference.
2	Concentrations of unmeasured elements can be calculated by stoichiometric relations to measured elements.
3	Matrix dilutant concentrations can be specified by the operator or calculated by difference.
4	The repetitive analysis of unknowns of the same type requires only the selection of an option number and entry of a sample identification name.
5	FPT accounts for absorbers in the primary and secondary x-ray paths such as: (a) air/vacuum/helium path; (b) mylar films on sample cups; and (c) pre- and post-filters.
6	FPT will predict intensities for operator defined concentrations.
7	Upon request, FPT will output all computational details for diagnostic purposes.

FPT PERFORMANCE

Stainless steel alloys have been selected for detailed illustration of the performance of the FPT program because they represent a very difficult task in spectral analysis. In addition, the interelement effects caused by absorption and secondary fluorescence are severe. Table 3 shows the results from analyzing a Cartech 80C standard as an unknown. Two excitation conditions were used. The standard error on the total composition reported in this and subsequent tables is computed as the square root of the sum of the squares of the absolute errors on the individual elements. Although this number doesn't have a completely rigorous statistical basis, it is a convenient figure of merit for comparing overall performance under different conditions*.

Table 5 compares the use of standards to standardless analysis. Generally the analytical accuracy improves as the standard becomes more similar to the unknown. Table 6 demonstrates that the FPT accuracy using a similar standard can be as good as the accuracy of a calibration curve method. ATAC (a more recent version of the FLINT program[20]) is an automated empirical calibration curve program with interelement corrections for matrix effects. Its accuracy on the

*The lack of rigor arises because the elemental concentration errors are not strictly independent random variables. There is some cross-correlation because of the interdependences of weight fractions in solving equations (1) to (5).

Table 5. NBS 1185 Steel Analysis with the FPT Program

Element	Certified Percent Concentration	Measured % Concentration with FPT Program using: No Standards	Pure Titanium Reference Standard	80C Brammer Steel Reference Standard
Si	0.40	0.57	0.55	0.53
Cr	17.09	18.88	18.08	17.12
Mn	1.22	1.29	1.17	1.20
Fe + remainder	(66.09)*	64.59	65.09	66.09
Ni	13.18	12.19	13.07	13.22
Cu	0.0067	0.10	0.01	0.02
Mo	2.01	2.37	2.03	1.83
Total	100.00	99.99	100.00	100.01
Standard Error on Total		2.6	1.4	0.23

*Fe concentration not certified. (66.09) value obtained by difference.
100 seconds live time, 25 kV, 15 μA, Rh anode, broadband excitation.

Table 6. FPT vs. ATAC Accuracy on Four Stainless Steels†

Element	Percent Concentration: Range	Average	Average Error: FPT‡	ATAC
Si	0.33 - 0.50	0.41	0.042	0.023
S	0.005 - 0.016	0.0074	0.0072	0.003
Cr	17.09 - 22.81	18.53	0.088	0.22
Mn	0.47 - 1.73	1.22	0.11	0.073
Fe + remainder*	61.38 - 81.15	68.27	0.18	0.18
Ni	0.29 - 13.18	9.91	0.068	0.11
Cu	0.0067- 0.39	0.16	0.015	0.015
Mo	0.15 - 2.34	1.25	0.066	0.015
Standard Error on Total			0.25	0.32

* Fe not certified; Fe concentration computed by difference for standard value and ATAC results.

† Standards: 1) 969 Cartech; 2) 93C Brammer; 3) 82A Brammer; and 4) NBS 1185.

‡ Brammer 80C used as a calibration reference standard for FPT with two excitation conditions.

300 and 400 series stainless steels has been well documented[20]. The similarity in accuracy between FPT and ATAC in this case is outstanding considering ATAC used 16 standards for the 300 series and 10 standards for the 400 series calibration while FPT used only one similar standard.

In order to get a picture of the accuracy of the FPT program over a wide range of sample types, the summaries in Figures 3, 4, and 5 have been compiled. The points in the graphs represent 297 elemental determinations from a variety of sample types including: NBS cements as pressed powders; stainless steel alloys; solder alloys; aluminum alloy; minerals; rocks and ores fused in $Li_2B_4O_7$ glass; and biological materials. Both K and L lines are involved in the analyses, and line energies range from 1 to 30 keV. For each determination, the concentration measured by FPT on a TEFA III is plotted against the certified concentration of the element or oxide in the sample. Scatter about the diagonal line indicates the relative errors in the measured concentrations. As is typical of XRF, the relative errors are smallest at high concentrations and degrade as the parts per million concentration range is approached. This trend is primarily a result of counting statistics, spectral background, and overlapping of low concentration peaks by larger peaks from high concentration elements.

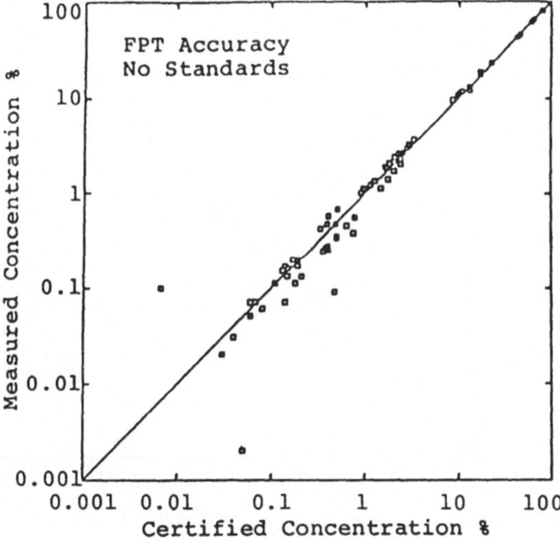

Fig. 3 The concentration measured by FPT versus the certified concentration on 91 elemental determinations illustrating the accuracy using no standards.

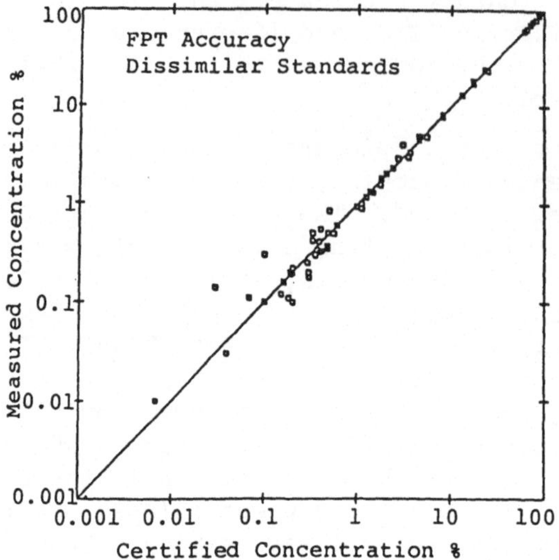

Fig. 4 The concentration measured by FPT versus the certified concentration on 75 elemental determinations illustrating the accuracy using dissimilar standards.

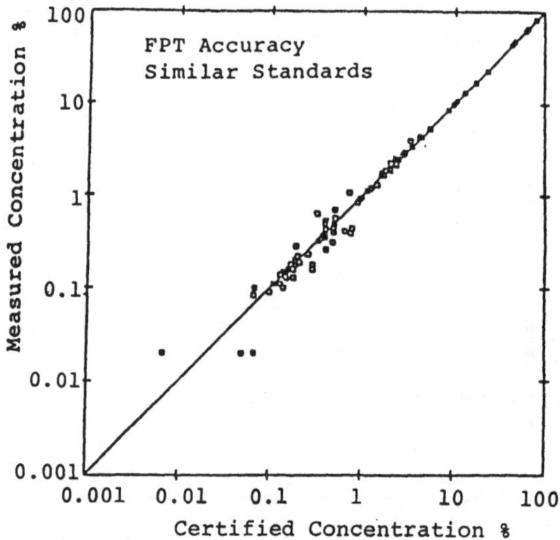

Fig. 5 The concentration measured by FPT versus the certified concentration on 131 elemental determinations illustrating the accuracy using similar standards.

Table 7. Average % Relative Errors Obtained with FPT*

% Concentration Range	Average % Relative Error		
	No Standards	Dissimilar Standards	Similar Standards
100 - 10	4.2	2.4	0.47
10 - 1	12	7.6	4.9
1 - 0.1	26	22	18
0.1 - 0.01	31	84	52

*Summary of 297 elemental determinations (91 no standards, 75 dissimilar standards, 131 similar standards).

Table 7 summarizes the percent relative errors from Figures 3, 4, and 5. For concentrations in the 1 to 100 percent range, the accuracy is improved by using standards. This is the range where the analytical accuracy is controlled by the quality of the fundamental parameters model. For concentrations less than 1%, the use of standards has very little impact on analytical accuracy. This occurs because the basic limit at low concentrations is the uncertainty in the net peak intensities.

CONCLUSIONS

The FPT fundamental parameters technique is a productive solution when it is impractical to obtain an adequate suite of similar standards for the calibration curve method. This is typically the case when a unique unknown must be analyzed. Using standardless analysis or dissimilar standards to solve this problem yields relative accuracies in the 2% to 5% range on major elements. Turnaround time for complete elemental analysis of a unique unknown is typically under one hour. This time includes sample preparation, qualitative analysis, quantitative analysis, and sample handling.

FPT is also useful where only a few, good quality, similar standards are available. Using a similar standard, 0.2% to 1% relative accuracies can be achieved on major elements. The program can be utilized to study sample excitation physics or to predict interelement effects for use with calibration curve methods.

REFERENCES

1. E. Gillam and H. T. Heal, Brit. J. Appl. Phys. 37:353 (1952).
2. J. Sherman, Spectrochim. Acta 7:283 (1955).
3. T. Shiraiwa and N. Fujino, Japan J. Appl. Phys. 5:886 (1966).
4. C. J. Sparks, Jr., Adv. X-ray Anal. 19:19 (1976).
5. J. W. Criss, Adv. X-ray Anal. 23:93 (1980).
6. J. W. Criss, L. S. Birks, and J. V. Gilfrich, Anal. Chem. 50:33 (1978).

7. R. W. Gould and S. R. Bates, X-ray Spectrom 1:29 (1972)
8. J. W. Criss and L. S. Birks, Anal. Chem. 40:1080 (1968).
9. Donald A. Stephenson, Anal. Chem. 43:1761 (1971).
10. Michael F. Ciccarelli, Anal. Chem. 49:345 (1977).
11. J. W. Otvos, G. E. A. Wyld, and T. C. Yao, Adv. X-ray Anal. 20:217 (1977).
12. R. B. Shen and J. C. Russ, X-ray Spectrom 6:56 (1977).
13. Ron Jenkins, R. W. Gould, Dale Gedcke, "Quantitative X-ray Spectrometry", Marcel Dekker, Inc., New York (1981), sections 2.9 and 10.7.
14. D. A. Gedcke, L. G. Byars, and W. H. Hardy, SEM/1982, in press.
15. W. H. McMaster, N. Kerr Del Grande, J. H. Mallett, and J. H. Hubbell, "Compilation of X-ray Cross Sections", UCRL-50174, National Technical Information Service, Springfield, VA (1969).
16. Walter Bambynek, Bernd Crasemann, R. W. Fink, H. U. Freund, Hans Mark, C. D. Swift, R. E. Price, P. Venugopala Rao, Rev. Mod. Phys. 44:716 (1972).
17. M. O. Krause, C. W. Nester, Jr., C. J. Sparks, Jr., and E. Ricci, "X-ray Fluorescence Cross Sections for K and L X-rays of the Elements, ORNL-5399", Oak Ridge National Laboratory, Oak Ridge, TN (1978).
18. J. A. Bearden, Rev. Mod. Phys. 39:78 (1967).
19. R. B. Shen, J. Criss, J. C. Russ, A. O. Sandborg, Adv. X-ray Anal. 23:99 (1980).
20. Bradner D. Wheeler and Nancy Jacobus, Adv. X-ray Anal. 22:395 (1979). See also the preprint published by EG&G ORTEC for additional details.

A COMPARISON OF THE XRF11 AND EXACT FUNDAMENTAL PARAMETERS PROGRAMS WHEN USING FILTERED DIRECT AND SECONDARY TARGET EXCITATION IN EDXRF

Ronald A. Vane

Kevex Corporation
1101 Chess Drive
Foster City, CA 94404

INTRODUCTION

The XRF11 program[1] by John Criss and the EXACT[2] program are two commercially available fundamental parameters programs for energy dispersive x-ray fluorescence spectrometry (EDXRF). These programs are both based on the same underlying equations[3] but use different approaches to the calculations. The EXACT program assumes monochromatic excitation, and the XRF11 program models polychromatic excitation sources. There are also great differences in how the two programs approach the iterations in the calculations and in how the data from standards are used in the two programs for calibration.

To produce the monochromatic excitation needed by the EXACT program, secondary targets have been the preferred method. But it is also possible to approximate monochromatic excitation by using filtered direct excitation.[4] The purpose of this study is to compare the data obtained from both secondary targets and direct filtered excitation as processed through both XRF11 and EXACT.

OPERATION OF EXACT AND XRF11

EXACT: This program uses the assumption that monochromatic excitation is used to excite the samples and that each analyte element is excited by the same primary radiation spectrum. The program is calibrated by the use of a single calibration coefficient for each element, which is found by measuring a standard of known concentration. Only one standard is needed for each element. The standard used may be either a single element standard or a type standard of the same type as the unknown, containing all of the elements.

The calibration constant contains those terms in the fundamental parameters equation which pertain to the x-ray flux intensity and instrumental factors. By doing a calibration on a type standard some factors which might otherwise cause systematic errors are calibrated out.

Filtered direct excitation works in EXACT if all of the exciting radiation lies above the absorption edges of the analyte elements. If this is the case, all of the exciting radiation may be approximated by a single effective energy for all of the analyte elements. Furthermore it does not matter what value is chosen as the effective energy, as long as it is within the range of the real primary spectrum. This can be easily shown without doing calculations. If a group of elements in a multielement sample is chosen and excited by different secondary targets, the absolute intensities of the peaks will change, but the peak ratios will remain nearly the same. Similarly when filtered direct radiation in the same energy range as the secondary targets is used on the same sample, the peak ratios will remain the same. Within reasonable limits any excitation condition which has all of its x-ray energies above the absorption edges of the elements of interest will excite these elements to give the same peak ratios. The only change is in the absolute intensity of the lines, and this is taken care of by the EXACT calibration constants.

XRF11: The XRF11 program has the ability to calculate the polychromatic spectrum or use monochromatic excitation for the exciting radiation. Alpha coefficients are calculated based on the theoretical intensities of lines, and these are used to speed the iteration process. More than one standard may be used for each element, and the program will adjust its calibration over matrix space to be correct at each standard.

When using XRF11, monochromatic excitation was specified when the secondary target data were used. When filtered direct excitation was used, the primary spectrum was calculated instead of making the monochromatic assumption.

USING FILTERS TO CREATE NARROW BAND POLYCHROMATIC RADIATION

As previously shown[4] "white" filters of various densities combined with varying kV setting on the x-ray tube can be used to create a narrow band polychromatic hump of radiation almost anywhere in the x-ray spectrum. The filter removes all x-rays below the cutoff energy, and the kV settings on the x-ray tube place a limit on the high-kV end. For this study white filters were chosen which gave cutoff energies at about the energy of the secondary target they were trying to mimic. Thus the cutoff energy was above the highest absorption edge in the group of analyte elements excited.

It does not seem to be important if there are absorption edges and lines of nonanalyte elements contained in the energy region of the bremsstrahlung hump. Even with secondary targets there is often scatter off the target which excites these high-energy lines with little effect on the results. The use of a type standard for calibration helps remove systematic errors from this source.

EXPERIMENTAL

Certain rules were established at the begining of the study to insure the comparability of the data. These rules were:

1. The same x-ray intensity data would be used by EXACT and XRF11.
2. The filters would be chosen to mimic the secondary targets. When using the data from filtered direct excitation, the source entries used in the EXACT program would be the same as the targets used for the secondary excitation. This avoids any differences in the results between secondary and direct excitation due to any differences in the absorption coefficients used by the program.
3. Only one standard from each set of samples would be used for calibration. This standard would not be included in the study of analytical accuracy done on the other standards.
4. The samples would not be moved on the tray between excitations. This would remove any random error contribution from removing and replacing the samples on the tray.
5. The excitation conditions would be selected to give 50% deadtime. All groups of samples would be counted at an optimum system count rate.

Two different matrices were selected for this study: NBS cements and bronze alloys. These represented two very different matrix types and both required multiple excitations. The cement standards were prepared as pressed pellets and were stored in a desiccator when not in use. The bronze alloys were the BNF (British National Foundry) C71 Gunmetals and the BNF C50 leaded bronzes. The surfaces of these samples were used as prepared by BNF.

The analyses were performed on a Kevex 0700 XRF subsystem which has the capability of both filtered direct and secondary excitation. The close coupling of the x-ray optics and power range of the x-ray tube in this system allow the system to be operated at the optimum count rates in both excitation modes.

Excitation conditions:

	BRONZE	
Elements	Secondary Target	Filtered Direct
Sn Sb	Gd, 60 kV, 1.5 mA	.76 mm Cu, 60 kV, .23 mA
Pb Zn Cu Ni Fe	Ag, 40 kV, .98 mA	.25 mm Cu, 40 kV, .40 mA*

*Also used for single direct excitation data.
Livetime: 100 seconds.

PORTLAND CEMENT

Elements	Secondary Target	Filtered Direct
Mg, Al, Si, P, S	none	no filter, 6 kV, .15 mA
K, Ca, Ti, Mn, Fe	Ge, 20 kV, 1.5 mA	.51 mm Al, 20 kV, .13 mA
Sr	Ag, 35 kV, 1.9 mA	.25 mm Cu, 35 kV, .80 mA

Livetime: 300 seconds.
All data from Kevex 0700 at 50% deadtime.

Unfiltered direct excitation was used for exciting the light elements (Mg, Al, Si, P, and S) in cement. The low fluorescence yields of these light elements and of low-energy secondary targets make direct excitation better for these elements. Successful calculations can be be made in EXACT by specifying Cl as the nominal target type. The Cl K lines are at the same energy as the Rh L lines from the x-ray tube. This approximation works because of the calibration procedure used. There is a rapid drop in fluorescence yield as Z drops in this group of elements, which overwhelms the errors caused by the differences between continuum excitation and monochromatic excitation. Because these errors are now small in comparison, the EXACT calibration procedure is able to correct for them over a larger area of matrix space than possible for higher Z elements.

TABLE 1
Count Rate Comparison
Secondary and Filtered Direct Excitation

Portland Cement NBS 633

Element	Concentration % oxide	Secondary Target (Ge)	Filtered Direct (20 kV, .508mm Al)
K	.17	3.803 cps	1.317 cps
Ca	64.50	2421	975.1
Ti	.24	7.843	2.883
Mn	.04	6.483	2.203
Fe	4.20	913.8	419.7
		(Ag)	(35 kV, .254mm Cu)
Sr	.31	305.7 cps	113.0 cps

Bronzes

Element	Concentration %	Secondary Target (Gd)	Filtered Direct (60 kV, .762mm Cu)
Sn	9.16	1644. cps	2560. cps
Sb	.51	103.4	163.2
		(Ag)	(40 kV, .254mm Cu)
Fe	.22	8.813 cps	5.200 cps
Ni	1.91	89.10	53.99
Zn	1.07	54.68	38.63
Pb	10.80	260.2	151.5

All at 50% deadtime.

RESULTS

For both sets of matrices, representative count rates from the filtered direct excitation and the secondary targets are compared in Table 1. Except for the excitation of Sn and Sb in bronze, the count rates in the peaks at 50% deadtime are greatest with the secondary targets. For the case of Sn and Sb in bronze, the exciting hump of radiation reaches down lower in energy (cutoff 35 keV) than the Gd secondary target line (43 keV). Thus the filtered direct excitation provided better excitation to the Sn and Sb lines than the Gd target in this case. For the other lines in cements and bronzes the secondary targets provided better excitation than the filters.

TABLE 2
Data Comparison

% Sn in Bronze

Sample	C50.02	C50.03	C50.04	C71.31	C71.32	C71.33	C71.34
List	10.5	8.53	11.1	4.60	6.10	5.1	8.2
Secondary Targets							
EXACT	10.59	8.60	10.74	4.20	6.15	4.85	7.70
XRF11	10.49	8.51	10.63	4.11	6.00	4.74	7.43
Filtered Direct - 2 Excitations							
EXACT	10.80	8.72	10.91	4.50	6.12	5.08	8.16
XRF11	10.73	8.62	10.83	4.39	5.95	4.94	7.87
Filtered Direct - 1 Excitation							
EXACT	10.76	9.03	10.86	4.98	6.96	5.82	9.27
XRF11	10.62	8.58	10.67	4.33	5.95	4.93	7.86

%CaO in Portland Cement

Sample	633	634	635	636	637
List	64.50	62.58	59.83	63.54	66.04
Secondary Target Excitation					
EXACT	65.19	(STD)	60.57	63.89	65.02
XRF11	65.54		60.97	64.22	65.35
Filtered Direct Excitation					
EXACT	65.13		60.04	63.06	64.58
XRF11	65.54		60.79	63.77	65.46

Table 2 shows the results for the calculation of Sn in bronze and Ca in cement by both methods. In order to more conveniently compare results for the major elements in each sample, an RMS (root mean square) deviation was calculated from the differences between the list and calculated values for each element.

Table 3 lists these RMS values for the major elements. These RMS values show that, overall, EXACT gave slightly better results than XRF11 from the same data for secondary target and filtered direct excitation.

TABLE 3
RMS Deviations - Bronze

Element	Sn	Pb	Zn	Ni
Secondary Target Excitation				
EXACT	.297	.128	.228	.130
XRF11	.413	.247	.403	.110
Filtered Direct - 2 Excitations				
EXACT	.158	.127	.145	.063
XRF11	.217	.258	.129	.057
Filtered Direct - 1 Excitation				
EXACT	.646	.472	.145	.078
XRF11	.251	.448	.195	.071

RMS Deviations - Portland Cement

Compound	CaO	SiO_2	Al_2O_3	Fe_2O_3	SO_3
Secondary Target Excitation					
EXACT	.717	.88	.30	.120	.32
XRF11	.911	.94	.37	.116	.17
Filtered Direct Excitation					
EXACT	.837	.93	.27	.116	.20
XRF11	.773	.82	.31	.125	.20

Also seen in the RMS error data was that filtered direct excitation (two excitations) gave slightly better results than the secondary target excitation for the bronzes. For the cements the two excitation methods gave about equal results. There is no easy explanation of why the filtered direct excitation gave better results in the bronzes in this set of data. Previous analyses on the bronze standards had given much better RMS results for Sn with secondary targets.

The monochromatic excitation assumption was broken to test EXACT and XRF11 by using a single filter excitation for bronze which placed the Sn peak on the excitation hump. The data using a single excitation on the bronzes (40 kV, 40 mA, .25 mm Cu filter) was processed by EXACT (Ba source specified) and XRF11 (polychromatic). The Sn and Sb lines lie in the middle of the excitation hump in this condition. The accuracy of the Sn and Sb results fell in EXACT and gave for Sn an RMS deviation of .646. In XRF11 the accuracy was still acceptable. This demonstrates the breakdown of the EXACT program assumptions when the primary radiation hump crosses the absorption edge of a major analyte element.

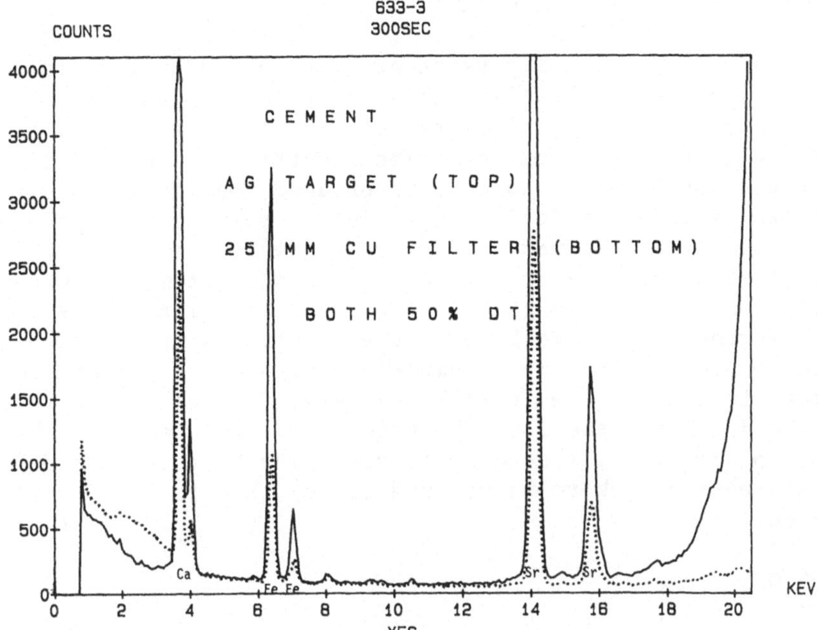

Figure I. Portland cement spectra showing the difference in peak intensity for Sr, Fe, and Ca between Ag secondary target or filtered direct excitation. Both are at 50% deadtime.

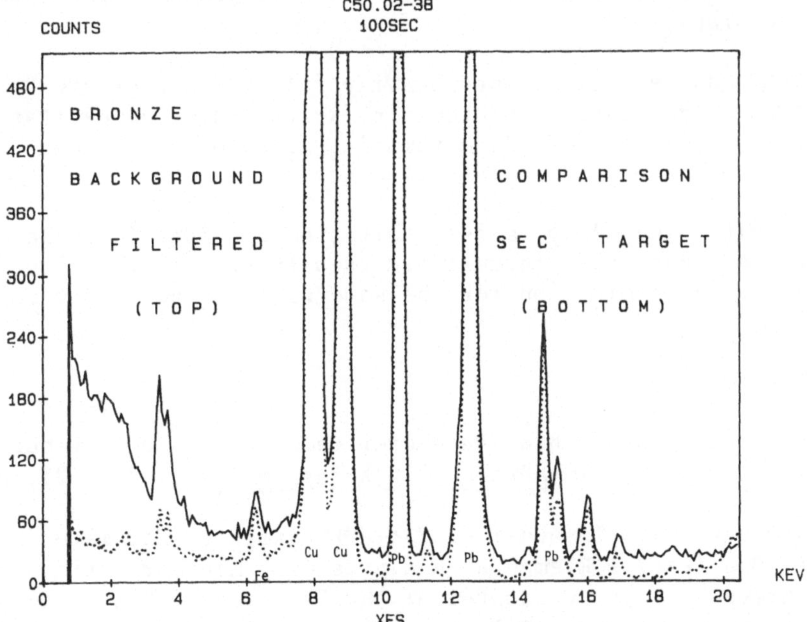

Figure II. Bronze spectra showing difference in background using Ag secondary target and equivalent filter with equal peak sizes. Deadtimes are not equal.

SECONDARY VERSUS DIRECT EXCITATION

The advantage of using secondary excitation over filtered direct excitation is shown in the data in Table 1 and the spectra in the figures. The secondary target is able to place all of its excitation flux at the most efficient location in the spectrum. For a given total system count rate or deadtime, the secondary target excitation provides the most counts per second in the peaks of interest. There is more scattered radiation in the spectrum from filtered direct excitation than from the optimum secondary targets when the analyte peaks are excited at a fixed rate. This gives a further advantage because the background under the peaks is not due to scatter, but to incomplete charge collection in the detector (tailing). The fewer high-energy scatter x-rays that reach the detector, the less the background will be under the peaks. To summarize, secondary targets give better peak count rates and better peak-to-background ratios than filtered direct radiation.

CONCLUSIONS

1. In the materials tested, EXACT gives better accuracy given the same data than XRF11 when monochromatic or narrow band polychromatic excitation is used and when only a single type standard is used for calibration. When polychromatic excitation crosses the absorption edge of an analyte element, XRF11 produces the expected better accuracy.

2. EXACT may be applied when heavily filtered direct excitation is used to approximate monochromatic excitation. The excitation must be a narrow band of polychromatic radiation above the absorption edges of the analyte elements.

3. Optimized secondary target excitation provides lower backgrounds and higher peak intensities (sensitivities) than does filtered direct excitation for the same total system count rate or deadtime.

REFERENCES

1. John Criss, "Fundamental-Parameter Calculations on a Laboratory Microcomputer," in Advances in X-ray Analysis, Vol. 23, p 93 (1980).
2. J.W. Otvos, G. Wyld and T.C. Yao, "Fundamental Parameter Method for Qualitative Elemental Analysis with Monochromatic X-ray Sources." Paper presented at the 25th annual Denver X-ray Conference, Denver, Colorado, 1976.
3. J.W. Criss and L.S. Birks, Anal. Chem., 40:1080 (1968).
4. R.A. Vane and W.D. Stewart, "The Effective Use of Filters with Direct Excitation in EDXRF," in Advances in X-ray Analysis, Vol. 23, p 231 (1980).

A GENERALIZED MATRIX CORRECTION APPROACH FOR ENERGY-DISPERSIVE X-RAY FLUORESCENCE ANALYSIS OF PAINT USING FUNDAMENTAL PARAMETERS AND SCATTERED SILVER Kα PEAKS

Leif Højslet Christensen

Isotope Division, Risø National Laboratory
DK-4000 Roskilde, Denmark

and

Iver Drabæk

Danish Isotope Centre, Skelbækgade 2
DK-1717 Copenhagen V, Denmark

ABSTRACT

An energy-dispersive x-ray fluorescence method has been developed for the direct determination of major and minor elements in infinitely thick samples of paint. Matrix absorption and enhancement corrections are iteratively calculated from a knowledge of tabulated fundamental parameters and the unknown weight fractions. An estimate of the significant light element fraction of the bulk sample required for the calculation of matrix attenuation is obtained using the scatter peaks of the silver secondary target. Relative elemental calibration constants and calibration factors for the coherent and incoherent peaks are determined experimentally using either thin-film standards or standards of known total composition. For routine analysis only one absolute standard is required. The method has been applied to different types of paint with a relative standard deviation better than 5% provided the counting statistics are not the limiting factor. The accuracy has been tested by comparing own results with those obtained either from the formulation or from instrumental neutron activation analysis.

INTRODUCTION

Potential applications of x-ray methods to coating analysis have recently been reviewed by Kamarchik, Jr. and Cunningham[1]. However, as pointed out by the authors, only a limited number of applications are described in the literature. This may partly be due to the problem of matrix effects complicating the quantitative conversion of peak intensities to elemental weight percentages. Unless the sample by some means can be presented to the spectrometer as an infinitely thin film, matrix absorption as well as enhancement effects are significant. For a paint sample containing 10% Ti and 3% Ca the enhancement of the Ca Kα line due to Ti is 30%.

A number of papers have dealt with different experimental procedures to compensate for matrix effects in paint samples. Chung et al.[2] have described a method for preparing a thin film containing an internal standard, thereby alleviating matrix effects. However, linear calibration curves have to be established for each element of interest, although the same curves can be used for various types of paint.

Using a dedicated nondispersive analyzer for the direct determination of lead in trade sales paints, Cunningham[3] established linear calibration curves by means of standard addition. In this case a calibration curve is needed for each element in each type of paint.

More recently, Kuntz and Towns[4] outlined a quantification method for lead using what they called the fundamental parameters "pure element behind a thin sample" technique. This method, however, requires two measurements and does not easily lend itself to automation.

The aim of the present work was to develop a universal, automated method for the direct determination of major and minor elements in different types of paint. This was accomplished using energy-dispersive x-ray fluorescence (EDXRF) and fundamental parameters.

THEORY

The heavy elements, which in this context are considered to be those with atomic numbers above 19, are quantified by means of their characteristic fluorescence peaks after correction for sample absorption and possible enhancement. The latter two corrections are iteratively calculated from a knowledge of tabulated fundamental physical parameters, e.g. elemental mass attenuation coefficients, photoelectric cross sections, fluorescence yields, etc.[5-8], and of all the unknown weight fractions. The computation of the absorption correction obviously requires an estimate of the significant light element fraction usually encountered in samples of paint. This is accomplished by using the method proposed by Nielson[9]. The scatter peaks of the silver secondary target after correcting for the absorption and the scatter contribution from the heavy elements are used for estimating the weight fractions of two light elements representative of the light element fraction of the bulk sample. This calculation is part of the iteration process.

The basic equations relating the net peak intensities and the elemental weight fractions for the case of infinitely thick samples and monochromatic excitation have been discussed in detail by several workers[10-11]. Nielson[9] and Van Dyck and Van Grieken[12] have previously outlined the equations for coherent and incoherent scatter as well as the assumptions inherent in these. We therefore confine ourselves to summarizing the equations.

For samples of infinite thickness the relationship between the fluorescence intensity, I_i, and the elemental weight fraction, W_i, is given by equation 1 in Table I. $K_{A,s}$ and $K_{R,i}$ are the absolute and relative calibration constants, respectively. $K_{A,s}$, which for a fixed irradiation geometry depends only on the kV/mA setting of the high-voltage generator, is determined experimentally using only one standard or flux monitor for each batch of samples. The experimentally determined relative elemental calibration constants, $K_{R,i}$, accounts for both the excitation and detection probability.

Table I. Basic Equations for the Fundamental Parameter Approach

Fluorescence equation:

$$I_i = K_{A,s} \cdot K_{R,i} \cdot A_i \cdot (1 + H_i^O) \cdot W_i \quad (1)$$

Scatter equations:

$$I_C = G_C \cdot A_C \cdot (\Sigma_i W_i \cdot \sigma_{iC}) \quad (2)$$

$$I_I = G_I \cdot A_I \cdot (\Sigma_i W_i \cdot \sigma_{iI}) \quad (3)$$

Simultaneous equations:

$$W_a \cdot \sigma_{aC} + W_b \cdot \sigma_{bC} = I_C \cdot (G_C \cdot A_C)^{-1} - \Sigma_j W_j \cdot \sigma_{jC} \quad (4)$$

$$W_a \cdot \sigma_{aI} + W_b \cdot \sigma_{bI} = I_I \cdot (G_I \cdot A_I)^{-1} - \Sigma_j W_j \cdot \sigma_{jI} \quad (5)$$

The terms A_i and H_i^o in equation 1 represent the matrix absorption and enhancement corrections, respectively.

The observed coherent and incoherent scattered Ag Kα secondary target peak intensities, I_C and I_I respectively, corrected for absorption in the sample matrix, A_C and A_I, are given by equations 2 and 3 in Table I. G_C and G_I are geometry-dependent proportionality constants, which can be determined by measuring samples of known total composition. The cross sections, σ_{iC} and σ_{iI}, are those of coherent and incoherent scattering, respectively, tabulated by McMaster et al.[5]

The weight fractions of two representative light elements, W_a and W_b, can now be estimated by solving the simultaneous equations 4 and 5. In these equations Σ_j denotes a summation over all the elements determined by means of their fluorescence peaks.

As mentioned by Nielson, several light element pairs satisfy equations 4 and 5. For that reason the pair is chosen whose incoherent-to-coherent scatter cross section ratio lies immediately on either side of the ratio of the observed scatter attributable to the light elements.

EXPERIMENTAL

Instrumentation

All measurements were carried out using a fully automatic Kevex 0700 energy-dispersive spectrometer connected to a Nuclear Data ND680 multichannel analyzer/microcomputer system. The spectrometer makes use of a low-powered rhodium anode x-ray tube and six switch-selectable secondary targets in a close geometry.

The x-ray tube was operated at 50 kV and 0.1 mA. Only one of the targets, i.e. Ag, was used for the analyses. The energy resolution of the Si(Li) detector is 155 eV at 5.90 keV.

Spectrum analysis and fundamental parameter approach calculations

The elemental peak intensities were determined using a non-linear least-squares fitting routine supplied by Nuclear Data. The intensities of the coherent and incoherent peaks were determined with the best precision using fixed energy windows.

Quantitative calculations based on the fundamental parameter approach outlined above were performed using the program MATRIX initially developed at the University of Aarhus. The program is coded in Algol and runs on a Burroughs B7800 computer. However, all the calculations might just as well have been performed by the ND680 microcomputer, a 16-bit LSI-11/2 with 56-kilobyte programmable memory.

Calibration standards and sample preparation

The calibration of the spectrometer was performed using either thin-film standards purchased from Micromatter, Eastsound, W.A., U.S.A. or analytical grade compounds of known total composition presented to the spectrometer as pressed pellets. A Zn thin-film standard was used as the flux-monitor.

Samples of paint are thoroughly stirred and poured into disposable spectrocups having a thin Mylar window. Batches of fifteen samples are then automatically processed by the spectrometer. In order to obtain adequate counting statistics for as many elements as possible a counting time of 2000 s was chosen.

RESULTS AND DISCUSSION

The method has been evaluated by measuring two different types of paint, i.e. a $ZnCrO_4$ primer and an emulsion paint. Concerning the latter type both a laboratory made and an ordinary paint were analyzed.

Table II shows the mean values and standard deviations obtained from the analysis of the $ZnCrO_4$ primer. Ten samples were prepared from the same batch of paint and subsequently measured. Besides Zn and Cr, Ti, Fe, and Pb were determined. The precision expressed as the standard deviation on a single determination was found to be as low as 0.7% for Zn, while it amounted to 6.3% for Pb mainly reflecting a higher variation in the determination of the Pb Lα intensity.

Comparative analysis was performed using instrumental neutron activation analysis (INAA). Cr, Zn, and Ti were determined using single element standards. Due to lack of time determination of Fe was omitted. Mean values together with standard deviations on single determinations are shown in Table II.

Using an approximative Student t-test no significant difference was found between the mean values obtained for Cr and Zn by the two analytical techniques. For Ti the difference is probably significant thus requiring further investigations.

Table II: Analysis Results for the $ZnCrO_4$ Primer

	EDXRF N = 10	INAA N = 4
Zn,%	16.59 ± 0.11	16.21 ± 0.53
Cr,%	4.17 ± 0.04	4.27 ± 0.16
Ti,%	4.33 ± 0.05	4.42 ± 0.05
Fe,%	0.69 ± 0.01	-
Pb,ppm	126 ± 8	-

The results for the emulsion paint is given in Table III. The formulation of this paint was known to be 2.00% for Ca and 11.99% for Ti and as seen in the table our results are in excellent agreement with this. The precision was found to be 1.1% for Ti, 4.2% for Zn, and 8.0% for Ca. Again, the observed variation mainly reflected the variation in the intensity determination.

Three other laboratories have analyzed this paint using different methods, and their results are included in Table III. Comparative IR spectroscopy of mixings simulating the paint sample was used by laboratories 1 and 2 while laboratory 3 used an ASTM method. Except for the high Ca result of laboratory 2, all the results agree to within 11% of the formulation.

Table III: Analysis Results for Emulsion Paint I

	Form.	Lab. 1 1)	Lab. 2 1)	Lab. 3 2)	EDXRF N = 8
Ca,%	2.00	2.00	2.84	-	1.99 ± 0.16
Ti,%	11.99	12.0	11.0	10.73	11.91 ± 0.13
Zn,%	-	-	-	-	0.120 ± 0.005

1) IR spectroscopy of reference mixings
2) ASTM D 1394 - 76

Ca, Ti, and Fe were determined in the laboratory-made emulsion paint and the results for these elements are given in Table IV. Once again the precision was found to be limited by that of the intensity determination. The standard deviation of single determinations ranged from 1.3% for Ti to 6.7% for Fe. However, the accuracy in this case was far from that obtained for the other paints.

Table IV: Analysis Results for the laboratory-made Emulsion Paint II

	Formulation	EDXRF N = 10
Ca,%	3.26	2.71 ± 0.12
Ti,%	12.02	13.44 ± 0.18
Fe,%	0.12	0.15 ± 0.01

For Ti and Fe our results are significantly greater than those obtained from the formulation, whereas the opposite is the case for Ca. This laboratory-made paint, however, suffered from a visual inhomogeneity which might explain the observed discrepancies.

To summarize, the proposed method is suitable for direct determination of several important constituents of paints using only one calibration standard. To a great extent the precision of the method is limited by that of the intensity determination. The accuracy is typically within 2 to 5%. Furthermore, the method is applicable to smaller equipment using radioisotope excitation and a minicomputer as computation facility.

ACKNOWLEDGEMENTS

We are grateful to the Danish National Science Research Council and the Danish Technical Science Research Council for covering part of our travel expenses for the 31st Annual Denver Conference. We wish to thank K. Heydorn for his comments on the preparation of the paper, G. Christensen and K. Eng who supplied us with real samples, and N. Pind who wrote a major part of the program MATRIX.

REFERENCES

1. P. Kamarchik, Jr. and G.P. Cunningham, "Applications of X-Ray Techniques to Coatings Analysis", Progress in Organic Coatings 8 (1980) 81-107.
2. F.H. Chung, A.J. Lorentz and R.W. Scott, "A Versatile Thin Film Method for Quantitative X-Ray Emission Analysis", X-Ray Spectrom. 3 (1974) 172-175.
3. G.P. Cunningham, "Applications of X-Ray Techniques to Coatings of Lead in Paint using a Dedicated X-Ray Fluor-

escence Analyzer", Proceedings of Fifth International Conference in Organic Coatings Science and Technology, 1979, 261-271, Adv. Org. Coat. Sci. Technol. Ser. Bd. 5.

4. G.S. Kuntz and R.L.R. Towns, "Determination of Lead in Paint by Energy Dispersive X-Ray Fluorescence Spectrometry", J. Coatings Technol., 54 (1982) 63-69.
5. W.H. McMaster, N. Kerr del Grande, J.H. Mallett, and J.H. Hubbell, "Compilation of X-Ray Cross Sections", UCRL 50174, (Secs. 1 and 2) (Rev. 1); Lawrence Radiation Laboratory, University of California, Livermore (1969).
6. W. Bambynek, B. Crasemann, R.W. Fink, H.U. Freund, H. Mark, C.O. Swift, R.E. Prince, and P. Venugopala Rao, "X-Ray Fluorescence Yields, Auger, and Coster-Kronig Transition Probabilities", Rev. Mod. Phys. 44 (1972) 716-816.
7. Md. R. Khan and M. Karimi, "Kβ/Kα Ratios in Energy-dispersive X-Ray Emission Analysis", X-Ray Spectrom. 9 (1980) 32-35.
8. M.O. Krause, C.W. Nestor, Jr., C.J. Sparks, Jr., and E. Ricci, "X-Ray Fluorescence Cross Sections for K and L X-Rays of the Elements", ORNL-5399, Oak Ridge National Laboratory, 1978.
9. K.K. Nielson, "Matrix Corrections for Energy Dispersive X-ray Fluorescence Analysis of Environmental Samples with Coherent/Incoherent Scattered X-rays", Anal. Chem. 49 (1977) 641-648.
10. C.J. Sparks, Jr., "Quantitative X-Ray Fluorescence Analysis using Fundamental Parameters", Adv. X-Ray Anal. 19 (1976) 19-52.
11. L.H. Christensen and N. Pind, "The Application of Energy-dispersive X-Ray Fluorescence and the Fundamental Parameter Approach to the Analysis of Ni-Fe-Cr Alloys", X-Ray Spectrom. 10 (1981) 156-161.
12. P.M. Van Dyck and R.E. Van Grieken, "Absorption Correction via Scattered Radiation in Energy-Dispersive X-Ray Fluorescence Analysis for Samples of Variable Composition and Thickness", Anal. Chem. 52 (1980) 1859-1864.

MULTIELEMENT ANALYSIS OF UNWEIGHED BIOLOGICAL AND GEOLOGICAL SAMPLES USING BACKSCATTER AND FUNDAMENTAL PARAMETERS*

K.K. Nielson

Rogers and Associates Engineering Corp.
Salt Lake City, Utah 84107

R.W. Sanders

Pacific Northwest Laboratory
Richland, Washington 99352

INTRODUCTION

A new approach to fundamental parameter calculations has been devised which makes use of incoherent and coherent backscatter intensities from the excitation radiation. The backscatter results from all sample constituents, and thus provides information on total sample mass as well as on bulk sample composition. By appropriately using the scatter intensities in fundamental parameter matrix calculations, accurate analyses of "unknown" samples can be obtained without prior knowledge of the sample matrix. The backscatter with fundamental parameter (BFP) method is especially advantageous for biological, geological and environmental samples because of their bulk quantities of carbon, oxygen, and other light elements which cannot be explicitly determined for traditional fundamental parameter calculations.

This paper describes extended capabilities of the BFP method, which was originally applied only to well-defined samples over a more restricted range of compositions.[1] The BFP calculation method is shown to accurately compute both the effective thickness and the bulk composition of individual samples. It thereby provides accurate multielement analyses of biological and geological materials without similar standards. Analyses of loose powders as well as pelletized samples is also demonstrated.

* This work supported in part by U.S. DOE Contract DE-AC06-76RLO 1830.

THEORY

The calculation of element concentrations from XRF peak intensities is generally complicated by sample self-absorption and enhancement effects. These matrix effects have been handled both empirically and with fundamental parameter calculations. Empirical techniques such as matched matrix standards, detailed influence coefficient calibrations, and special sample preparation procedures are most common, and also most time-consuming. Fundamental parameter corrections for matrix effects offer simplified calibrations and avoid errors from poorly-chosen standards. However, the mathematical matrix corrections require knowledge of all major element concentrations, and have usually been limited to metal alloys or other samples where all major constituents are observed in the XRF spectrum or are otherwise of known concentration. Since carbon, oxygen, and other light elements are major variable components of biological and geological materials and are not observed in ordinary XRF analysis, they restrict the application of fundamental parameter methods for a wide segment of analytical applications. This restriction is overcome in the present BFP method by estimation of the light-element components of the sample from the incoherent and coherent backscatter intensities.

The light-element components of the sample matrix are represented in the BFP calculations by masses of two "representative" light elements. These elements and their masses are determined iteratively with the matrix corrections in the following manner. The incoherent and coherent scatter from the measured elements are first estimated by separately summing the products of the measured element masses (g/cm^2) and their scatter cross sections. These sums of the measured-element scatter are then subtracted from the observed incoherent and coherent scatter intensities, and the differences are attributed to the uncharacterized light element part of the sample matrix. Since incoherent and coherent scatter cross sections vary differently with atomic number, these two light-element scatter quantities describe the approximate atomic number as well as the total mass of the light elements.

The ratio of the incoherent/coherent scatter from the light elements is used to select two light elements with appropriate atomic numbers. Their masses (g/cm^2) are then calculated from two simultaneous equations which satisfy the light-element quantities of incoherent and coherent scatter.[1,2] The two light element masses are next included with the observed element masses (g/cm^2) to estimate the total sample mass as well as the appropriate matrix corrections. Corrected element masses are then used iteratively to re-determine the light elements and to revise the matrix corrections until a self-consistent set of corrections and element masses is found for a given XRF spectrum.

The BFP method is designed to determine and accommodate any sample thickness (g/cm^2) from infinitely thin (no absorption of low-energy x-rays) to infinitely thick (maximum absorption of backscattered x-rays). Since optimum sensitivities are obtained with intermediate thicknesses, the method is usually used in the intermediate range to determine sample thickness as well as bulk composition. Calculated sample thicknesses are generally as accurate as gravimetric measurements, and are especially helpful for irregular or powdered samples whose average effective thickness is difficult to measure. By defining the thickness and bulk composition of individual samples, the BFP method avoids much of the time-consuming sample preparation and calibration of traditional XRF analysis.

EXPERIMENTAL

The BFP method was evaluated by analyses of a variety of biological and geological standard reference materials. These included NBS coal (SRM 1632)[3], NBS orchard leaves (SRM 1571)[4], IAEA soil (SOIL5)[5], USGS rock (AGV-1)[6], and a PNL oil shale reference material (COS-1)[7]. The samples were each analyzed as self-supporting pressed pellets of five different thicknesses, generally covering the range of 25-250 mg/cm^2 to examine the effects of varying sample thickness. The sample pellets were each weighed to allow comparison of calculated sample thicknesses with actual sample thicknesses. The COS-1 shale was also analyzed as loose powder supported by thin plastic films to demonstrate the ability to analyze powdered solids. Five samples of varying powder thickness were analyzed.

XRF excitation utilized zirconium $K_{\alpha,\beta}$ x-rays from a secondary source, and detection utilized an energy-dispersive Si(Li) detector. The instrument allowed only the backscatter caused by the sample to reach the detector. Typical analysis livetimes were 50 minutes. A multielement, thin-film sensitivity curve[2] was used as the only elemental calibration for the analyses. The backscatter peaks were both calibrated with sensitivity factors (intensity per unit mass per unit scatter cross section) as described elsewhere.[2] Data analysis utilized the SAP3 computer code[2] with a modification to use K_β backscatter peaks instead of K_α peaks. The SAP3 code performed peak analysis as well as the BFP calculations, and printed measured element concentrations, calculated light element concentrations, peak analysis parameters and matrix correction parameters. Typical execution times were 30 seconds for determining 24 elements in a 1024-channel spectrum using a 64K PDP-11/34 computer operating under RSX-11M.

RESULTS AND DISCUSSION

The multielement analyses of the coal, orchard leaves, soil and rock are summarized in Table I in terms of mean element concentrations from the five analyses, reference concentrations reported previously, standard deviations among the five analyses, and standard deviation predicted for individual analyses from the peak counting statistics. Two important conclusions can be drawn from the data in Table I. First, the mean concentrations are in generally good agreement (1-2 std. dev.) with the reference values for both biological and geological materials, indicating an excellent accommodation of very different sample matrices. Second, the standard deviations among analyses covering a ten-fold variation in sample thickness are comparable with the precisions predicted by peak counting statistics. This indicates that the variations due to different sample thicknesses were small compared to the random errors normally associated with XRF analysis.

A more detailed examination of the pellet analyses was made to quantitatively evaluate possible systematic errors from calculated sample thicknesses. Calculated thicknesses were individually compared with actual thicknesses, and found to have a standard relative error of ten percent. The average relative bias in this comparison was found to be less than 0.01 percent,

TABLE I
ANALYSES OF PELLETIZED COAL, ORCHARD LEAF, SOIL5 AND AGV-1 STANDARDS WITH VARYING SAMPLE THICKNESSES

	Si	S	Cl	K	Ca	Ti	V	Cr	Mn	Fe	Ni	Cu	Zn	Ga	As	Se	Br	Rb
COAL Mean[a]	2.9[b]	1.23[b]	600	0.28[b]	0.43[b]	0.10[b]	25	25	40	0.83[b]	16	18	34	5.9	5.7	3.0	18.0	18
Ref.Conc.[c]	3.2		890	.28	.43	.11	35	20	40	.87	15	18	37		5.9	2.9	19.3	21
Std.Dev.[d]	.4	.09	100	.03	.02	.01	4	5	4	.04	2	1	2	.3	.6	.4	.7	1
S_{stat}[e]	.3	.09	200	.02	.02	.01	10	5	4	.04	1	1	2	.5	.5	.3	1.0	1
O.L. Mean	<0.7[b]	0.18[b]	600	1.43[b]	1.88[b]	47	<8	5	95	299	3.2	15	26	<.6	16	<0.4	9.6	11
Ref.Conc.		.23	700	1.47	2.09			2	91	300	1.3	12	25		14	.1	10.	12
Std.Dev.		.02	70	.05	.05	19		2	5	21	.2	1	1		1		.4	1
S_{stat}		.04	160	.07	.10	8		3	6	15	.5	1	1		1		.6	1
SOIL5 Mean	26.5[b]	<.19[b]	<800	2.03[b]	2.54[b]	0.51[b]	168	41	960	4.5[b]	15	87	412	20	121	1	6.4	135
Ref.Conc.	27.0			1.95	2.5	.52	130	38	900	4.8		80	370		110	2	6.0	120
Std.Dev.	.4			.06	.06	.02	37	10	30	.2	7	3	15	2	4		.5	4
S_{stat}	1.5			.11	.13	.03	30	17	50	.2	3	5	20	2	6		.8	7
AGV-1 Mean	26.9[b]	<.19[b]	<800	2.30[b]	3.45[b]	0.63[b]	107	<35	766	4.9[b]	19	60	86	20	4	<1	<1.0	69
Ref.Conc.	27.6		110	2.40	3.50	.62	125	12	763	4.7	19	60	84	20	1		0.5	67
Std.Dev.	.8			.13	.14	.02	41		38	.2	3	4	2	1	1			3
S_{stat}	1.6			.12	.18	.03	40		40	.2	3	4	5	2	1			4

[a]Mean of five analyses with varying sample thickness. [b]Weight percent concentration units; all others are in parts-per-million. [c]Reference concentrations in Ref. 3 for Coal; Ref. 4 for Orchard Leaves; Ref. 5 for Soil5; and Ref. 6 for AGV-1. [d]Standard deviation of the five analyses with varying sample thickness. [e]Standard deviation of a single determination from peak counting stats.

indicating no significant systematic error. Errors in element concentrations were found to be six times lower than errors in calculated thick-ness, due to the normalizing effect of dividing individual element masses (g/cm^2) by the total calculated sample mass (g/cm^2) to obtain concentration units (ppm or percent). Although the concentration error is reduced by normalizing to sample mass, it should be noted that this normalization does not make the BFP method a relative method (i.e., independent of live-time or excitation intensity). Incorrect excitation intensities or livetime estimates cause erroneous estimates of total sample mass and thereby bias the resulting concentrations.

Examination of the representative light elements chosen for the five reference materials indicates reproducible selections despite varying sample thicknesses. The respective light elements and average total light element concentrations calculated for coal, orchard leaves, SOIL5, AGV-1 and COS-1 were carbon and oxygen (92%); carbon and oxygen (95%); nitrogen and oxygen (57%); nitrogen and oxygen (55%) and nitrogen and oxygen (68%). These concentrations represent all elements not determined in the XRF analysis.

The analyses of the pelletized COS-1 samples are reported separately in Table II for comparison with the five replicate analyses in powder form. The separate mean concentrations from powder and pelletized samples are reported, and are found to compare well with reference concentrations. The absolute error between the XRF concentrations and the reference concentrations is listed, and is generally comparable to or smaller than the uncertainty in individual determinations estimated from peak counting statistics (S_{stat}).

TABLE II
ANALYSES OF OIL SHALE STANDARD COS-1
AS POWDERS AND IN PELLETS OF VARYING THICKNESS

	Si[a]	S[a]	K[a]	Ca[a]	Ti[a]	V	Cr	Mn	Fe[a]	Ni	Cu	Zn	Ga	As	Se	Rb	Sr	Pb
Pellet Mean[b]	14.4	0.60	1.55	10.3	0.18	85	37	328	1.96	33	40	69	9.4	43	2.4	72	660	24
Powder Mean[c]	14.9	.48	1.53	9.9	.15	72	36	318	2.08	30	38	64	9.2	40	2.0	67	616	22
Ref. Conc.[d]	14.4	.65	1.62	9.9	.17	87	34	313	2.01	25	38	67	8.9	44	2.0	72	679	22
Error[e]	+.2	-.11	-.08	+.2	.00	-8	+2	+10	+.01	+7	+1	-1	+.4	-3	+.2	-3	-41	+1
S_{stat}[f]	1.1	.09	.08	.5	.01	25	14	20	.11	4	3	4	1.1	2	.5	3	33	2
S_{treat}[g]	.7	.19	.02	.7	.04	20	2	15	.19	4	3	7	.3	5	.8	7	70	2
S_{rep}[h]	.8	.04	.04	.2	.01	23	8	14	.08	2	2	3	.6	1	.2	3	16	1

[a]Weight percent concentration units; all other concentrations are in parts-per-million. [b]Means from five analyses with varying pellet thickness. [c]Means from five analyses of powder samples of varying thickness. [d]Reference concentrations recommended in Ref.7. [e]Average of pellet and powder means minus reference concentration. [f]Standard deviation of a single determination estimated from peak counting statistics. [g]Standard deviation between sample treatments (pellet & powder) from analysis of variance, one degree of freedom. [h]Standard deviation among replicate analyses with varying thickness, from analysis of variance, eight degrees of freedom.

The significance of differences between the pellet and powder means was evaluated with one-way analyses of variance for each element in Table II. The analyses allowed partitioning the standard deviation among the two sample treatments (S_{treat}) from the standard deviation among replicate analyses (S_{rep}). The analyses indicated that there are significant differences between the powder and pellet means for S, Ca, Ti, Fe, Zn, As, Se, Rb and Sr at the $p<0.05$ significance level, but not for the other nine elements, which had equally powerful statistical tests. This suggests that the simple difference in physical sample form may not be important. Instead, different grain compositions, grain settling, or grain size effects could have contributed to the differences. In either case, the biases are small, particularly compared with the peak uncertainties for each element.

In summary, BFP calculations allow direct quantitative analysis of unknown, unweighed samples by calculation of sample bulk constituents and thickness from the backscatter peaks. Analysis of accuracy and precision indicates that bias is minimal, and that the major source of uncertainty is in the random errors from peak counting statistics. The method is ideal for analyzing unknown samples because it does not require similar standards. Instead, it utilizes the same general-purpose thin-film calibrations used for thin air-filter samples, and on the other extreme, the same calibrations can even accommodate thick metal alloy samples.[8] Besides reducing preparation and calibration costs, the BFP method avoids the potential contamination and dilution often associated with sample preparation because most materials can be directly analyzed as loose powders, solid lumps or pellets.

REFERENCES

1. K.K. Nielson, Matrix Corrections for Energy Dispersive X-ray Fluorescence Analysis of Environmental Samples with Coherent/Incoherent Scattered X-rays, Anal. Chem., 49:641-648 (1977).
2. K.K. Nielson and R.W. Sanders, The SAP3 Computer Program for Quantitative Multielement Analysis by Energy Dispersive X-ray Fluorescence, U.S. DOE Report PNL-4173 (1982).
3. J.M. Ondov et al, Anal. Chem., 47:1102-1109 (1975).
4. NBS Certificate of Analysis, SRM-1571 (1971).
5. R. Dybczy-Nski et al, Report on the Intercomparison Run Soil-5 for the Determination of Trace Elements in Soil, IAEA-RL-46 (1978).
6. F.J. Flanagan, Geochim. Cosmochim. Acta, 37:1189-1200 (1973).
7. C.L. Wilkerson et al, Interlaboratory Analysis of Major and Trace Elements in a Colorado Oil Shale Reference Material, Pacific Northwest Laboratory, PNL-SA-9236, June 1981.
8. K.K. Nielson, R.W. Sanders and J.C. Evans, Anal. Chem. 54 (1982), in press.

A CORRECTION METHOD FOR ABSORPTION IN THE ANALYSIS OF AEROSOLS BY EDX SPECTROMETRY

A.S.M. de Jesus, D.J. van der Bank and E.S. Wesolinski

Isotopes and Radiation Department
Nuclear Development Corporation of South Africa (Pty) Ltd.
Pelindaba, South Africa

ABSTRACT

Low-energy x-rays can undergo considerable absorption in the sample material when EDX spectrometry is used in multi-element analysis of aerosol-loaded filters. An expression which corrects for the effects of absorption, based on counting aerosol filters and standards on both sides, is derived. The method is experimentally verified by using the absorption of the 60 keV gamma-ray of ^{241}Am in aluminium and also by comparison with PIXE results for environmental filters.

INTRODUCTION

Energy dispersive x-ray fluorescence (EDX) spectrometry has been extensively used in the multi-elemental analysis of air pollution aerosols.[1-4] The method is sufficiently precise and accurate but suffers from the inconvenience that low-energy characteristic radiation may be significantly absorbed in the filter especially when cellulose filters are used. Since elemental distributions across the filter are not known, absorption corrections become difficult to apply. Several methods for dealing with the problem have been proposed by various authors,[5-8] with various degrees of success.

The work reported here, based on double-side counting,[6-7] suggests a first-order approximation to correct for absorption in the filter. Some of the assumptions made are known not to be strictly correct, but this is considered to be warranted by the simplicity of the method as seen against the accuracy obtained. The main assumptions made are that the primary x-rays undergo negligible attenuation in the filter and aerosol material, which is the case with Mo K x-rays, frequently used in EDX spectrometry; particle size effects are negligible[9]; and the correction factor due to absorption in the aerosol is small for normal aerosol loads (200-500 μg/cm^2)[10].

The problem is approached by defining an infinitely thin layer L (see Fig 1) in the filter, where all the elements being analysed are assumed to be concentrated. The postulation of this imaginary layer is related to the concept of equivalent depth used by Adams and Van Grieken[6] but its position is such that the generated characteristic x-rays are attenuated by a layer of effective thickness x or y, depending whether side U or D is facing the detector. The effective thickness t of the filter is taken as a constant and is given by

$$t \simeq x_s + y_s \simeq x_f + y_f \qquad (1)$$

where the symbols s and f refer to standard and filter to be analysed, respectively. This approximation would appear to be contentious but it seems justified if the filters are not too thick. A numerical integration performed for an exponential, a linear and a constant aerosol concentration distribution across the filter thickness confirmed that a maximum error of approximately 10 % is made for a maximum value of t equal to two half-thicknesses.

For a standard filter with known concentration of a specific element one may write

$$N_s(U) = A_s\ e^{-\mu x_s} \text{ and } N_s(D) = A_s\ e^{-\mu y_s} \qquad (2)$$

where A_s is the characteristic x-ray intensity without absorption in the filter; $N_s(U)$ and $N_s(D)$ represent the characteristic x-ray intensity after absorption in layers x_s and y_s, respectively; and μ is the mass absorption coefficient of the filter for the characteristic radiation of interest. Similar equations may be written for the filter to be analysed.

By using equation (1) and defining $R = \frac{N(U)}{N(D)}$, expression

$$A_f = A_s \cdot \frac{N_f(D)}{N_s(D)} \cdot \left(\frac{R_f}{R_s}\right)^{\frac{1}{2}} \qquad (3)$$

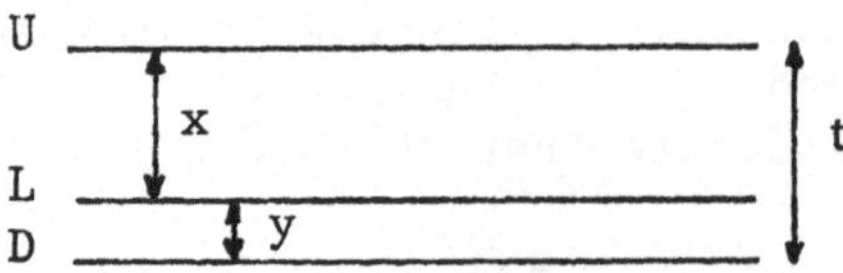

Fig. 1. Schematic representation of filter profile

can be derived, where the factor $\left(\frac{R_f}{R_s}\right)^{\frac{1}{2}}$ corrects for the effects of absorption in the standard and in the unknown. Since A_s can easily be related to the known elemental concentration in the standard and all the other quantities are measurable, the elemental concentration in the unknown filter can readily be determined.

EXPERIMENTAL

It is virtually impossible to prepare standards with known specific elemental distributions across the filter thickness. Stacks of aluminium plates were therefore used to simulate filters, and ^{241}Am (it emits 60 keV gamma rays), appropriately deposited on each plate, served to create different distributions of radiation sources across the thickness of the stacks. Because it was assumed that the primary radiation underwent negligible absorption in the filter, this is equivalent to generating fluorescent radiation at various depths in the "filter" but under much more stringent conditions seeing that considerably more material is involved than in the case of true filters.

Constant, linear and exponential distributions of activity across the stacks were considered. Three different thicknesses of stacks, namely 3.5, 12 and 66 mm were used so as to cover a wide absorption range. The stacks used as standards were those having a constant distribution. The results obtained are shown in Table 1 from where it is evident that, after applying the derived correction, most results are within 5 % of the true value of 124 μCi. Even in the case of the 66 mm stacks, which correspond to approximately six half-thicknesses, the maximum variation from the true value was, in the worst case, only 20 %.

Table 1. Comparison of True and Measured ^{241}Am Activity in Aluminium Stacks

Stack Thickness (mm)	Distribution of Activity	True Activity (μCi)	Measured Activity (μCi)		Correction Factor (%)
			Before Correction	After Correction	
3.5	Linear	124 ± 1	125 ± 1	120 ± 2	3.9
	Exponential	124 ± 1	132 ± 1	124 ± 2	7.0
12	Linear	124 ± 1	134 ± 2	121 ± 2	9.7
	Exponential	124 ± 1	149 ± 2	120 ± 2	20
66	Linear	124 ± 1	167 ± 3	115 ± 3	31
	Exponential	124 ± 1	249 ± 4	103 ± 8	59

Table 2. Comparison of PIXE and EDX Results for Aerosol Collected in Air Filter

Element	PIXE (μg/cm^2)	EDX (μg/cm^2)	
		Without Correction	With Correction
S	11.1 ± 0.5	92.7 ± 1.9	31.9 ± 1.4
Cl	1.8 ± 0.2	12.6 ± 0.9	14.3 ± 0.6
K	14.1 ± 0.3	24.0 ± 1.0	16.1 ± 0.6
Ca	41.7 ± 0.5	65.1 ± 1.2	43.2 ± 0.8

Comparison of results obtained by PIXE with those obtained with the present method when applied to environmental samples, an example of which is shown in Table 2, indicates that concentrations of S and Cℓ can be considerably underestimated in the case of PIXE.

CONCLUSION

Because the correction which has been derived is largely independent of elemental distribution across the filter, cellulose filter standards can easily be prepared. The method is easy to apply and constitutes a viable alternative for correcting for absorption when cellulose filters are used, or whenever there is reason to believe that the elements to be analysed are distributed across the filter thickness.

REFERENCES

1. S.K. Perry and F.P. Brady. Nucl. Instr. Methods 108 (1973) 389.
2. J.R. Rhodes. IEET Trans. Nucl. Sci. 21 (1974) 608.
3. D.C. Camp, J.A. Cooper and J.R. Rhodes. X-ray Spectrom. 3 (1974) 47.
4. P. Van Espen and F. Adams. Anal. Chim. Acta 75 (1974) 61.
5. T.G. Dzubay and R.O. Nelson. Advances in X-ray Analysis, 18 (1973) 619.
6. F.C. Adams and R.E. Van Grieken. Anal. Chem. 47 (1975) 1767.
7. R.E. Van Grieken and F.C. Adams. X-ray Spectrom. 5 (1976) 61.
8. B.H. O'Connor, G.C. Kerrigan, W.W. Thomas and R. Gasseng. X-ray Spectrom. 4 (1975) 190.
9. J.R. Rhodes and C.B. Hunter. X-ray Spectrom. 1 (1972) 113.
10. P. Verbeke, P. Van Espen and F. Adams. Anal. Chim. Acta 100 (1978) 31.

XRF ANALYSIS BY COMBINING THE STANDARD ADDITION METHOD WITH MATRIX-CORRECTION MODELS

Peter B. De Groot

Celanese Chemical Co., Inc.
P. O. Box 9077
Corpus Christi, TX 78408

ABSTRACT

The standard addition method is often useful for single-element analyses in matrices of unknown composition or those difficult to reproduce. Results are good at low concentrations where concentration vs. intensity is approximately linear, but serious errors occur at higher concentrations. A method is shown here for greatly extending the range of the standard addition approach by combining it with with matrix correction models. Expressions incorporating the well-known alpha-or beta-correction coefficients are derived for the observed intensities before and after addition of the analyte. K-ratio measurements on the sample plus a single standard addition allow the calculation of the correction coefficient and analyte concentration. Several standard addition levels can also be used and concentration obtained from non-linear regression analysis. A computer program has been developed to perform these calculations. Analyses of CuO on silica and of four metals in a metal oxide catalyst over a wide range of concentrations gave results with a relative accuracy generally within ±5%.

INTRODUCTION

Recent advances have been made in fundamental parameters X-Ray Fluorescence (XRF) analysis methods which have reduced the number of standards needed and relaxed the requirement that the unknown matrix be duplicated closely. Nevertheless, an occasional sample is found for which it is impractical or difficult to produce suitable standards. Examples from our laboratory include licensed catalysts which we are not allowed to analyze except for specific contaminants, and natural materials and corrosion deposits which have complicated multi-element compositions.

In such cases the conventional standard addition method usually works well if the analyte concentration is below about 1% (1). The conventional method calculates the unknown concentration from a linear extrapolation of the intensity vs. concentration slope to the origin. At higher concentrations, intensity vs. concentration is generally quite non-linear, and the conventional standard addition method gives very large errors. However, if the real form of the relationship between intensity and concentration is known, the standard addition method can be modified to produce a non-linear extrapolation to zero and hence the true unknown concentration. This has been done in this work, and the resulting method tested on some samples of known composition.

CALCULATION METHOD

Matrix correction models such as those of Raspberry and Heinrich (2) have been used to describe the relationship of intensity or k-ratio (observed intensity/pure element intensity). The matrix and the standard compound added can each be considered one component of a two-component system, which the models describe quite accurately. The k-ratio in this case becomes the intensity relative to the pure compound added. The model equation is then re-written in terms of the usual k-ratio and the concentration, C, of the analyte element before standard addition. For the alpha-coefficient model used when absorbtion effects predominate, the relationship is:

$$k = \frac{I}{I_0} = \frac{C + A/ZW}{(1+\alpha)(1+A/W)-\alpha(C+AZ/W)/Z} \tag{1}$$

where W = Wt. sample before addition
A = Wt. standard compound added
Z = Wt. fraction analyte in A

A corresponding expression for the beta-coefficient model used when fluorescence effects predominate for the analyte is:

$$k = \frac{I}{I_0} = \frac{(C+AZ/W)\left[1 + \frac{(C+AZ/W)}{Z(1+A/W)}\right]}{(1+\beta)(1+A/W)+(1-\beta)(C+AZ/W)/Z} \tag{2}$$

These equations are non-linear equations in the dependent variable k, the parameters C and alpha (or beta), and the independent variables Z, A and W. The equation for the unknown (A = 0) and one other with any level of added standard can be solved for C and alpha. More than one level of addition can also be treated by non-linear regression analysis to determine the best- fit values of the parameters. Many computer programs exist which are suitable for this procedure. Listings of the programs used in this work are available from the author.

EXPERIMENTAL METHODS

Sample preparation

All solids were ground in a mortar and pestle to -100 mesh. Known concentrations of CuO on silica were prepared by slurrying a weighed amount of SiO_2 in water and adding a measured volume of standard $Cu(NO_3)_2$ solution. The pH was then adjusted to 6.0-7.0, and the material dried under vacuum on a rotary evaporator. The resulting solid was then calcined in air at 500°C overnight. Standard additions were made by adding additional volumes of $Cu(NO_3)_2$ solution to this material, then evaporating and calcining as above.

Standard additions were made to the mixed metal oxide catalyst by adding either MnO_2, H_2WO_4 or MoO_3 to the catalyst and wet-grinding the mixed solids. The material was then slurried with additional water, evaporated to dryness on a rotary evaporator, and dried in air at 120°C for several hours. In the case of vanadium, V_2O_5 was simply ground together with the catalyst in a Tekmar Model A10-S grinder.

All samples were re-ground to pass 100 mesh and pressed into planchettes for XRF analysis.

X-Ray Intensity Measurement

X-Ray intensities were measured with a Philips Universal Vacuum Spectrograph equipped with Cr and W target x-ray tubes, LiF and EDDT analyzing crystals, and flow-proportional and scintillation detectors. The Lα intensity was measured for W and the Kα was used for all other elements.

The pure element intensity for the calculation of k-ratios was obtained from measurements on oxides of Mn, V, and W, and from the pure elements as well as oxides for Cu and Mo. All intensities were measured in duplicate on each of two duplicate planchettes.

Experimental Strategy

To obtain the best results from a standard addition method, a compromise must be made. On the one hand, the addition of the analyte-containing compound must be sufficiently large to produce an acceptable statistical error in the change in intensity measured. On the other, the addition of analyte should be kept small, so the sample plus added standard will approximate the original matrix as closely as possible.

The determining factor in the precision of the calculated concentration via the standard addition method is the relative error in the difference in intensity before and after addition. Using values of 0.5% for the precision of intensity measurement, and a goal of no more than ±5% error in the relative intensity difference, it can be shown that the ratio of intensity after standard addition to that before must be >=1.15. For estimating the amount to be added, the amount originally present is assumed equal to the k-ratio.

RESULTS AND DISCUSSION

Results

Results of XRF analyses by the non-linear standard addition method described here are given in the table. Two different levels of standard addition were used for each of the elemental analyses. These are designated in the table by the element symbol, followed by a number, e.g., Cu1, Cu2, etc. The concentration was calculated in two ways.

First, data from each of the addition standards were paired with data from the unknown and the explicit solution for the two parameters alpha (or beta) and C obtained. Second, all of the data were used in a non-linear regression fit for the best values of C and and alpha or beta. Results are compared with the known concentrations of the elements in the two test samples. The true analyte concentrations for the Cu/SiO_2 samples were established as described above in "Sample Preparation." The true values for the mixed metal oxide catalyst are the result of analyses by different methods by our laboratory and another industrial laboratory.

Effect of Multiple Standards

Surprisingly, the concentrations calculated from the least-squares fit of the unknown plus two or more levels of standard addition are essentially the same as those calculated from the highest addition level and the unknown alone. All the data were weighted equally in the non-linear regression program used. An appropriate weighting of the data might change the domination of

Table 1. Concentration Calculated by Non-linear Standard Addition Method

Sample	elem.	Stds. Used	Istd / Iunk	True C, %	Calculated C, %	coeff*	rel. error,%
CuO/ Silica	Cu	Cu1	1.15	4.99	4.63	-0.6363a	-7.2
		Cu2	1.27		4.94	-0.6101a	-1.0
		Both			4.95	-0.6109a	-0.8
CuO/ Silica	Cu	Cu3	1.10	12.5	12.3	-0.5906a	-1.6
		Cu4	1.19		13.3	-0.5498a	+6.4
		Both			13.3	-0.5520a	+6.4
Catalyst	Mo	Mo1	1.08	45.6	45.4	0.3078a	-0.4
		Mo2	1.15		46.4	0.4016a	+1.8
		Both			46.2	0.3793a	+1.3
	W	W1	1.34	8.83	8.45	-0.4402a	-4.3
		W2	1.64		8.83	-0.4142a	0.0
		Both			8.82	-0.4121a	-0.1
	Mn	Mn1	1.20	6.41	6.81	1.852a	+6.2
		Mn2	1.35		6.40	1.670a	+0.2
		Both			6.40	1.659a	+0.2
	V	V1	1.21	6.10	6.02	1.242a	-1.3
		V2	1.44		5.86	1.245a	-3.9
		Both			5.86	1.245a	-3.9
	V	V1	1.21	6.10	6.21	1.550b	+1.8
		V2	1.44		6.27	1.681b	+2.8
		Both			6.27	1.584b	+2.8

* a= alpha, b= beta

the fit by the highest standard, but this has not yet been investigated. The best overall result from the present data would be the average of the two explicit solutions rather than the non-linear regression result.

Absorbance vs. Fluorescence Model

The present samples are not suitable for adequately testing the standard addition/fluorescence model method. However, vanadium would experience the strongest fluorescence effects because its x-ray absorbtion edge is just below the Mn Ka x-ray emission energy. Both absorbance and fluorescence model results for vanadium are given in the table. The two models produced about equally accurate results. For all the other elements the absorbance model produced a more accurate result, as expected.

Analysis Without Pure Elements

With two or more levels of standard addition, it is possible to rewrite equations (1) and (2) so that I is the dependent variable, and solve for I_0 along with alpha or beta and C. This eliminates the need for a pure element. This procedure resulted in less accurate values (10-15% relative errors) of C except for tungsten, where the result was about the same as before. In addition, the values obtained for alpha or beta and Io were very unrealistic in the cases of V and Mn. The calculation of these coefficients is very sensitive to small variations in measured intensities.

Accuracy and Range of Applicability

Accuracy does not seem to depend on concentration in the unknown to any great extent. The method generally gave results accurate to within about 5% relative over a concentration range of about 5-45%. The determining factor (outside of sample preparation) in the accuracy is probably the magnitude of the alpha or beta coefficient; that is, the deviation from linearity that must be corrected for. The data given here are insufficient to confirm this experimentally, however.

CONCLUSION

Combining the standard addition method with empirical matrix correction models yields a moderately accurate single element analysis which can be useful in cases where standards required. for other methods are difficult or impractical to obtain. Only a single standard addition and the pure element or oxide are required.

ACKNOWLEDGEMENT

I thank Celanese Chemical Company, Inc., for permission to publish this work, and I especially thank Joyce E. Purdy for diligent and careful sample preparation and XRF analyses.

REFERENCES

1. E. F. Kaelble, Handbook of X-Rays, McGraw-Hill, New York (1967).

2. S. D. Rasberry and K. F. J. Heinrich. Anal Chem 46:81-89 (1974).

ACCURATE GEOCHEMICAL ANALYSIS OF SAMPLES OF UNKNOWN COMPOSITION

J. Kikkert

Philips Application Laboratories
X-Ray and Nuclear Analysis
Almelo, The Netherlands

INTRODUCTION

Considering all fields of analysis by X-ray fluorescence spectrometry geochemical analysis is probably the most complex. The geochemical analyst is often requested to analyse samples of unknown composition and origin. While these samples need to be analysed for a number of elements of economic or geochemical significance it is often impossible or uneconomic to analyse for all elements that could possibly be present in the sample.

To obtain meaningful results the geochemical analyst often has to adapt his procedures to build in safeguards for the unexpected. Firstly the geochemical analyst has to obtain a measured intensity for the element of interest, this measurement must be free from interference from other elements, and have an optimal peak to background ratio. This is achieved by means of:

- Optimal excitation of the element of interest through the choice of anode material and the use of 100 kV excitation;
- Choice of high resolution conditions, such as the use of the LiF 220 crystal to reduce line overlap to a minimum;
- Careful selection of backgrounds which are free of line overlap positions, which accurately predict the background underneath the peak. The optimal background positions for the analysis of geochemical elements using a Philips PW 1400 spectrometer are given in [1];
- Detection of unsuspected line overlap by correlating measured peak and background intensitites to the intensity of Compton scattered tube lines.

An extended version of this part of the paper is available as an Philips internal publication [1]. Once reliable net peak intensities have been obtained these have to be converted into concentrations via matrix corrections that compensate for all elements present in the sample although most of these have not been analysed.

MATRIX CORRECTIONS USING COMPTON SCATTERED TUBE LINES

Matrix corrections based on Compton scatter are well documented e.g. [2,3].

This method works well if matrix variations are limited and no significant absorption edges occur between the Compton line and the region of primary absorption or the fluorescent line. Provided the concentration of the elements giving rise to these edges is less than 2000 ppm, the effect of these edges can be ignored when performing routine geochemical analysis.
Where a more generally applicable method is required to cope with samples varying over a wide range of compositions then this simple method has to be extended. In that case the solution to the problem can be split in two stages:

a. Estimation of mass absorption at the Compton wavelength.
Compton scatter is the incoherent scatter of X-rays by the electrons of the atoms.
All elements except Hydrogen have almost the same ratio of electrons to atomic weight and thus scatter X-rays with almost the same efficiency. The measured Compton scattered tube line is attenuated by the absorption that takes place as the radiation enters and leaves the sample after scatter.
If all elements present in the sample, scatter the tube line with equal efficiency, this then leads to:

$$\mu_c = \frac{A}{R_c} \quad . \qquad (1)$$

μ_c = massabsorption at the Compton wavelength
A = constant
R_c = count rate of the Compton scattered tube line

Fig. 1 shows the scatter cross-sections for common rock forming oxides based on the data of Mc. Masters [5].
The scatter cross-sections show a decrease, corresponding to the decrease in the ratio number of electrons to atomic weight with increasing atomic number.

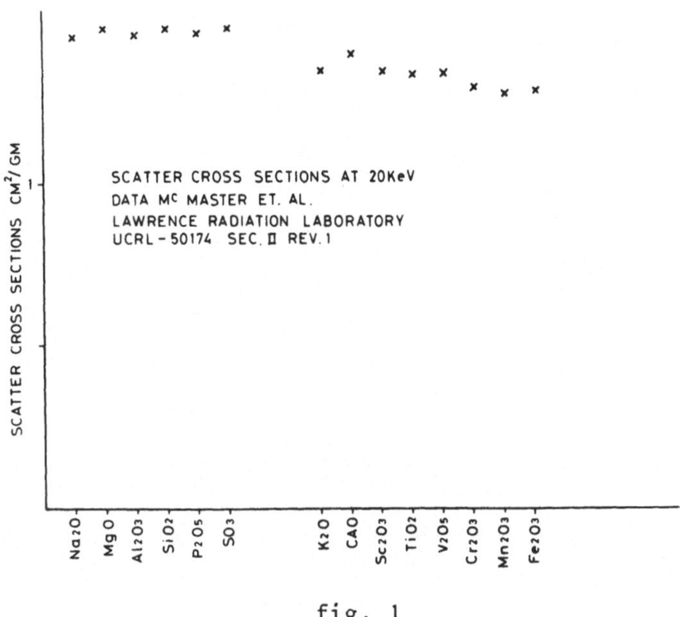

fig. 1

As a result, if Compton scatter is used to correct for the mass absorption of samples containing high concentrations of heavier elements, such as Ca or Fe, then this method overestimates the mass absorption by 10 to 15 percent compared to samples consisting mainly of silicates.

In practice it can be shown that a more correct form of eq. (1) which takes into account the scatter efficiency of the sample is:

$$\mu_c = A_1 \frac{\sum_j S_j C_j}{R_c} \tag{2}$$

S_j = scatter cross-section of elements
A_1 = constant

The decrease in scatter cross section is almost a linear function of atomic number. A simple correction for this can be made by adding a constant (B) to the Compton count in the denominator of equation 1.

b. Extrapolation of mass absorption to the wavelength of primary and secondary absorption.
The ratio between the mass absorption of different geochemical samples remains constant over a wide range of wavelengths provided that no significant absorption edges occur in the region of interest.

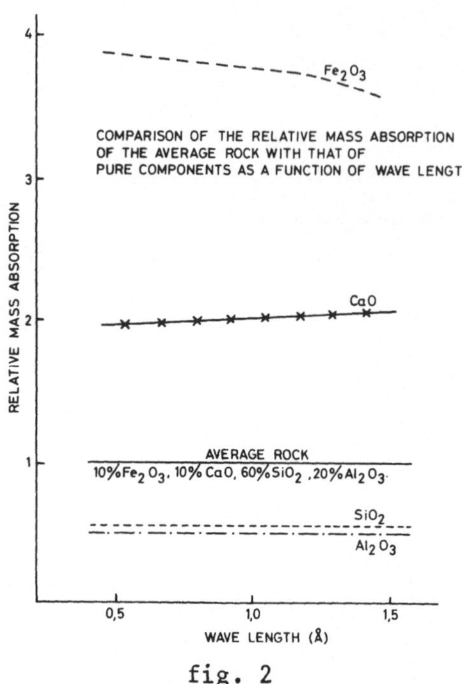

fig. 2

Fig. 2 shows the calculated massabsorption of pure Fe_2O_3, CaO, SiO_2 and Al_2O_3 relative to the massabsorption of a sample selected to represent the average rock. The composition of this sample was chosen as 10% Fe_2O_3, 10 % CaO, 60% SiO_2 and 20% Al_2O_3. This sample is shown with a relative massabsorption of 1. As can be seen the relative massabsorption remains almost constant over a wide range of wavelengths, even for these extreme cases. The relative mass absorption of the unknown with respect to the standards can thus be determined at the Compton wavelength and applied at the wavelength region of primary absorption and the wavelength of the characteristic line. Should significant absorption edges occur between the Compton wavelength and the wavelength at which absorption occurs, then a correction needs to be made. This leads to the following equation:

$$C_i = E_i R_i \quad (\mu_c + \sum_j J_j C_j) \qquad (3)$$

μ_c = relative massabsorption at the Compton wavelength.
E_i = calibration constant
J_j = correction for the absorption edge due to element j.

The summation is over all elements that have an absorption edge in the region of interest. As these elements are usually trace elements the contribution of most of these is extremely small. Only elements present in concentrations above 2000 ppm give rise to corrections greater than 1% relative.

If the simplest form of relation between Compton and mass absorption is used, this leads to the equation as introduced by Kikkert [4].

$$C_i = E_i R_i \left(\frac{A_o}{R_c} + \sum_j J_j C_j \right) \tag{4}$$

A_0 = a constant, usually the Compton intensity of a mass absorption reference standard.

In practice two types of cases can be distinguished. Elements measured on the low energy side of the Compton wavelength have a number of absorption edges between the Compton and the characteristic line. This is shown for the analysis of Zn in the presence of significant concentrations of As in fig. 3.
If the relative massabsorption is determined at the Compton wavelength then corrections have to be applied for both the Zn and As absorption edge. For elements determined on the high wavelength side of the Compton line this correction is negative. For the average silicate rock this correction is 7 - 10 percent relative for each percent of the element giving rise to the absorption edge.

For elements determined, using lines on the high energy side of the Compton line, a positive correction needs to be made, which for the average silicate rock is of the order of 6 - 10 percent relative for each percent of the element giving rise to the absorption edge.

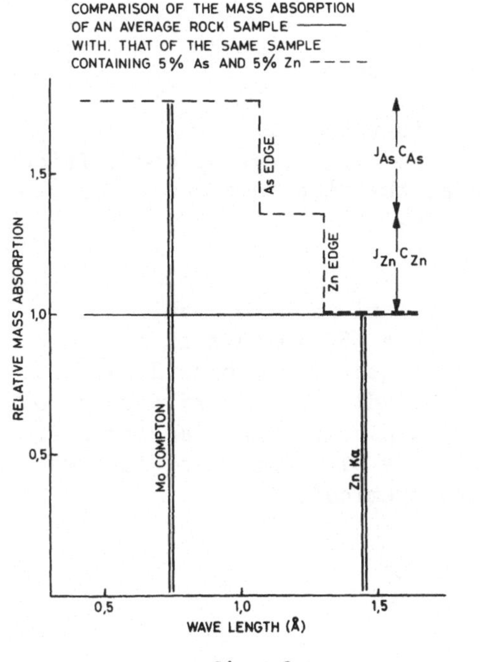

fig. 3

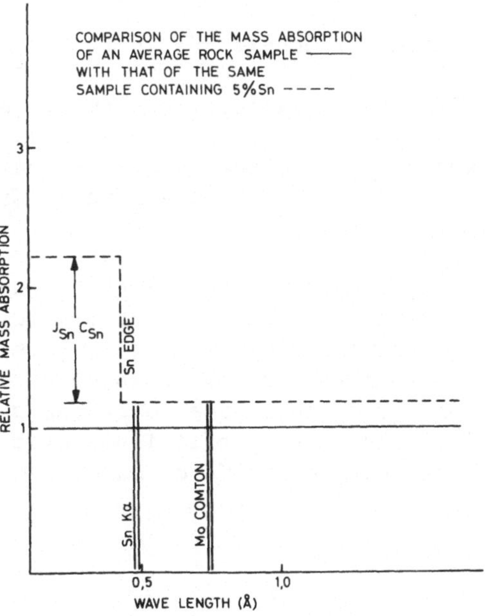

fig. 4

An example of a correction for an absorption edge between the Compton and the region of primary absorption is the analysis of Sn where Compton scatter is used for matrix correction [6]. As shown in fig. 4, the secondary absorption at Sn Ka can be predicted without a correction for the Sn absorption edge but such a correction is necessary to calculate the primary absorption.

Thirtyeight tin metalurgical products were analysed. The composition of the samples varied in the range:

Sn	0.3 - 52%
S	0.5 - 34%
Fe	6 - 52%
As	0 - 2%

The samples were diluted ten fold with a 1:1 mixture of KCl and quartz. This reduces matrix effects to the extend that equation (4) can be applied successfully. In this case the equation becomes

$$C_{Sn} = E_{Sn}\, R_{Sn} \left(\frac{A_o}{R_c} + 0.069\, \frac{C_{Sn}}{10} \right) \qquad (5)$$

A_0= Compton scatter of a mass absorption reference standard

The factor ten is due to the ten fold dilution.

The analytical results are shown in table 1. The method corrects successfully for the wide range of compositions encountered in the samples - including the high Sn concentrates - without having to determine S, Si or Fe.

In recent work [7] these jump corrections have been succesfully related to theoretical alphas [8] so that these corrections can now be calculated from theoretical considerations, taking into account enhancement and including a primary absorption correction integrated over the whole wavelength range applicable. The calculated coefficients agree closely with empirical factors derived from standards made from pure chemicals.

Table 1
Analysis of Sn products

SAMPLE NO	%Sn CHEM	%Sn XRF	SAMPLE NO	%Sn CHEM	%Sn XRF
1	39.3	39.0	21	0.76	0.77
2	42.0	42.1	22	0.83	0.85
3	31.3	30.9	23	0.35	0.36
4	12.2	12.0	24	0.49	0.49
5	30.9	30.5	25	0.30	0.31
6	41.6	41.7	26	0.46	0.48
7	41.2	41.3	27	0.73	0.73
8	52.1	52.0	28	0.40	0.41
9	13.2	13.1	29	0.72	0.71
10	26.4	26.6	30	0.59	0.58
11	26.6	26.8	31	0.67	0.67
12	20.7	20.8	32	0.46	0.45
13	20.0	20.0	33	2.66	2.62
14	25.6	25.7	34	1.12	1.09
15	14.1	14.1	35	0.35	0.35
16	24.0	24.3	36	0.12	0.11
17	0.36	0.38	37	0.14	0.14
18	0.47	0.47	38	0.21	0.21
19	0.56	0.56	quartz blank 1	----	<0.01
20	0.64	0.64	quartz blank 2	----	<0.01

Chemical Analysis by Renison Ltd, Tasmania.

REFERENCES

1. Philips' Application Report "Accurate Geochemical Analysis of samples of unknown composition and origin"
2. R.C. Reynolds, Am. Mineral 52, 1493 (1967)
3. G. Alderman and J.W. Kemp, Anal. Chem. 34, 812 (1962)
4. J.N. Kikkert, 4th Analytical Symposium Melbourne, 1974
5. McMaster et al. Lawrence Rad. Lab. UCRL - 50174
6. J.N. Kikkert, "XRF application on Tin Ores and Concentrates", Philips' Application Report (1978)
7. J.N. Kikkert and W.K. de Jongh, X-ray Spectrometry, to be published
8. W.K. de Jongh, X-ray Spectromety 2, 151 (1973)

XRF ANALYSIS OF VEGETATION SAMPLES AND ITS APPLICATION TO MINERAL EXPLORATION

T. K. Smith and T. K. Ball

Institute of Geological Sciences
London and Keyworth, United Kingdom

ANALYSIS OF VEGETATION SAMPLES

Many analysts use a wet or dry oxidation treatment prior to determination of minor and trace elements in non-fossilised vegetation. This stage is often unnecessary for XRFS since the technique is sufficiently sensitive for many purposes to permit the analysis of unashed material. Indeed, inaccurate results may be obtained if volatiles are lost, if no correction is made for loss on ignition, or if such correction is of large magnitude.

During ashing the organic fraction of a sample is largely or completely removed by oxidation. This process may be simple dry ashing in air at several hundreds of degrees centigrade, wet ashing in an oxidising acid at a lower temperature, or a combination of both. Variations of these standard methods have included rapid destruction of organic matter at 390°C by a twentyfold excess of molten equimolar sodium and potassium nitrates in a borosilicate beaker[1]. Ashing may have a benefit in concentration of an analyte if the sensitivity of the analytical technique is inadequate for the original material. Interference by organic substances may also be removed. A number of problems, however, may arise. In the case of dry ashing cross contamination may occur in the furnace through convection currents. If the ash is to be used directly for analysis without dilution, a large and varying factor is introduced for conversion of analyte concentrations in ash to those in the original sample. Perhaps most importantly, it may cause concentration changes through loss of volatiles, or retention or contribution by the ashing vessel. Many papers have been published on this latter subject. For example, Strohal et al.[2] investigated volatilisation of trace elements during dry ashing of a mollusc and concluded that wet ashing was to be recommended. Gleit and Holland[3] however devised a method whereby a stream of oxygen excited by a radiofrequency discharge enabled dry ashing to take place below 100°C. Wet ashing also has some problems in that there is a risk of explosion and that impurities may be introduced through the reagents. In addition to the avoidance of such difficulties the requirement for analysis of relatively large numbers of samples, such as occur in geochemical reconnaissance, is an incentive to analyse the material in its unashed form. Jenkins et al.[4] did so in 1966 and this was also the approach more recently employed by Norrish and Hutton[5].

The procedure adopted in this study was simply to dry the samples in a fan-assisted oven at 85°C for approximately six hours, to comminute them in a commercial blender, and then to press 15 grams at 25 tons load in a 40mm die. A vibratory disc mill is unsuitable for some of the more resinous conifers because of adherence to the pot. Attempts to improve the coherence of some broadleaf pellets by the use of a urea-formaldehyde resin of especially low viscosity and rapid hardening at room temperature were frustrated by corrosion of the die caused by increase in the chemical activity of the resin at high pressure. Further contraindications were extended briquetting time and contamination by trace elements. The largely cellulosic form of the samples exerts a beneficial, approximately constant dominance that ashing would remove. A further benefit is that standards, including those of low concentration, may accurately be made by adding a measured volume of solution of known analyte concentration to cellulose powder. After drying and disintegration in a coffee mill the cellulose may readily be pressed to form discs very similar in X-ray response to that of the unknowns.

A model for calculation of the element concentration from fluorescent X-ray intensity may take the form used by de Jongh[6] and using the same notation:

$$C_i = (D_i+E_iR_i)(1+\Sigma\alpha_{ij}C_j)$$

The set of simultaneous equations may be reduced to one by putting j equal to i.

$$C_i = (D_i+E_iR_i)(1+\alpha_{ii}C_i)$$

This is possible in determination of individual elements present in minor or trace concentration if R_i is replaced by

$$\frac{R_i^x}{R_i^m} \cdot \frac{R_b^m}{R_b^x}$$

where x is the sample, m is the monitor standard, and b is a background wavelength chosen such that its intensity is approximately inversely proportional to the mass absorption coefficient of the sample for the analytical line in question. This will provide matrix correction for the other elements present. Although this procedure is not necessary for pure vegetation samples it is desirable for mulls, where there may be an admixture of mineral grains. Some standards containing a proportion of silica were therefore prepared to simulate this sample type.

APPLICATION TO MINERAL EXPLORATION

The successful introduction of any technique of prospecting into a well-established scheme must show overall advantages either in that the same basic information is obtained but the technique is simplified or that the novel method provides useful data not readily obtained by its competitors. Pedogeochemistry is the major method used for tactical prospecting and its high success rate at locating mineral occurrences attests to the general validity of the procedure. Soil samples are typically collected at depths of up to 1m. Tree roots can sample soil substrates to a much greater depth and the root system is often as large as the crown. The trees should reflect the composition of the media sampled by the roots and if this is an orebody, the occurrence should be reflected in the composition of the crown. This is particularly important when the orebody is covered by barren overburden such as glacial drift or raised beach deposits.

The Coed-y-Brenin, Gwynedd copper porphyry has been eroded by glacial action and is characterised by being overlain by a mixture of thin residual soil and a thick, hydromorphically enriched glacial till[7]. The area was newly planted with trees from 1953 onwards and in most cases it is possible to choose species represented inside and outside the copper rich area. A traverse was selected, the first criterion being to choose an area underlain by similar rock types but containing variable amounts of copper. The line crosses a part of extensive areas of impeded drainage with a formerly thick copper-bearing peat underlain by glacial drift. The major part of the traverse however crosses steep, well-drained ground with thin residual soil cover. A difference was found between surface soil and vegetation profiles. The soils have high Cu contrast and content but reflect hydromorphic and soil creep anomalies; the vegetation has relatively low contrast and content but indicates the underlying orebody.

The Long Rake, Derbyshire is a vein system extending for some 6 km consisting of a number of orebodies of fluorite and barite with minor galena and sphalerite, the combined sulphides reaching a maximum of about 11%. The rake is a normal fault and the orebodies consist mainly of fault infill but where splay faulting occurs extensive replacement orebodies are formed in the acute angle. At Haddon Fields one such location has a managed woodland consisting mainly of conifers (Picea yezzoensis) but also contains a substantial population of the common elder (Sambucus nigra). In this area the lode is underlain by about 2.5m of glacial drift with a high loess content.

The response of S. nigra along a traverse at a high angle to the orebody is illustrated in Fig. 1a,b,c. Here Ca, Ba and Pb concentration in live twigs all accurately reflect the position of the orebody. Dead twigs exhibit a reduction in the concentration of Ca and Ba compared with live growth, but the reverse is observed for Pb. In dead twigs the concentration of Pb is significantly higher than in live growth but only the latter positively identifies the position of the orebody. With P. yezzoensis the Ca and Ba concentrations in both dead twigs and live growth accurately reflect the position of the vein. However in contrast to S. nigra the dead twig Pb content is a good indicator of the presence of the vein whereas for the live growth the relationship is obscure. Fig. 1d shows the merged data giving determinations of Pb in all of the trees collected along the traverse irrespective of species or of whether the samples were of live or dead growth. Although the highest value is still located over the orebody the contrast is substantially reduced. While not illustrated the Ca and Ba results show only marginal improvement.

DISCUSSION

It can be shown that the analytical data from easily accessible parts of trees such as side branches and twigs can locate orebodies successfully and there is an abundant literature on the subject[8,9]. For the Coed-y-Brenin deposit the biogeochemical method shows an advantage over standard soil geochemical techniques in that somewhat smoother and more useful data are obtained. At Haddon Fields the biogeochemical data gives a precise and accurate location for the orebody in an area of foreign overburden affected by considerable hydromorphic transport of ore elements. Ca concentrations are anomalously high over the vein even though the vein is hosted in limestone, presumably reflecting the greater solubility of calcium fluoride. Different species of trees tolerate different concentrations of both nutrient and toxic elements and the response must be interpreted on traverses which are culture specific. Even very high grade orebodies do not affect the whole of the overlying flora in the same manner. Great care must also be taken that the same type of sample is selected from the trees, mixing of dead and live twigs being contraindicated. Whereas there appears to be little difference in response to e.g. nutrient Ca, toxic Pb behaves in a totally different manner between dead and live tissue in P. yezzoensis and S. nigra. The latter appears to be a useful species for further research because of its wide distribution and its tolerance for toxic metals, as evidenced by its apparently healthy growth in heavily polluted industrial environments.

One of the major difficulties in biogeochemical prospecting

has been that although sample collection has been easier than with almost all other media, sample preparation has been in contrast rather tedious. The described technique differs in that preparation is no more difficult than for soil and that detection limits by XRFS are adequate for most prospecting purposes. Trees can reflect the composition of the underlying substrate; in the instances outlined above so did the composition of the soils, the difference being that the tree anomalies referred only to the orebodies, whereas soil samples also reflected hydromorphic and soil creep anomalies. Silvigeochemistry, if interpreted with care, can thus provide a useful contribution to the increasingly broad spectrum of techniques available to the prospecting geologist.

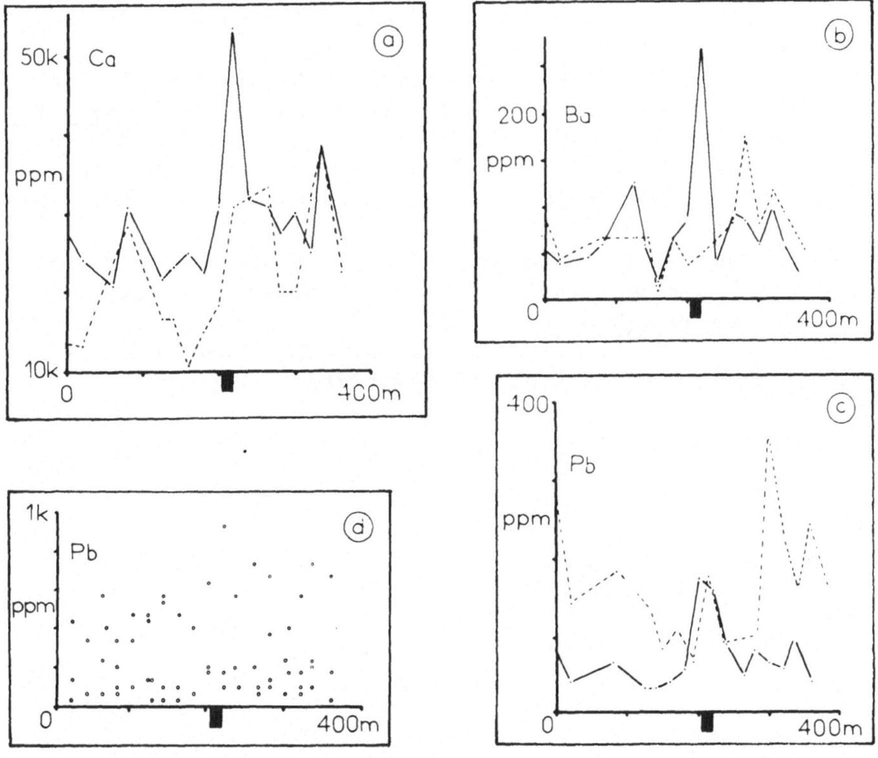

Fig. 1. Profile over Long Rake barite/fluorite/galena/sphalerite orebody (solid block) (a) Ca (b) Ba (c) Pb in Sambucus nigra (Solid lines: live twigs; dashed lines: dead twigs) (d) Pb in all species including both live and dead growth.

ACKNOWLEDGEMENT

This paper is published with the approval of the Director, Institute of Geological Sciences (NERC).

REFERENCES

1. J. M. Bowen, Use of sodium and potassium nitrates for decomposing organic samples for elementary analysis. Anal. Chem. 40:969 (1968).
2. P. Strohal, S. Lucic and O. Jelisavcic, The loss of cerium, cobalt, manganese, protactinium, ruthenium and zinc during dry ashing of biological material. Analyst 94:678 (1969).
3. C. E. Gleit and W. D. Holland, Use of electrically excited oxygen for the low temperature decomposition of organic substances. Anal. Chem. 34:1454 (1962).
4. R. Jenkins, P. W. Hurley and V. M. Shorrocks, Plant mineral analysis by X-ray fluorescence spectrometry. Analyst 91:935 (1966).
5. K. Norrish and J. T. Hutton, Plant analyses by X-ray spectrometry. X-Ray Spectrometry 6:6 (1977).
6. W. K. de Jongh, X-ray fluorescence analysis applying theoretical matrix corrections. Stainless steel. X-Ray Spectrometry 2:151 (1973).
7. R. Rice and G. J. Sharpe, Copper mineralisation in the forest of Coed-y-Brenin, North Wales, Trans. Instn. Min. Metall. 85:B1 (1976).
8. C. E. Dunn, The biogeochemical expression of deeply buried uranium mineralisation in Saskatchewan, Canada. J. Geochem. Explor. 15:437 (1981).
9. A. W. Rose, H. E. Hawkes and J. S. Webb, "Geochemistry in Mineral Exploration," Academic Press, London (1979).

APPLICATION OF XRF TO MEASURE STRONTIUM IN HUMAN BONE *IN VIVO*

L. Wielopolski, D. Vartsky, S. Yasumura, and S.H. Cohn

Brookhaven National Laboratory, Medical Research Center
Upton, New York 11973

INTRODUCTION

Strontium appears in the earth's crust at a concentration of about 400 ppm and in sea water about 8.1 ppm. Consequently, all living forms have evolved in the presence of this alkaline earth and have incorporated it in their tissues. The exact role of Sr in the human body remains ambiguous (1), even though Sr has been extensively studied with reference to the radiation hazard from ^{90}Sr (2).

Metabolic behavior of Sr, in some aspects, is similar to that of Ca (3), but substantial dissimilarities between these two are also well recognized (1,4). It has been suggested that Sr stimulates the formation of bone matrix. At higher concentrations, however, it interferes with calcification mechanism of the bone matrix (5).

Strontium has occasionally been put to therapeutic use. Therapeutic effects of Sr have been reported in an adult case of the Fanconi Syndrome (6), and Sr lactate appears to have some value in the treatment of osteoporosis (7). There is also some evidence that Sr can harden bone and teeth, and hence is considered to be an anticariogenic agent in man (8).

As a basis for better understanding the role that Sr fulfills in human body, it is desirable to measure directly the main Sr store in human body. Although strontium is omnipresent in human tissues, 99% is stored in the mineral portion of the bone (9). In the present study x-ray fluorescence (XRF) was applied to measure the strontium content of the tibial shaft *in vivo*. The feasibility studies showed that normal levels of stable strontium in the bone can be measured successfully.

METHOD

X-ray fluorescence spectroscopy was utilized for the measurement. In the present study either ^{125}I (Te x-rays) or ^{109}Cd (Ag x-rays) was used to stimulate 14.1 keV Sr K x-rays. The emitted x-rays were detected by a Si(Li) detector and processed by standard nuclear spectroscopy electronics.

Because of the low energy of the characteristic Sr x-rays, only superficial bones could be considered for the measurement of Sr. In the present work, the tibial shaft was selected as the measurement site. The geometrical configuration of the source, the detector and subject's leg is shown in Fig. 1.

The radiation dose delivered to the skin surface (approximately 1 cm^2) is one rem, and the dose delivered to the bone marrow subjacent to the exposed surface of the tibia is about 50 mrem.

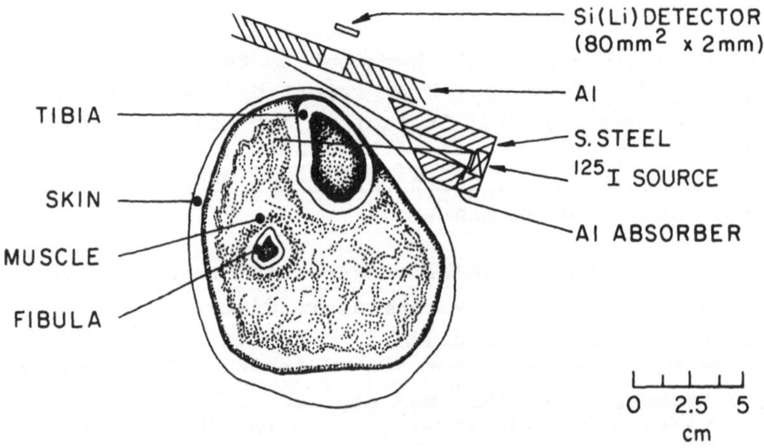

Fig. 1. Schematic diagram of the source, tibia, and detector configuration.

RESULTS

A typical spectrum acquired from a dry human tibial shaft is shown in Fig. 2. The Sr photopeak is clearly visible above the background. Two lead peaks are also visible, as well as Zn (in trace levels) (10) and Ca, the main constituent of the bone. The spectrum from the measurement of a small area of the tibia in an intact leg from a cadaver is shown in the upper part of Fig. 3. A Sr peak is clearly visible. To confirm that the measured Sr was located in the bone, a second measurement was taken from the mid-calf tissue of the same cadaver, see lower part of Fig. 3. It is apparent that the Sr level in the skin and in the soft tissue is considerably below the detection limit of the present system. The Fe and Cu peaks originated in the collimator materials.

The results of all the measurements are summarized in Table 1. Listed are the source used to induce Sr x-rays, the net number of counts in the Sr peak together with the error of the measurement, and how well the Sr peak is defined above the background. The net number of counts were normalized to an exposure dose of 1 rem.

Quantitative calibration of the Sr measurement was performed by assaying bone specimen, from the measurement site, by atomic absorption. The calibration for the post mortem measurements only, is shown in Fig. 4. It should be noted that the value of the point 64 in Table 1 is corrected in Fig. 4 for increased attenuation corresponding to 1mm of overlying tissue. The corrected value is 1533 instead of 1224. The calibration line in Fig. 4 represents,

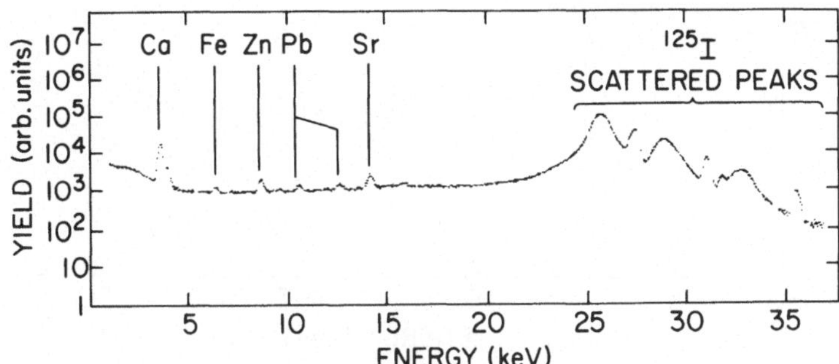

Fig. 2. L x-ray spectrum taken from a dry bone, induced by ^{125}I source. The peaks above channel 500 are backscattered peaks.

therefore, the slope of the Sr signal for an exposure of 1 rem and overlying tissue thickness of 2.5 ± 0.5 mm. The lower detection limit of Sr (defined as three standard deviation above the background) is about 15 μg Sr/g wet bone, and the reproducibility of the system is about 5%.

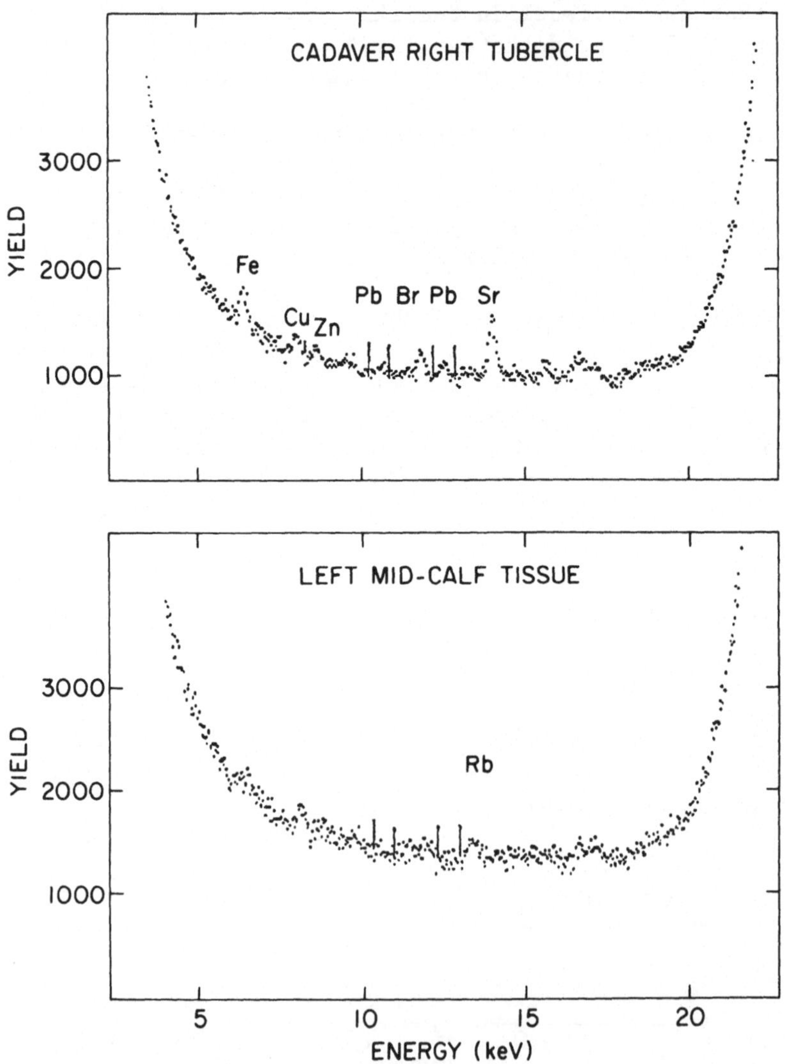

Fig. 3. L x-ray spectrum measured on intact cadaver leg irradiated by ^{125}I source. 1 rem skin dose.

TABLE 1

STRONTIUM MEASUREMENTS ON INTACT TIBIAL SHAFTS, EXPOSURE 1 REM

Sample	Source	Net counts* ± SD	σ's** Above Bkg.
(1) P.M.			
59	I-125	1481 ± 128	17.1
60	"	1895 ± 133	21.3
61	"	1124 ± 175	12.1
64	"	1274 ± 170	14.4
65	"	1497 ± 139	15.3
67	Cd-109	2986 ± 151	29.7
(2) A.L.			
1610	Cd-109	1004 ± 155	9.0
1610	I-125	1363 ± 175	11.3
2109	Cd-109	888 ± 193	6.7
3188	"	1363 ± 202	9.7
4442	"	2363 ± 191	19.3
(3) N.			
H.S.	"	1784 ± 179	13.5
M.S.	"	840 ± 176	6.4
S.M.	"	607 ± 170	5.1
K.E.	"	2089 ± 180	16.8
S.H.C.	"	1497 ± 170	12.7
R.S.	"	1609 ± 187	13.1
D.V.	"	1660 ± 170	13.3

(1) P.M. - Postmortem measurement
(2) A.L. - Amputated legs measurement from diabetic patients
(3) N - Normal volunteers
* Uncorrected for overlying tissue thickness
** $\sigma = \sqrt{\text{Background}}$

DISCUSSION

The preliminary results demonstrate the feasibility of measuring normal levels of Sr in the human body *in vivo*. Although the method is limited to superficial bones, the intensity of the measured Sr signal is satisfactory. The present set-up is also used to measure lead in the bone *in vivo* (11,12). The chief limitation of the method is the attenuation of the signal due to overlying tissue. It is important, therefore, in particular, for epidemiological studies, that the measured Sr signal be normalized per unit of overlying tissue thickness. In the present study, the thickness of the

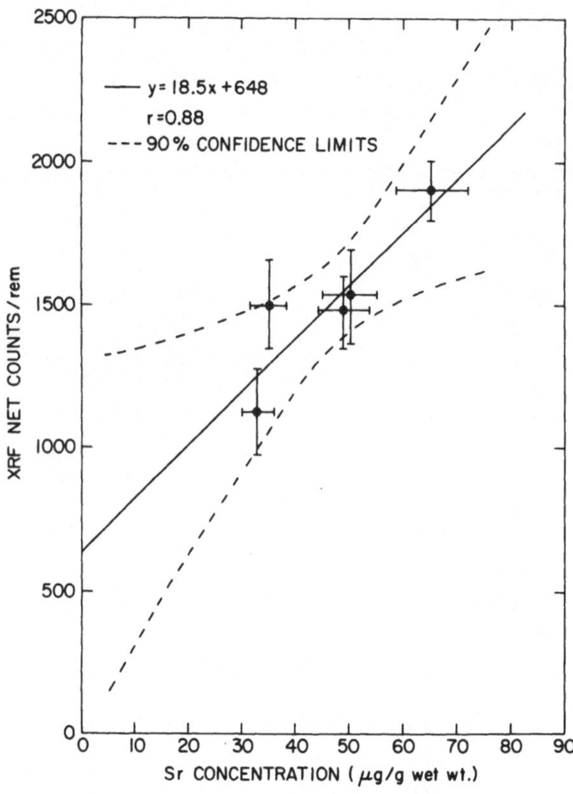

Fig. 4.

Strontium calibration. XRF net count/rem at 2.5 mm overlying tissue thickness versus Sr concentration in the bone determined by atomic absorption.

overlying tissue in cadaver legs has been determined directly by excision of the tissue. In vivo, the overlying tissue thickness can be determined ultrasonically with an accuracy of 0.3 mm with the use of a 7 MHz ultrasound transducer (12). Alternatively, the measured signal can be normalized with respect to coherently and incoherently scattered radiation (13). An improved response to the Sr signal can be attained by using polarized radiation to excite the Sr characteristic radiation (12). ^{125}I and ^{109}Cd were found to be equally effective sources for the Sr excitation.

Since most of the work reported in the literature deals with Sr transport and balance in human body, a noninvasive method to measure the steady state condition of Sr in the bone facilitate a new insight into the role of Sr in metabolic bone diseases, and epidemiological studies of lower levels of caries prevalent in populations with higher levels of Sr in drinking water.

We thank Dr. J. Rosen for preparing the bone samples and we acknowledge the assistance of Mr. G. Boykin and Ms. A. Kronenberg.

REFERENCES

1. H. A. Schroeder, I. H. Tipton, and A. P. Nason, Trace metals in man: strontium and barium, J. Chron. Dis. 25, 491-517 (1972).
2. Gi-Ichino Tanaka, Hisao Kawamura, and Etsuko Nomura, Distribution of Strontium in the skeleton and in the mass of mineralized bone, Health Phys. 40, 601-614 (1981).
3. G. Mazzuoli, E. Biagi, and G. Coen, Comparison of intravenous infusion of stable strontium with the Calcium Tolerance Test, Acta Medica Scan. 170, 21-30 (1961).
4. S. H. Cohn, S. W. Lippincott, E. A. Gusmano, J. S. Robertson, Comparative kinetics of Ca-47 and Sr-85 in man, Rad. Research 119, 104-119 (1963).
5. G. E. W. Wolstenholme and C. M. O'Connor, Bone structure and metabolism, p^{252}, J. & A. Churchill Ltd., London (1956).
6. G. Gothoni, Successful strontium therapy in an adult case of the Fanconi Syndrome, Acta Medica Scan. 170, 111-116 (1961).
7. F. E. McCaslin and J. M. Jones, Effects of stable strontium in treatment of Osteoporosis, Ed. S. C. Skoryna, Handbook of Stable Strontium, pp 563-579, Plenum Press (1981).
8. M. E. J. Curzon and P. C. Spector, Strontium in human dental enamel, pp 581-592, ibid.
9. J. A. Spadaro and R. D. Becker, The distribution of trace metal ions in bone and tendon, Calc. Tiss. Res. 6, 49-54 (1970).
10. R. O. Becker, J. A. Spadaro, and E. W. Berg, The trace elements of human bone, J. Bone and Joint Surg. 50, 326-334 (1968).
11. L. Wielopolski, D. N. Slatkin, D. Vartsky, K. J. Ellis and S. H. Cohn, Feasibility study for the in vivo measurement of lead in bone using L x-ray fluorescence, IEEE Trans. Nucl. Science, 28, 114-116 (1981).
12. L. Wielopolski, J. F. Rosen, D. N. Slatkin, D. Vartsky, K. J. Ellis, and S. H. Cohn, Feasibility of non-invasive analysis of lead in the human tibia by soft x-ray fluorescence, submitted for publication in Medical Physics.
13. R. E. Snyder and D. C. Second, The in situ measurement of strontium content in bone using x-ray fluorescence analysis, Phys. Med. Biol., 27, 515-529 (1982).

DETERMINATION OF BORON OXIDE IN GLASS BY X-RAY FLUORESCENCE ANALYSIS

T. Arai and T. Sohmura

Rigaku Industrial Corp., Osaka Japan

H. Tamenori

Nippon Sheet Glass Co., LTD., Hyogo, Japan

INTRODUCTION

In the last few years, at the Denver X-Ray Analytical Conference, the author and co-workers presented two papers which described the principle and applications of carbon analysis by X-ray fluorescence based upon a monochromatization technique consisting of total reflection and filtering. Instead of the wavelength dispersive method based on Bragg reflection, this monochromatization, combining total reflection by a selected mirror and an appropriate filter, offered an alternative approach for the purpose of increasing measured X-ray intensity.[1] The analytical performance of quantitative determination of carbon content in steel, cast iron and coal were reported.[2,3]

In this report, as the extension of X-ray fluorescence analysis for soft and ultrasoft X-rays, the principle and instrumental developments for monochromatization of B-K X-rays and the analytical application to boron oxide in glass are described.

The determination of boron oxide in glass is important as an analytical task for the glass industry and as a development in soft X-ray instrumentation. Since the boron oxide content can have a marked influence on a number of properties, (i.e. hardness, strength, heat and chemical-resistance, small thermal expansion, phase-separation, etc.) the quantitative analysis of boron oxide content is quite important for production control and quality of glass. Because it is not easy to estimate the content of boron oxide during the melting process because of the high volatility of boron oxide from molton glass, rapid analysis of glass is required in place of regular chemical analysis, which has higher sensitivity and is more accurate than the X-ray fluorescence analysis. However, X-ray fluorescence analysis is simple and fast and can be applied simultaneously to the determination of other constituents in glass.

Fundamentals established at the beginning of this study were: First, because the emitted X-ray intensity by electron or proton excitation, which has high excitation efficiency, is strongly influenced by the surface condition of an analyzing sample, X-ray excitation was adopted. Second, the X-ray excitation method is more suitable to the analysis of low atomic number elements because the analyzing depth of the X-ray excitation method is much deeper than that of electron and proton excitation techniques. Third, a regular sealed-off X-ray tube should be used for convenience in routine laboratory applications.

In order to compensate for the intensity reduction which arises from the low efficiency of the X-ray excitation method, a high efficiency monochromator was adopted, since specular reflection of characteristic X-rays can become nearly total external reflection at a small grazing incidence angle and its reflected intensity is very high. A total reflection mirror and an appropriate filter have been studied as a monochromator for B-K X-ray measurement and the analysis for boron oxide in glass was performed.

PRINCIPLE OF MONOCHROMATIZATION

In order to design a total reflection monochromator, the selection of a mirror material for B-K X-rays and the determination of a setting angle of the selected mirror were made based upon the totally-reflected intensity curve versus glancing angle (abbreviated to I-ϕ curve), which was calculated with the Fresnel equation derived by Henke,[4] using physical constants of bulk materials. In Fig. (1) the calculated I-ϕ curves of various mirror materials are shown for B-K X-rays and in Fig. (2) for C-K X-rays. It is shown that, when a mirror material has a small absorption coefficient for irradiating X-rays, the intensity of total reflection becomes high and, when it has a large absorption coefficient, the intensity is low. Since these I-ϕ curves are strongly dependent upon the linear absorption coefficient, the linear absorption coefficients of mirror materials for B-K X-rays are listed in Table (1) for preliminary selection of mirror materials.[5,6] Since the mirror material to be used should have a small absorption coefficient for B-K X-rays and a large absorption coefficient for C-K X-rays, which are the interfering X-rays, it is shown in Fig. (1) and (2) that the mirror material can be selected from Mo-metal, Nb-metal, boron and graphite. For the purpose of selecting mirror materials more precisely, the ratio of dispersing reflection angle (abbreviated to RDA) was introduced, which was discussed in a previous paper.[3] After looking for the mirror material which has a small RDA, a molybdenum mirror (Mo-mirror)

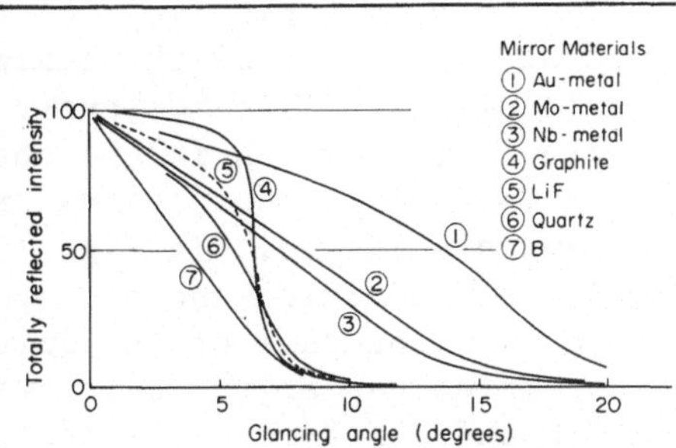

Figure 1 — Calculated I-ϕ curves of various mirror materials for C-K X-rays

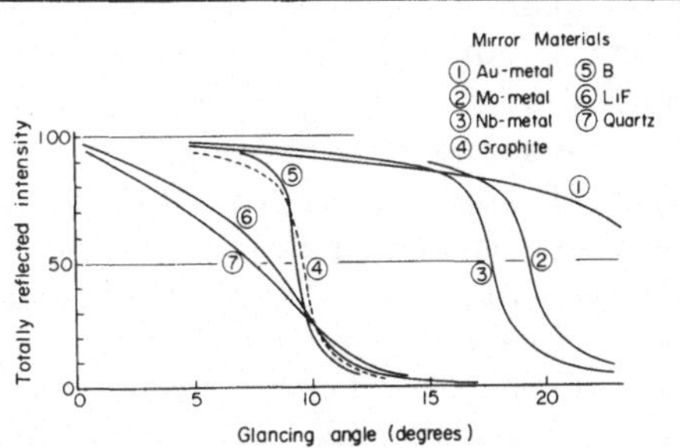

Figure 2 — Calculated I-ϕ curves of various mirror materials for B-K X-rays

Table 1 — Calculated fundamental parameters of mirror materials for total reflection of B-K X-rays

Mirror materials	Linear absorption coeficient for B-Kα	RDA	$\emptyset_{50}$		$\emptyset_{SA}$
			$\emptyset_{max}$	$\emptyset_{min}$*	
B	7.75×10^3	0.030	9.39	3.88	6.63
Graphite	14.6	0.047	9.64	6.39	8.01
Nb-metal	37.8	0.028	17.69	7.37	12.53
Mo-metal	48.5	0.025	19.29	8.18	13.73
LiF	68.5	0.21	8.42	6.28	7.35
Quartz	115.	0.32	7.51	5.64	6.58
Au-metal	173.	0.068	24.55	13.84	19.20

* $\emptyset_{min}$ is devided from C-Kα

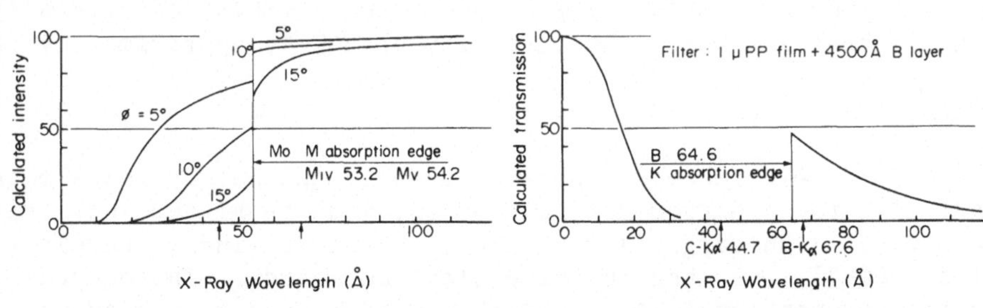

Figure 3 — Total reflection and transmitted intensity versus X-ray wavelength

Table 2 — Comparison of 50 percent intensity and RDA of B-K X-rays as calculated, measured and reported

Mirror materials	$\emptyset_{50}$ (degrees) Calculated	Measured	Reported	RDA Calculated	Measured	$\emptyset_{SA}$ (degrees) Calculated	Measured
Nb-metal	17.69	7.35		0.028	0.16	12.53	5.45
Mo-metal	19.29	7.85		0.025	0.17	13.73	5.03
Glass	6.84*	6.25		0.32*	0.29	5.99	6.00
Quartz	7.51	5.90		0.32	0.31	6.58	5.88
LiF	8.42	6.60	5.4	0.21	0.38	7.35	6.15
Graphite	9.64	—	6.3	0.054	—	8.01	—
B	9.39	2.65		0.030	0.28	6.63	3.35
Au-metal	24.55	4.18	5.6	0.068	0.61	19.20	3.36

* Fused SiO_2 ** after Lukixskii (1964)

was adopted as monochromator medium. The setting angle (ϕ_{SA}) of the Mo-mirror for B-K X-rays was determined at the center of the maximum glancing angle (ϕ_{max}) which is the angle of 50% of the I-ϕ curve of B-K X-rays and the minimum glancing angle (ϕ_{min}) which is the angle of 50% of the I-ϕ of the C-K X-rays. In Table (1), the setting angles of various mirror materials for B-K X-rays are listed. In Fig. (3), the calculated total reflection intensity of Mo-mirror and a composite filter versus X-ray wavelength are shown. In the case of the intensity curve of the Mo-mirror, there is an absorption gap at the X-ray wavelength of about 54Å, which is between C-K and B-K X-rays. A filtering technique can be applied to eliminate longer and shorter wavelength X-rays than those of B-K. The filter used was the combination of a one micron polypropylene film and a boron layer of 4500Å which resulted in intensity reductions of 50% for B-K and 98.3% for C-K X-rays. In Fig. (3) the calculated transmission of a composite filter is shown and the influence of the absorption gap of boron located at 64.6Å is illustrated clearly, Combining the total reflection and transmission of a filtering film, it is possible to detect X-rays of about 20Å wavelength but most of these X-rays can be discriminated by our means of pulse height selection based upon the energy difference betweeen B-K and 20Å X-rays. When interfering X-rays have the same wavelength as the measuring X-rays, strongly transmitted X-rays may be observed because of the small absorption of the filtering materials. In the case of interfering X-rays, which have longer wavelength than that of the measuring X-rays, a reduction of interfering effects may be obtained by the use of a filter because of the difference in absorption between B-K and the interfering X-rays. The synthesized intensity of B-K is about 40% while the intensity of C-K is approximately 1%.

EXPERIMENTAL AND RESULTS

The X-ray fluorescent apparatus used in this study was a regular sequential X-ray spectrometer, the S/Max, made by Rigaku Industrial Corporation. A Rh target X-ray tube with a thin window, namely a Machlett OEG-75, was used as the excitation source. The detector was a gas flow proportional counter with a window made from a one micron polypropylene film covered with an evaporated layer of aluminum, coupled with a low noise preamplifier. Because of the broadness of the pulse height distribution of soft X-rays, monochromatization using pulse height selection was not effective.

Testing mirrors for total reflection of various X-ray wavelengths, which were flat glass plates coated with selected metals and specially prepared solid plates, were investigated as the dispersing medium. A scanning goniometer was used for the measurement of the I-ϕ curves of various materials, using a narrow beam of parallel X-rays which were emitted from pure materials.

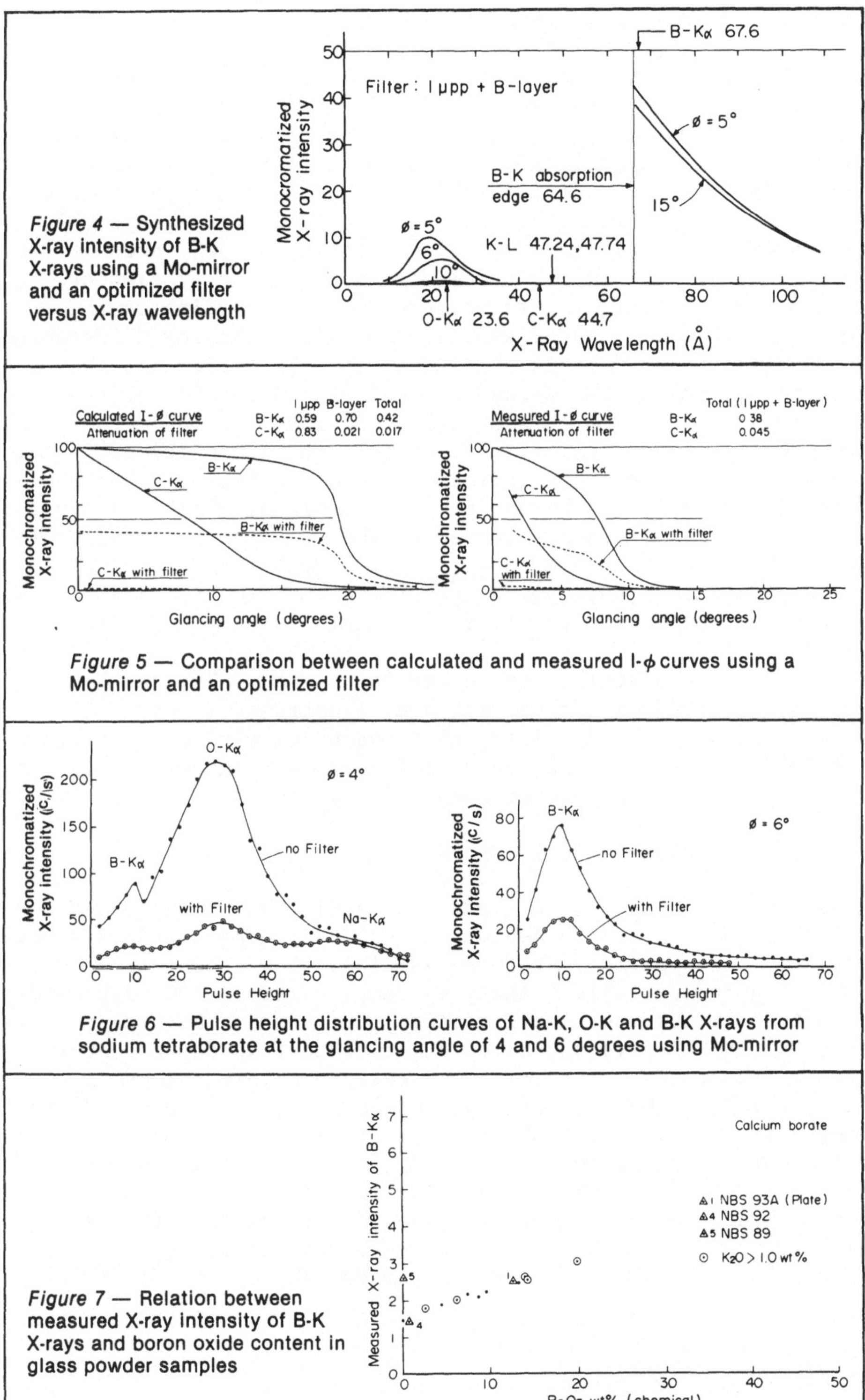

Figure 4 — Synthesized X-ray intensity of B-K X-rays using a Mo-mirror and an optimized filter versus X-ray wavelength

Figure 5 — Comparison between calculated and measured I-ϕ curves using a Mo-mirror and an optimized filter

Figure 6 — Pulse height distribution curves of Na-K, O-K and B-K X-rays from sodium tetraborate at the glancing angle of 4 and 6 degrees using Mo-mirror

Figure 7 — Relation between measured X-ray intensity of B-K X-rays and boron oxide content in glass powder samples

EXPERIMENTAL RESULTS AND DISCUSSIONS

Synthesized I-ϕ curves of B-K and C-K rays are shown in the upper part of Fig. (4) and measured I-ϕ curves in the lower part of Fig. (4), using Mo-mirrors. The I-ϕ curves of no-filter measurements are indicated with a solid line and of with-filter measurements with a dotted line. A large difference was found between synthesized and measured I-ϕ curves, which may have been caused by the physical condition of the Mo-mirror surface. In Table (2) 50% intensity angles (ϕ_{50}) are almost the same, but calculated ϕ_{50} of heavy element materials are much larger than measured ϕ_{50}. According to the difference of the calculated and measured ϕ_{50}, similar differences in RDA and ϕ_{SA} occur. It is concluded from the difference that in the case of Mo-mirror for B-K X-rays the calculated ϕ_{SA} is 13.73 degrees, but the measured ϕ_{SA}, which should be used for B-Kα monochromatization is 5.03 degrees. When the filter is used, the intensity of C-Kα is very low because of the absorption of the boron filter. For the practical application of boron analysis in glass, which contained potassium oxide as a minor component, the large setting angle of 6 degrees rather than measured ϕ_{SA} of 5.03 degrees was adopted in order to reduce the overlapping interference of the L series X-rays of potassium, whose is very similar to that of the C-K X-rays.

Pulse height distribution curves for Na-Kα, O-Kα, and B-Kα generated from sodium tetroborate are illustrated in Fig. (5) at glancing angles of the Mo-mirror of 4 and 6 degrees, without and with filter. At a glancing angle of 6 degrees, monochromatization for B-K X-rays is accomplished when a filter is used.

APPLICATION OF BORON OXIDE IN GLASS

Boron oxide analysis in glass was carried out using NBS standard samples (89, 92 and 93 A-sheet glass), Japanese standard samples (R501, R-1) and specially prepared samples, the content of which was determined chemically. Analyzed samples were < 200 mesh powders, pressed into briquettes. In Fig. (6), the relation between measured intensity of B-K X-rays and boron oxide content is shown. At the content of about 43%, there are two measured points for calcium borated, which is the raw material for borosilicate glass. A small influence from the L series X-rays of potassium was found, notwithstanding the use of boron filters. The increase in measured intensity of NBS-89 results from the overlapping interference of the N series X-rays of lead. A correction for these interferences was made using Eq. (1); a,b and d for K_2O were decided by the least square method and d for PbO was calculated from experimental data to fit Eq. (2).

$$wt\% = Ia + b \sum_j dj\ Wj \quad (1)$$

Where wt% : X-Ray Content
I : Measured X-Ray Intensity
Wj : Interfering Element Content (wt%) of j component
lj : Correcting coefficient of j component
a and b : Experimentally Determined Constant

$$B_2O_3\ wt\% - 0.1218\ I - 17.19 - 0.125W_{K_2O} - 0.69W_{PbO} \quad (2)$$

Table (2) gives the analytical results for glass samples showing chemically determined contents of boron oxide, measured X-ray intensities and analytical accuracy. The measured X-ray precisions for low content and 19.0% samples were 0.2 and 0.42% respectively. The measured sample precision for calcium borate was the same as the measured X-ray precision which was obtained from measuring ten samples prepared from the well mixed powder. The analytical accuracies varied between 0.5 and 1% for boron oxide content from 1 to 20%, respectively, depending on the laboratory.

Table 3 — X-ray analytical content of boron oxide in glass samples

Sample No.	Kinds of glass	B_2O_3 (chemical)	X - Ray intensity (I)	B_2O_3 (X - Ray)	Δ (chemi-X-ray)	K_2O	PbO
R - 501	High boron borosilicate	13.2 wt%	245.7 c/s	12.65 wt%	-0.55 wt%	0.30 wt%	wt %
S - 2	E glass for fiber	8.6	207.3	8.01	-0.59	0.20	
3	for optical	19.9	302.4	18.92	-0.98	2.8	
4	for thermos	13.9	262.3	14.47	+0.57	1.1	
5	for microwave oven	14.2	255.4	13.88	-0.32	0.1	
6	for watch cover	2.5	178.4	2.58	+0.08	7.7	
R - 1	Soda-Lime for sheet	0.0	145.1	0.26	+0.26	0.84	
NBS - 92	Low boron borosilicate	0.7	143.6	0.16	-0.54	(0.55)	
93A	High boron borosilicate	12.56	249.7	13.21	+0.65	0.01	
89	Lead - Barium	0.0	257.7	—	—	8.40	17.50
NP - 20	Borosilicate	4.38	186.6	5.33	+0.95	0.80	
21	Borosilicate	8.39	216.1	9.03	+0.64	0.35	
22	Borosilicate	6.21	201.2	7.04	+0.83	(1.08)	
23	Borosilicate	9.54	219.7	9.48	-0.06	0.33	
30	Calcium - Borate	43.6	729.2		σ = ±0.65		
31	Calcium - Borate	45.5	719.7				

Calculating Formula $wt\% = 0.1218\ I - 17.19 - 0.25\ W_{K_2O} - 0.69\ W_{PbO}$

CONCLUSION

An X-ray fluorescence spectrometer using a total reflection monochromator for the detection of boron K series has been developed on the basis of commercially available instrumentation. The combination of total reflection mirror and appropriate filter for boron analysis was investigated. Boron oxide in glass in the range of 1.0 to 20% can be analyzed with a precision of 0.2% at the lower content and 0.42% at 19.0%; the analytical accuracy was 0.65% after correcting, for K_2O and PbO. The main problem is the difference between X-ray precision and analytical accuracy which should be improved in the future to better meet the analytical requirements in glass industry applications.

ACKNOWLEDGEMENTS

The authors express their greatest gratitude for kind consideration rendered by Mr. Hikaru P. Shimura, President of Rigaku Ind. Corporation and Mr. A. Umayahura, Manager of Laboratory of Nippon Sheet Glass Company, LTD.

REFERENCES

1) T. Asai, Japanese J. Appl. Phys. vol. 21 (1982) p. 1347

2) M. J. Janiak, T. Utaka and T. Arai, 28th Denver X-Ray Conference(1979

3) M. J. Janiak, M. Funahashi and T. Arai, 29th Denver X-Ray Conference (1980)

4) B. L. Henke, Phys. Rev. vol. 6 (1972),p. 94

5) B. L. Henke and E. S. Ebisu, Advances in X-Ray Analysis vol. 17 (1974), p. 150

6) B. L. Henke and M. L. Schattenburg, Advances in X-Ray Analysis, vol. 19 (1976), p. 749

7) A. P. Lukirskii, E. P. Svinov, O. A. Ershov and Yu. F. Shepelev, Opt. & Spectrosk (*USSR) vol. 16 (1964), p. 186

MEASUREMENT OF COMPOSITION AND THICKNESS FOR SINGLE LAYER COATING WITH ENERGY DISPERSIVE XRF ANALYSIS

Robert Shen and Alan Sandborg

EDAX Laboratories
EDAX International, Inc.
Prairie View, IL 60069

INTRODUCTION

Methods for rapid and accurate analysis of coating composition and thickness are of major importance for the semiconductor industry. Typical coating problems are Al, Au or Pt coatings on metal and Cr/Ni alloy or SiO_2/P_2O_5 glass on Si wafers. Due to the nature of the coatings and the variety of elements, a software package, COAT95 has been developed to cover the following three possible situations:

1. Single element coating for thickness only.

2. Compound, elements A and B, coating on substrate element S, for composition and thickness.

3. Compound, elements A and S, coating on substrate element S, for composition and thickness.

The main goal of the COAT95 software is to have all possible models for the single layer coating problems of the semiconductor industry. Within each model, the user can still select several different options to fit his particular requirement. Of course, the COAT95 software can also apply to other coating problems. The software will run on an EDAX PV9500 system with 32K DEC LSI-11 computer and dual floppy disk drives.

EQUATION AND METHOD

1. Single element coating for thickness only:

The user can select one of the two models, which are as follows:

$$T = - a * \log (I_1 / I_2) + b \quad (1)$$

$$T = - a * \log (1 - I_3 / I_2) + b \quad (2)$$

Here T and I are thickness and intensity. a and b are slope and intercept. In normal cases, I_1 is the intensity of the element in the substrate and I_3 is the intensity of the element in the coating. I_2 is the intensity of the corresponding pure element. However, in some special cases, for example a compound coating, as long as the composition is fixed, I_1, I_2 and I_3 may be any elements as long as the curve fit nicely. The user also can choose to use the equations with or without the log factor.

2. Compound A and B coating on S substrate:

$$C_A = a * \log (I_A / I_B) + b \quad (3)$$

$$T = \frac{a}{C_A + C_B\mu_B/\mu_A} * \log (I_1 / I_2) + b \quad (4)$$

Here, C and μ are concentration and mass absorption coefficients at the energy of the emission line for the substrate element. In EQ(3), the intensity ratio is used. So the composition calculation will be independent of thickness variation. Other notations are defined same as in Model 1.

3. Compound A and S coating on S substrate:

$$C_A = \frac{I_A}{F * (a_1 + a_2F + a_3F^2 + . . .)} \quad (5)$$

$$T = a * \log (1 - I_1 / I_2) + b \quad (6)$$

Here F term can be coating thickness (T) or intensity ratio of the substrate element to its pure element (I_1/I_2). The terms in parenthesis in EQ(5) are the polynomial expression of (1-EXP(-$\mu\rho$T)) term.

In the normal case, the thickness should be used. However, if the given thickness values in standards are not accurate, then intensity ratio (I_1/I_2) may be used. In EQ(6), the composition of the coating is assumed not change too much. EQ(6) gives a relatively poor fit to the experimental data based on our experience.

With each model, a set of standards are needed to calculate the coefficients using linear least square fit method. Then the standards or unknowns are recalculated using the coefficients determined from the fit.

EXPERIMENT AND RESULTS

The standard samples used here come from a leading semiconductor company. There is no confirmation for the given thickness and concentration values. A standard EDAX PV9500 system with Rh target x-ray tube was used in this study.

1. Single element coating for thickness only:

A set of standards for aluminum coating on silicon wafers was used. The range of coating thickness was 0.5μm to 1.7μm. Since the peak to background ratio was very high, gross intensities were used. Two models of calculation were tried to fit the data. One was using EQ(1) and the other was, in EQ(1), replacing I_1/I_2 with I_3/I_1, intensity ratio of Al to Si. The results are listed in Table 1 with the absolute average deviation 0.03μm and 0.017μm, respectively. Due to the coating thickness and/or the sample shape of the coating, the user can select any intensity of element or intensity ratio of elements to fit the data in this software.

2. Compound A and B coating on S substrate:

A few Cr/Ni compound coatings on Si wafer were used here. The typical coating thickness was only about 0.01μm. There is not much intensity for Cr and Ni. The measured gross intensities (counts per second) are listed in Table 2. The results calculated from the EQ(3) and (4) are also listed in Table 2. The absolute average deviation for the Cr concentration is 0.5% and the thickness is 0.0006μm. Here sample #C13 was omitted in the thickness calculation. Due to the limited samples, further study may be needed.

3. Compound A and S coating on S substrate:

One set of SiO_2/P_2O_5 glasses on Si wafer were used in this study. A simple peak strip method was used to get the net peak intensity. The absolute average deviation for the calculated P_2O_5 concentration was 0.14%. In the P_2O_5 calculation, the Si ratios were used instead of thickness data. The absolute average

TABLE 1: Al-COATING ON Si WAFERS

SAMPLE	MEASURED GROSS INTE(C/S)		THICKNESS (KÅ)#		
	Al	Si	GIVEN	CALC-1*	CALC-2*
A1	3510.69	2222.38	10.80	10.49	10.79
A2	2054.50	5680.90	5.00	5.33	4.98
A3	4393.15	918.03	17.00	**	**
A4	3153.31	2530.80	9.80	9.78	9.85
A5	3096.86	2982.01	9.60	8.88	9.14
A6	3690.38	1726.15	11.60	11.88	11.98
A7	4538.95	786.40	16.00	16.20	15.90
A8	2351.09	4816.70	6.00	6.24	6.16
AVE DEV				0.30	0.17

*CALC-1: I_1/I_2 is used
*CALC-2: I_1/I_3 is used
**Omitted in the CALC, # 1KÅ = 0.1μm

TABLE 2: Cr/Ni COATING ON Si WAFTER

MEASURED GROSS INTENSITIES (C/S):

SAMPLE	Si	Cr	Ni
C13	6748.21	67.96	27.64
C14	6720.29	80.46	24.24
C15	6685.66	45.96	43.06
C16	6761.03	84.24	20.60

GIVEN & CALC CONC(%) & THICKNESS (100Å):

	GIVEN		CALC.	
SAMPLE	CR(%)	(100Å)	CR(%)	(100Å)
C13	61.00	0.80	60.03	*
C14	67.00	1.40	67.30	1.32
C15	39.50	1.00	39.82	0.99
C16	72.00	1.20	72.35	1.29
AVE DEV			0.49	0.06

*OMITTED IN THE CALC.
ANALYSIS CONDITION: 20KV, 75μA, 200SEC

TABLE 3: SiO_2/P_2O_5 COATING ON Si WAFERS

SAMPLE	MEASURED NET INTE(C/S)		GIVEN P_2O_5	GIVEN THICKNESS
#	Si	P	(%)	(KÅ)
H01	7722.04	31.88	2.4	6.00
H02	7653.81	46.48	3.2	6.10
H03	7593.61	68.78	4.2	6.50
H04	7550.97	88.72	5.2	6.60
H05	7526.15	115.10	6.3	6.70
H06	7467.19	130.23	7.3	6.75
H07	7447.68	137.39	7.9	6.78
H08	7448.31	148.08	8.5	6.70
H09	7382.78	204.23	11.1	6.70
H10	7479.75	184.31	10.7	6.40

SAMPLE	P_2O_5 CONCENTRATION(%)		THICKNESS (KÅ)	
	CALC	DIFF	CALC	DIFF
H01	2.31	0.09	6.06	-0.06
H02	3.14	0.06	6.24	-0.14
H03	4.38	-0.18	6.36	0.14
H04	5.44	0.24	6.48	0.12
H05	*(6.91)		6.53	0.17
H06	7.45	-0.15	6.66	0.09
H07	7.74	0.16	6.69	0.09
H08	8.35	0.15	6.69	0.01
H09	10.96	0.14	6.82	-0.17
H10	10.65	0.05	6.63	-0.23
AVE DEV		0.14		0.12

*OMITTED IN THE CALC. & TREATED IT AS AN UNKNOWN
PURE Si INTENSITY: 8621.26 (C/S)
ANALYSIS CONDITION: 10KV, 300μA, 100SEC

deviation for the coating thickness was 0.12(KÅ). The calculated results were also listed in Table 3. Plots of given and calculated P_2O_5 and thickness from the data are examined. The P_2O_5 plot is good, but thickness plot is poor.

For the precision study, one sample has been analyzed for ten times without moving the sample. Then the same sample has been rotated 45 degrees before each analysis. The results show the relative standard deviation for the P_2O_5(%) is 1.9% and 2.2% for both with and without sample movement, respectively. The difference of the results do not seem affected significantly by the sample orientation.

DISCUSSION

In the single element coating, the users may select I_1, I_2 or I_3 for any element in EQ(1) or EQ(2). With different exciting voltages, the user may find that different elements should be used even for a same set of samples.

In compound A and B on S coating, the intensity ratio is used in EQ(3). That means the concentration calculation of the coating will be independent of the coating thickness, sample shape or system instability.

In the SiO_2/P_2O_5 on Si wafer study, the results for the thickness calculation are relatively poor. A better model will be needed for the thickness calculation. Fortunately, another non-EDS method is available now for the thickness measurement only.

The precision study shows that the variation of P_2O_5(%) is mainly due to x-ray counting error of the small P peak.

DETERMINATION OF LIGHT ELEMENTS ON THE CHEM-X MULTICHANNEL SPECTROMETER

Y.M. Gurvich, A. Buman, and I. Lokshin

Bausch & Lomb
9545 Wentworth St.
Sunland, CA 91040

For X-ray fluorescence determination of light elements, vacuum X-ray spectrometers are usually used to increase the transmission of the optical path for the soft X-radiation. A vacuum of about 50μHg is required for the determination of elements with atomic numbers down to 9(F).

A vacuum system however, complicates the sample presentation mechanism of a spectrometer; it requires vacuum seals, additional fastening of the flow detector window and, as a results, increases the price of the spectrometer. Besides, a vacuum has certain disadvantages for some applications:

- analysis of liquids requires a sophisticated sample cell or special sample preparation techniques that complicated the analytical procedure and are not always efficient;

- distension of the liquid cell window under vacuum may lead to its erruption and contamination of the X-ray chamber or, at least, to additional errors due to changes of the geometry of the system;

- volatile liquids are evaporated rapidly under vacuum causing inaccuracies in the X-ray analysis;

- briquettes may outgas or even rupture in a vacuum spectrometer unless previously outgassed in a vacuum chamber or oven [1];

- analysis of moist briquettes (frequently common in production control) requires prolonged evacuation of the sample chamber, sometimes for several minutes;

- analysis of loose powders in vacuum is practically impossible.

In the Chem-X, Bausch & Lomb's new 8 channel X-ray wavelength-dispersive spectrometer with low power X-ray tube, the analysis of light elements in powder, liquid and solid samples is performed in a He atmosphere. The sample stage is separated from the X-ray chamber by a primary film window in order to maintain a He atmosphere inside spectrometer; therefore, during analysis the sample is situated in a air medium. This permits the analysis of loose powders and eliminates numerous problems of liquid's analysis under vacuum.

A comparison of the features of the Chem-X and conventional X-ray multichannel and sequential spectrometers is shown in Table 1. Introduced at the 30th Annual Denver X-Ray Conference [2], the Chem-X was able to determine elements from Al(Z=13) up. Further experiments demonstrated Chem-X's capabilities for the determination of Mg(Z=12); this will extend significantly the application of the instrument.

A TLAP analyzing crystal was chosen for Mg determination because of its high reflective efficiency: It is 2.5 times higher than for RAP, about 4 times higher than for KAP [3], and more than an order of magnitude higher than for ADP (101), the widely recommended crystal for Mg determination. Our experiments have shown that the TLAP crystal was stable in a He atmosphere at least for several weeks, while the long term stability and behavior of many acid phthalates under vacuum is highly dependent on both the material and the thickness of the detector window and the primary window of the spectrometer. The standard window of the Chem-X is made of polycarbonate film which is about two times less transparent for the soft radiation than polypropylene film. For the determination of Mg, the special primary window was made from a stretched polypropylene, 2 - 2.5 μ thick. For MgKa radiation, the transmission of such film is about 75% - 80%.

Regular daily maintenance of the Chem-X includes periodic changing of the primary window which is constantly exposed to X-ray radiation: Once per 8, 24, or 48 hours, depending on the workload. The thin polypropylene film used for Mg determination demonstrated good durability and reproducibility of results for two days; therefore, its utilization does not change the routine Chem-X maintenance.

In a flow proportional counter, the standard 2 μ aluminized polycarbon film (as a window) and P10 gas (Ar/CH_4 mixture) were used for Mg determination. This is the same detector type that is

TABLE 1 Characteristics of the Chem-X and conventional WDX spectrometer

CHARACTERISTICS	CHEM-X	CONVENTIONAL WDX	
		Multichannel	Sequential
Range of Elements	Al - U	(C) F - U	F - U
Number of Monochromators	Up to 8	Up to 30	1
Power on X-ray tube	Up to 200 W	Up to 3 kW	Up to 2-3 kW
Water Cooling	Not Required	Required	
X-ray Path	He, air	Vacuum, air (He optional)	
X-ray tube / sample position	Sample is above	Tube is above	Tube is above or below
Sample types	Solid, liquid, loose powder	Solid, briquetted powder (liquid inconvenient)	
Replacement or adjustment of monochromators	By operator	By manufacturer	
Number of samples loaded	1	1-7 (250-300 optional)	4-10
Type of crystals	Flat	Curved, flat	Flat
Detectors	Flow & sealed proportional	Flow & sealed proportional, scintillators	
Integration time (sec.)	50-100	30-50	20-50 per element
Detection limit	10-30 ppm	1-10 ppm	1-10 ppm
Computer	Yes	Yes	Yes
Matrix effects correction	Yes	Yes	Yes
Drift correction	Yes	Yes	Yes
Typical price	$60,000 and lower	$180,000 to 300,000	$100,000 to 180,000

used for Al and Si determination. The resolution of the detectors was 33% for MgKα which is equivalent to 15% for FeKα; this agrees with characteristics of flow detectors previously published [1].

The resolution of the detector is important for Mg determination in various materials. In Fig. 1, the energy distribution of the Mg monochromator is shown, recorded from NBS Portland Cement standard SRM-636, which contains 3.9% MgO. On the right side of the figure, the ascending slope of the CaKα peak, 3rd order reflection, is seen (the maximum of the peak is outside of the PHA voltage range); on the left side, the escape peak of ArK-series stands out, due to an incomplete absorption of CaKα fluorescent radiation in the detector medium. The position of the escape peak is exactly proportional to the difference between the energies of the CaKα and ArKα fluorescent radiation.

The window setting for Mg determination should separate the analytical peak from both CaKα, 3rd order, and ArK-escape peaks. In our experiments, the best results (optimum sensitivity and line contrast, or BEC) were obtained with a 2.5-3V window.

Determination of Mg is important for the Chem-X application in the cement industry. Multichannel X-ray spectrometers with conventional X-ray tubes are quite routine analytical tools for production control and quality assurance in modern cement manufacturing. They provide determination of all necessary oxides, starting with Na_2O. After the Mg monochromator is commercially available, the Chem-X could cover similar analytical tasks. Determination of Na, if necessary, can be made by an inexpensive flame photometer that requires approximately 5 - 10 minutes for an analysis.

Results of Mg determination in cement on Chem-X are demonstrated in Fig. 2. Seven NBS Portland Cement Standards (SRM 633-639) were measured. The concave shape of the curve is due to the absorption of the MgKα soft radiation in the heavier components of the cement matrix (with $Z \geq 13$). The sensitivity of the monochromator is about 15 cps/% MgO; the limit of detection is 0.05% MgO for an integration time of 200 seconds.

Experiments show that the accuracies of analysis achieved on the Chem-X and X-ray spectrometers with regular X-ray tubes (2-3kW) are quite comparable. In Figs. 3-6, the analytical results of NBS standards are shown for major cement components: Al_2O_3, SiO_2, CaO and Fe_2O_3. The intensities of the Chem-X and a conventional WDX spectrometer are plotted in count rates per watt power on the X-ray tubes. In spite of different count rates, especially for light elements, analytical results of both instruments are similar: Deviation of individual data from calibration lines (which defines the accuracy of analysis) is the same.

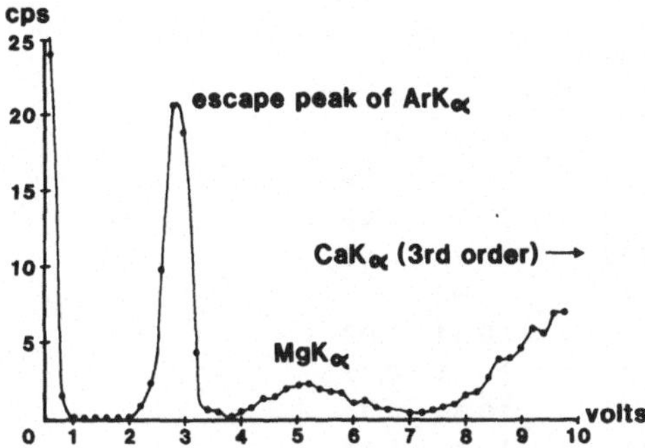

Fig. 1 Energy distribution of the Chem-X's Mg Monochromator

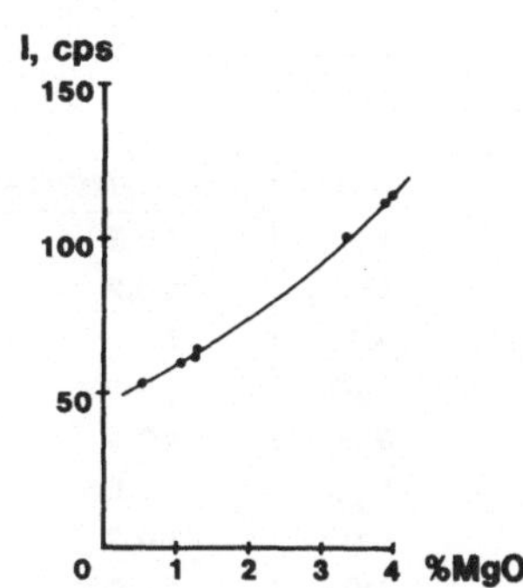

Fig. 2 Analysis of MgO in cement on the Chem-X

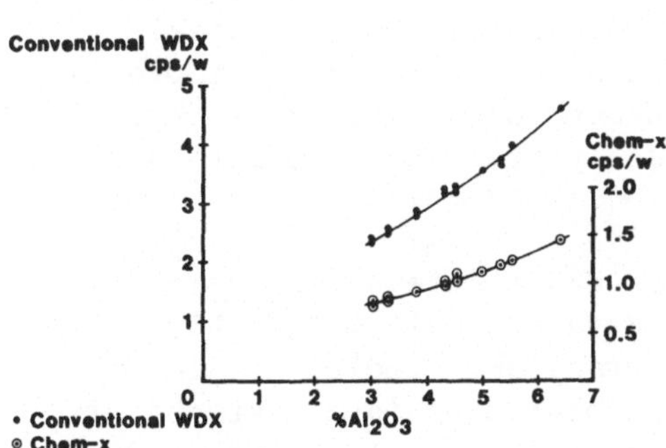

Fig. 3 X-Ray analysis of Al_2O_3 in cement

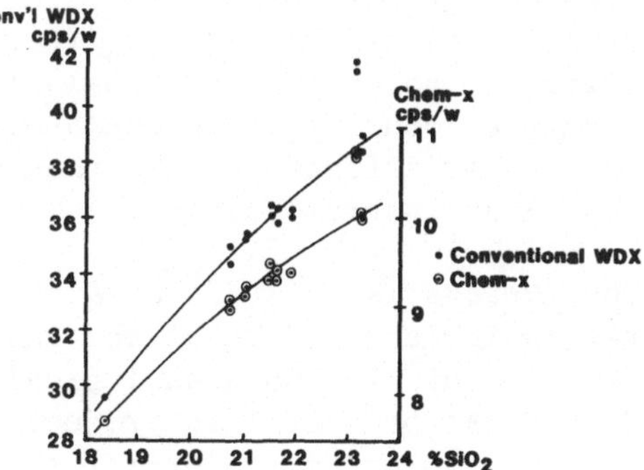

Fig. 4 X-Ray analysis of SiO_2 in cement

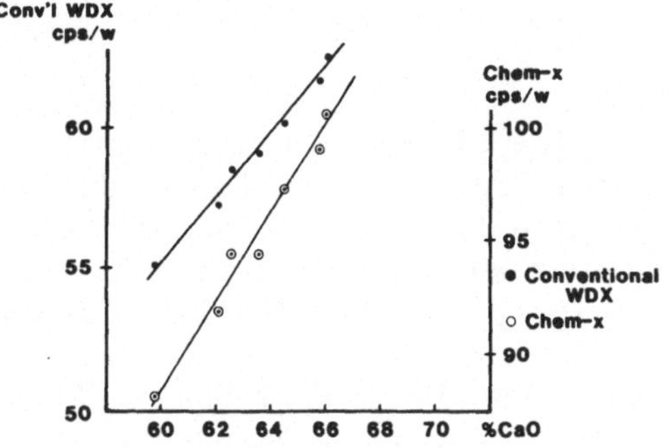

Fig. 5 X-Ray analysis of CaO in cement

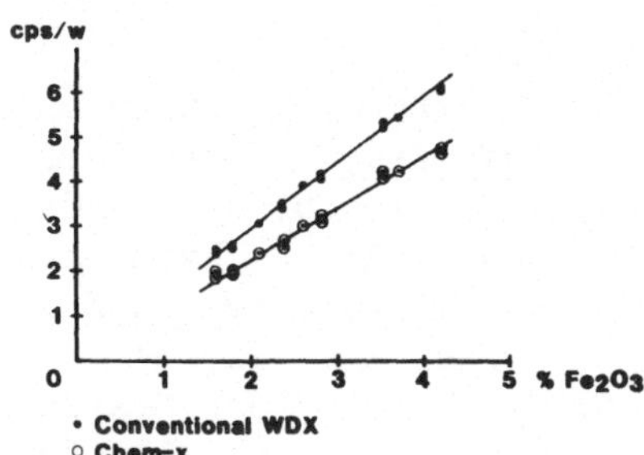

Fig. 6 X-Ray analysis of Fe_2O_3 in cement

TABLE 2 X-Ray analysis of NBS Portland Cement standards

	DEVIATION, % OXIDE				
		CHEM-X		CONV'L WDX SPECTROM.	
OXIDE	REQUIREMENTS	MAX.	σ	MAX.	σ
MgO	±0.2	±0.24	0.12	±0.18	0.09
Al_2O_3	±0.2	±0.07	0.04	±0.10	0.05
SiO_2	±0.2	±0.26	0.15	±0.26	0.15
P_2O_5	±0.03	±0.01	0.01	--	--
SO_3	±0.1	±0.10	0.07	±0.11	0.07
K_2O	±0.05	±0.03	0.02	±0.03	0.02
CaO	±0.3	±0.64	0.3	±0.39	0.3
TiO_2	±0.03	±0.02	0.01	±0.01	0.01
Fe_2O_3	±0.10	±0.07	0.05	±0.08	0.04

Even when the sample deviates obviously far from the calibration line, as in Fig. 4 for SiO_2 (perhaps due to the mineralogical effect), the errors are equal with both spectrometers. It proves that the accuracy of cement analysis depends on sample preparation and matrix effects, rather than on analytical characteristics of the X-ray instrumentation.

The average (standard) deviations for NBS standards are demonstrated in Table 2. Most of the Chem-X data are well within ASTM requirements for both analytical accuracy and precision. The Chem-X can be useful in production laboratories in all stages of cement manufacture: Certification of raw materials (limestone caly, iron ore, sand, coal, shale, slags, etc.), production control of raw mix, kiln feed, clinker, and quality assurance of finished cement.

References

1. E. P. Bertin, "Principles and Practice of X-Ray Spectrometric Analysis," Second Edition, Plenum Press, N.Y. 1079 pp. (1975).
2. J. Lucas-Tooth, B. W. Adamson, Y. M. Gurvich, "The Analysis of Copper Alloys by Chem-X, Low Power WDX Multichannel Spectrometer," Advances in X-Ray Analysis, V. 25" pp. 169-172, Plenum Press, N.Y. (1981).
3. A. J. Burek, "Long Grating Spacing Crystals," "Workshop on X-Ray Instrumentation for Synchrotron Radiation Research," Stanford Linear Accelerator Center, April 3-5, 1978; pp. III/8-16.

SIMULTANEOUS DETERMINATION OF 36 ELEMENTS BY X-RAY FLUORESCENCE SPECTROMETRY AS A PROSPECTING TOOL

Clive E. Feather and Fritz C. Baumgartner

Anglo American Research Laboratories
P.O. Box 106, Crown Mines 2025, South Africa

INTRODUCTION

In regional reconnaisance geochemical prospecting, it is desirable to determine as many elements as possible, in order to provide a comprehensive cover for the detection of most types of ore mineralization. Multi-element simultaneous instruments are ideally suited to this task, but the maximum number of wavelength dispersive channels (monochromators) which may be fitted to any one of the commercially available X-ray spectrometers is approximately thirty. Because some of these channels are required for the measurement of background, the practical limit for spectral lines is about twenty-five. Additional elements may be determined by inclusion of a scanning channel, but its sequential operation increases the analysis time per sample.

In what is believed to be a unique installation, Applied Research Laboratories (ARL - now Bausch and Lomb) were able to supply, to the authors' specifications, a computer-controlled ARL 72000S X-ray spectrometer which is fitted with twenty-eight monochromators, a Kevex 7000 gamma-X energy dispersive spectrometer (EDS) (in the position normally occupied by two monochromators), and a Herzog automatic sample feed with two interchangeable tables capable of delivering 600 samples per day. With this system, routinely thirty-six elements are determined simultaneously in pressed powder briquettes. The instrument is controlled by software custom-written by ARL, and data reduction is effected by programs written in BASIC by the authors at Anglo American Research Laboratories (AARL), in which all popular correction models are used. Instrument settings and a list of the elements which are determined, working ranges, limits of detection, and precisions obtainable are given in Table 1.

Table 1 : Instrument settings, elements determined, working range, theoretical lower limits of detection, and practical precision of the system.

Instrument : Combined ARL 72000S and Kevex 7000 gamma-X
Tube : End window OEG, rhodium target
Excitation : 60 kV, 40 mA

Element and line	Wavelength Å	Detector	Working Range		Precision (1 std. dev.)
			L.L.D. (3s)	Upper	
F Kα	18.31	Ar f.p.	0.5%	11%	0.3 % at 1.7%
Na Kα	11.91	Ar f.p.	0.1%	5%	0.2 % at 2.9%
Mg Kα	9.89	Ar s.p.	0.1%	10%	0.2 % at 2.0%
Al Kα	8.34	Si(Li)	0.5%	10%	1.1 % at 10.8%
Si Kα	7.13	Si(Li)	0.5%	30%	1.0 % at 30.0%
P Kα	6.16	Si(Li)	0.1%	5%	0.1 % at 1.0%
S Kα	5.37	Si(Li)	0.1%	5%	0.1 % at 1.0%
K Kα	3.74	Si(Li)	0.1%	4%	0.18% at 4.4%
Ca Kα	3.36	Si(Li)	0.1%	15%	0.07% at 0.8%
Ti Kα	2.750	Si(Li)	0.1%	8%	0.12% at 2.5%
V Kα	2.505	Si(Li)	0.1%	8%	0.10% at 2.3%
Cr Kα	2.291	Si(Li)	0.1%	8%	0.10% at 5.0%
Mn Kα	2.103	Si(Li)	0.1%	6%	0.10% at 1.0%
Fe Kα	1.937	Si(Li)	0.1%	15%	0.14% at 6.7%
Co Kα	1.790	Ne s.p.	5 ppm	1100 ppm	3.9 ppm at 100 ppm
Ni Kα	1.659	Ar s.p.	5 ppm	3100 ppm	1.6 ppm at 100 ppm
Cu Kα	1.542	Xe s.p.	5 ppm	1000 ppm	3.0 ppm at 100 ppm
Zn Kα	1.436	Xe s.p.	5 ppm	1000 ppm	6.1 ppm at 100 ppm
As Kβ	1.057	Xe s.p.	5 ppm	1000 ppm	5.1 ppm at 100 ppm
Se Kα	1.106	Xe s.p.	2 ppm	1000 ppm	1.3 ppm at 100 ppm
Rb Kα	0.927	Xe s.p.	2 ppm	1400 ppm	2.5 ppm at 100 ppm
Sr Kα	0.877	Kr s.p.	2 ppm	1000 ppm	0.9 ppm at 100 ppm
Y Kα	0.830	Xe s.p.	2 ppm	1000 ppm	1.7 ppm at 100 ppm
Zr Kα	0.787	Xe s.p.	2 ppm	5000 ppm	0.9 ppm at 100 ppm
Nb Kα	0.748	Xe s.p.	1 ppm	1000 ppm	1.0 ppm at 100 ppm
Mo Kα	0.711	Xe s.p.	1 ppm	1000 ppm	1.7 ppm at 100 ppm
Sn Kα	0.492	Xe s.p.	7 ppm	1000 ppm	7.0 ppm at 100 ppm
Sb Kα	0.472	Xe s.p.	7 ppm	1000 ppm	7.7 ppm at 100 ppm
Te Kα	0.453	Xe s.p.	7 ppm	1000 ppm	7.8 ppm at 100 ppm
Ba Lβ2	2.404	Kr s.p.	50 ppm	1100 ppm	38 ppm at 300 ppm
Ta Lα	1.522	Xe s.p.	5 ppm	1000 ppm	2.0 ppm at 100 ppm
W Lα	1.476	Xe s.p.	5 ppm	1300 ppm	3.0 ppm at 100 ppm
Pb Lβ	0.982	Xe s.p.	3 ppm	1000 ppm	3.9 ppm at 100 ppm
Bi Lα	1.144	Xe s.p.	3 ppm	1000 ppm	3.4 ppm at 100 ppm
Th Lα	0.956	Xe s.p.	2 ppm	1000 ppm	4.1 ppm at 100 ppm
U Lα	0.911	Xe s.p.	2 ppm	1000 ppm	2.0 ppm at 100 ppm

NOTE: f.p. = flow gas proportional counter (ARL 72000S)
s.p. = sealed gas proportional counter (ARL 72000S)
Si(Li) = 10 mm^2 lithium drifted silicon detector (Kevex)
Precision is calculated from data accumulated over six weeks by measurement of control standards at the end of each daily run.

SAMPLE PREPARATION

Sample preparation is simple and rapid. Geochemical prospecting loam samples are generally sieved through an 80 mesh (177 micrometres) screen, and the undersize is mixed with 9% styrene copolymer and 1% paraffin wax binder[1]. Rocks are pulverized before adding the binders. The mixture is milled for four minutes in 100 ml vessels in a six-vessel Siebtechnik vibro-rotary milling unit, and pressed at 30 tons into 30 mm (12 g) or 40 mm (16 g) diameter briquettes. Both standards and samples are prepared in the same manner. At least every twentieth sample is prepared in duplicate, and the replicates are run separately at the end of each run together with control standards.

DETERMINATION OF MAJOR AND MINOR ELEMENTS

Fourteen major and minor elements are determined semi-quantitatively by using a combination of the EDS (Al, Si, P, S, K, Ca, Ti, V, Cr, Mn and Fe) and monochromators (F, Na and Mg). Internationally accepted standards, chosen to cover a wide range of matrices, were used to generate inter-element correction factors by multiple regression analysis using the computer program XRF4 by H.E. Marr[2]. To some degree, these factors also compensate for mineralogical effects, well known to be a principal source in error when using powder briquettes at long wavelengths. For example, in soil samples, abundant silica is usually due to the presence of quartz, alumina relates to clay minerals, and Fe_2O_3 to hematite.

TRACE ELEMENTS

Twenty-two trace elements (Co, Ni, Cu, Zn, As, Se, Rb, Sr, Y, Nb, Zr, Mo, Sn, Sb, Te, Ba, Ta, W, Pb, Bi, Th, and U) are determined by using monochromators and the principles of Reynolds[3,4] to determine mass absorption coefficient, and by the method of Feather and Willis[5] in which interference-free background measurements are used to calculate the backgrounds beneath adjacent spectral peaks. In the latter method, by using oxide blanks, background at a chosen wavelength may be related to the reciprocal of mass absorption coefficients at that wavelength. The linear relationship obtained intercepts the intensity axis (Figure 1), implying that part of the background is not matrix dependent, and which Feather and Willis[5] have called the residual background.

Provided that a major element absorption edge does not lie between two spectral positions, the mass absorption coefficients at the two wavelengths are directly related. Thus, when comparing two background positions, the matrix dependent components are also directly related, and the intercept relates to the difference in the residual backgrounds at the two spectral positions (Figure 2).

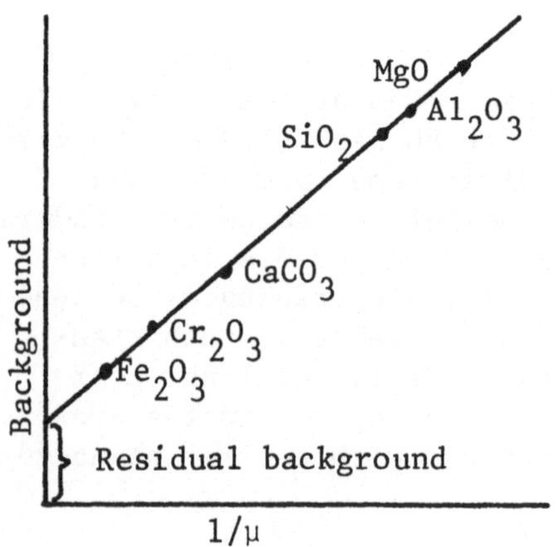

Figure 1. Background vs reciprocal of mass absorption coefficient (μ).

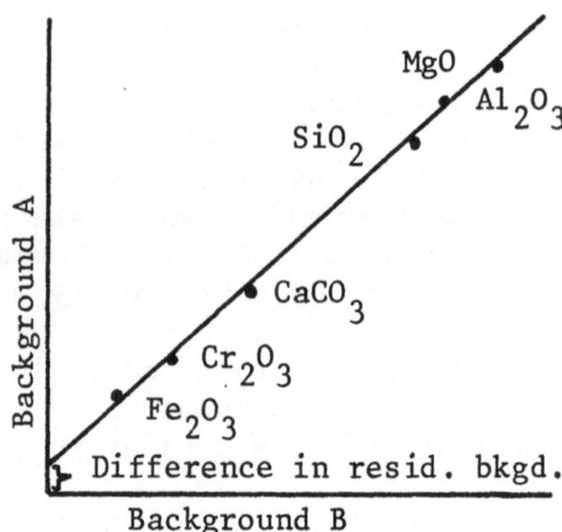

Figure 2. Relationship of backgrounds at differing wavelengths.

It is possible to calculate background beneath several spectral peaks from a single background monitoring position. Because of high intensity, a tube line Compton peak is normally chosen. In the present application Rh Kβ Compton was chosen, which is interference-free when analyzing geochemical prospecting samples. When dealing with simultaneous spectrometers, the individual channel performance differences (e.g. crystal reflectance, detector efficiency) are constants in the background relationship.

VARIATIONS IN BACKGROUND

When examined in detail, the background relationships are not linear, and some of the plotted points for the oxide blanks do not lie on the best fit regression lines, despite which instrument is used. In addition to the ARL 72000S and Kevex EDS, tests have been carried out in Philips PW1220, PW1270, PW1400 and Siemens SRS100 spectrometers, giving similar results. Scatter was also shown when using a comprehensive set of well analysed natural samples including igneous rocks (from dunite to granite), and some sedimentary rocks. The spectral positions which were chosen were BiLα, WLα, TeKα, SbKα, and SeKα. These elements occur at levels below detection limits in ordinary rocks and soils. Deviations from the best fit regression lines of background vs Compton peak, when expressed in terms of concentration, were shown to be most significant indeed (Table 2). The average absolute relative error for Bi was found to be 14 ppm, 14 ppm for W, 36 ppm for Te, 32 ppm for Sb, and 3 ppm for Se. Further investigation showed that the scatter in the background relationship was absent if glass discs of the blanks were prepared, e.g. by fusion with lithium tetraborate. It was thus clear that the scatter was due to the crystalline character of the individual constituents in the briquettes.

Table 2 : Summary of the comparison of results obtained for bismuth, tungsten, tellurium, antimony and selenium with and without corrections to calculated background using empirically determined major element influence factors. For all the elements in these samples the recommended value is <1 ppm. The average absolute errors of the corrected values is generally similar to the theoretical limits of detection.

Sample No.	Rock type	Bi (ppm)		W (ppm)		Te (ppm)		Sb (ppm)		Se (ppm)	
		no corr.	with corr.	no corr.	with corr.	no corr.	with corr.	no corr.	with corr.	no corr.	with corr.
C- 18 Sy-2	Syenite	18.9	0.3	- 5.5	4.4	- 76.9	19.3	-57.2	- 3.8	5.2	0.1
C-333 MRG-1	Gabbro	14.6	0.5	-15.0	1.5	- 41.2	- 3.2	-45.9	- 2.1	3.5	0.3
C-335 NIM-D	Dunite	0.7	0.4	-22.4	7.8	- 49.5	-12.3	-44.2	- 8.6	0.7	0.4
C-336 NIM-G	Granite	14.8	0.0	- 5.9	- 2.3	- 33.5	-13.0	-24.8	- 4.7	3.0	0.2
C-338 NIM-N	Norite	12.1	- 0.3	-24.2	- 5.5	- 41.6	8.1	-35.3	3.1	0.8	- 0.5
C-339 NIM-P	Pyroxene	5.6	- 0.6	-19.7	1.1	- 50.3	- 1.0	-44.8	- 1.4	0.4	0.0
C-340 NIM-S	Syenite	20.4	0.7	-18.7	- 0.9	- 24.6	5.9	-25.3	0.3	3.2	0.1
AARL 1	Granite	18.3	0.5	- 9.9	4.7	8.2	40.8	- 7.6	18.3	2.9	0.0
AARL 2	Granite	17.0	- 0.2	-12.4	- 8.4	- 10.7	19.3	-22.4	6.6	3.0	0.6
AARL 3	Granite	16.8	- 0.3	-11.0	0.0	- 23.5	13.3	-23.7	4.4	3.5	0.4
AARL 5	Syenite	31.6	1.2	- 6.1	- 3.3	-102.4	- 8.3	-85.2	- 5.8	6.2	0.1
AARL 6	Pyroxenite	10.5	- 1.1	-27.6	- 3.8	- 45,7	21.0	-42.1	8.0	2.2	- 0.1
AARL 7	Peridotite	3.1	0.2	20.9	- 2.7	- 6.8	4.5	13.1	2.8	0.4	- 0.2
AARL 8	Plagioclase	13.7	- 0.3	-18.0	3.5	0.4	4.0	3.8	3.2	0.0	- 0.2
AARL 9	Slate	10.8	0.2	- 4.3	0.7	-35.4	- 2.7	-29.5	- 3.8	0.8	- 0.1
AARL 10	Mica Sand	17.6	0.9	- 8.7	1.0	18.7	- 7.4	5.5	- 6.2	2.6	- 0.1
Theoretical limit of detection		3		5		7		7		2	

The major element concentrations of the natural samples were known, and it was decided to carry out multiple linear regression analysis in order to obtain factors which expressed the influence of the major elements on the actual background. The factors would also relate indirectly to the mineralogy of the samples. The results show a considerable improvement on the uncorrected results. Average absolute errors were reduced to about 1 ppm for Bi, 4 ppm for W, 12 ppm for Te, 8 ppm for Sb, and less than 1 ppm for Se, which approximate to the theoretical limits of detection (Table 2). When the samples being treated were limited to more narrow suites, e.g. soils, the errors in background determination were even further reduced. The method requires that the concentration of major elements should be accurately known, however, and exhaustive tests have shown that Kevex EDS, combined with powder briquettes, gives results which are not sufficiently accurate.

SOURCE OF BACKGROUND VARIATIONS - DIFFRACTION EFFECTS

By virtue of the very fine particle size of the components in a briquette (90 per cent less than 75 micrometres) the overall surface area of the sample of immense. These fine particles have a completely random orientation in the surface of the briquette, although plate-like minerals have been known to show a preferred orientation parallel to the pressing die surface. The Bragg equation will be frequently satified, as shown in the sketch in Figure 3, in which continuum of a wavelength for which the monochromator is set is diffracted into the detector. A large range of incident angles is possible. On average (62.5° for ARL 72000S), for monochromators set for elements with wavelengths shorter than FeKα(1.94Å), direct diffraction of the tube continuum into the monochromators will occur only from d-spacings of less than about 1.1 Å.

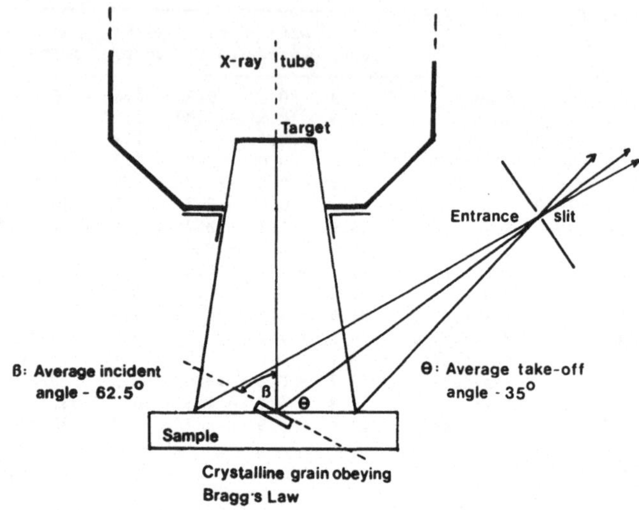

Figure 3. Example of diffraction by a particle in the surface of a pressed powder briquette.

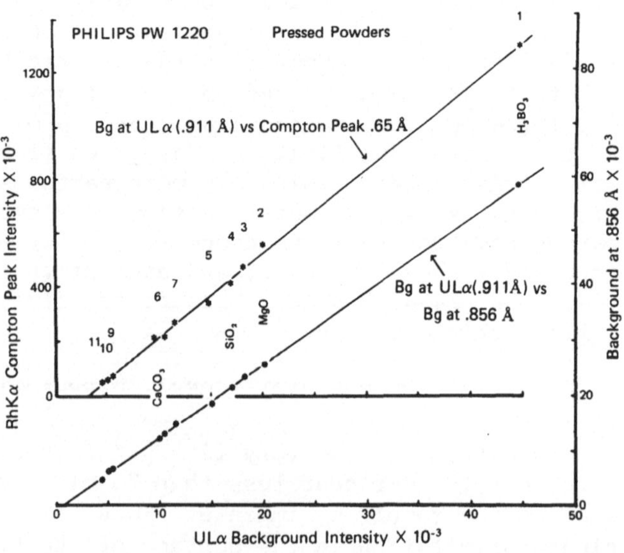

Figure 4. Comparison of background relationships
- U Lα (.911 Å) vs .856 Å
- U Lα (.911 Å) vs .650 Å

By returning to the set of oxide blanks in the original investigation (Figure 2), it was confirmed that if the reagents used to prepare these blanks gave rise to X-ray diffraction peaks from d-spacings of less than 1.1 Å, the background was enhanced.

When comparing the background positions which are very close and where all other instrumental conditions are kept constant, the relationship is good, because the contribution from the crystalline characterisitics of the briquette components will be practically the same at both background positions. For example, a better relationship is obtained when the ULα (0.911 Å) background is related to nearby background at 0.856 Å using the Philips PW1220, as shown in Figure 4, and may be compared in the same Figure with the relationship with the somewhat distant RhKβ Compton peak (0.65 Å).

It is impractical to measure background adjacent to every peak of interest. In the case of multi-channel spectrometers, with fixed wavelength channels and differing take-off angles, analyzing crystals and detectors, it becomes impossible to measure adjacent background positions under the same instrumental conditions. Consequently, the problem has to be overcome by the mathematical model described above, or by a simple direct approach of matching blanks which the anticipated mineralogy. It has been found, for soils from South Africa, that a range of blanks from quartz through 50/50 quartz/$CaCO_3$ to 50/50 quartz/Fe_2O_3 approximates to their typical mineralogy. In the determination of U and Th in ores from Witwatersrand gold mines, a quartz-pyrite binary is most successful. In calcines of sulphide concentrates, a quartz-Fe_2O_3 binary gives excellent results.

By using the latter approach, the system has been shown to produce highly precise results with excellent lower limits of detection (Table 1). Results obtained for internationally accepted standards are in good agreement with recommended values.

DATA INTERPRETATION

The combination instrument has been in operation for over two years, and experience has shown that the practical production is between 7 500 and 10 000 samples per month, mainly stream sediments and soil reconnaisance geochemical prospecting samples. This gives rise to a formidable volume of data; between 270 000 and 360 000 element determinations per month. To facilitate interpretation of these data, the output is automatically transferred by an off-line PDP 11/24 computer to a geochemical data-base and archiving system running in an HP 3000/44 computer. The latter facility provides, among other services, computer plotted concentration contour maps which may be used as overlays on geological maps of the same scale.

CONCLUSIONS

36 elements are determined simultaneously in up to 10 000 samples per month by using a unique combination of an ARL 72000S spectrometer with 28 monochromators, a Kevex EDS, and Herzog sample changers. Errors in determination of background for trace elements were found to be due to the diffraction of X-rays from the crystalline phases in the surface of the pressed powder briquettes. Correction factors based on major element contents of the samples have proved successful in overcoming the problem, but efficiency of this approach was found to depend upon an accurate determination of major elements. In general, it was found that the EDS, in combination with pressed powder briquettes, could not provide a sufficiently accurate determination of major elements.

An alternative approach of matching the blanks with the expected range of mineralogy of the unknowns has proved successful, and is in daily use. The output is in machine readable form, and is automatically transfered to a geochemical data-base for further processing.

ACKNOWLEDGEMENTS

The assistance and advice of Mr. D. Sermin of Applied Research Laboratories (Bausch and Lomb), Switzerland, and of Prof. J.P. Willis, Department of Geochemistry, University of Cape Town, is gratefully acknowledged. Also, the authors would like to thank their colleagues for advice and encouragement during the design and commissioning of the equipment, the Management of the Metallurgical Laboratory of Anglo-Transvaal Consolidated Investment Co. Ltd. for permission to use their Philips PW1400 spectrometer, and the Management of Anglo American Research Laboratories for support and permission to publish this paper.

REFERENCES

1. C. van Zyl, Rapid Preparation of Robust Pressed Powder Briquettes Containing a Styrene and Wax Mixture as Binder, X-ray Spectrometry, 11:29 (1982).
2. H.E. Marr, XRF4 - Computer Programming for X-ray Analysis, Bureau of Mines, I.C. 8712, U.S. Dept. of Interior (1976).
3. R.C. Reynolds, Matrix Corrections in Trace Element Analysis by X-ray Fluorescence: Estimation of Mass Absorption Coefficients by Compton Scattering, Am. Mineral., 48:1133 (1963).
4. R.C. Reynolds, Estimation of Mass Absorption Coefficients by Compton Scattering: Improvements and Extensions to the Method, Am. Mineral., 52:1493 (1967).
5. C.E. Feather and J.P. Willis, A Simple Method of Background and Matrix Correction of Spectral Peaks in Trace Element Analysis by X-ray Fluorescence Spectrometry, X-ray Spectrometry, 5:41 (1976).

ELEMENTAL ANALYSIS OF GEOLOGICAL SAMPLES USING A MULTICHANNEL, SIMULTANEOUS X-RAY SPECTROMETER

J. B. Cross and L. V. Wilson

Phillips Petroleum Company

Bartlesville, OK 74004

INTRODUCTION

X-ray spectrometry has been used successfully for major and trace element analysis of geological samples (e.g. reference 1). Its advantage is providing accurate and precise results in a rapid manner: a key factor in selecting analytical methods for mineralogical studies involving large numbers of samples. All elements with atomic numbers greater than 9 (fluorine), except noble gases, can be determined with sensitivities ranging down to ppm levels.

The analytical method described in this report is for the determination of the major elements (i.e. Na, Mg, Al, Si, P, K, Ca, Ti, Mn and Fe) in geological samples. Commercially available automated fusion devices are used for sample preparation. Subsequent rapid and precise sample measurements were made with a multichannel, simultaneous, wave-length dispersive x-ray spectrometer. The method provides a high throughput of samples without loss of accuracy and precision. Described are details on sample fusions, standard preparations, computation methods and customer analytical reports. The method is evaluated through the analysis of well-documented geological standards. Applications to samples collected by Phillips geologists provide additional method evaluation.

EXPERIMENTAL

All measurements were made on a Siemens MRS-400 simultaneous, multichannel x-ray fluorescence spectrometer. It is equipped with 23 fixed channels (individual spectrometers) and a scanning channel, all under vacuum. A PDP 11/04 Digital computer/RX01 floppy disk system controls the spectrometer and is used for data storage and analytical calculations. The analytical measurements (x-ray intensities) for all 10 elements of interest are made simultaneously in a period of about 100 seconds per sample. Each channel is equipped with a logarithmically curved crystal for increased sensitivity and resolution. Both gas-flow proportional counters and NaI scintillation detectors are used, depending upon the element measured.

Samples were prepared by first grinding 6 to 8 grams to 1-20 microns with a Retsch automated grinder. Soil samples were dried at 105°C for 16 hours and rock samples at 140°C for 3 hours to remove absorbed moisture. The dried sample (0.8g) was fused in 6 grams of lithium tetraborate ($Li_2B_4O_7$) and 1 gram ammonium nitrate. The fusions were done on an Angstrom Inc. automated fusion device ("Puff") in platinum/gold alloy crucibles. The "Puff" heats the samples to 550°C for 8 minutes, then for 12 minutes at 1100°C. The crucible and burner are then tilted to $\sim$ 45° and rotated at 3 rpm, while the burner is maintained at 1100°C, for 20 minutes. The fused materials were cooled in the crucibles and the resulting glass disc measured without further preparation. (The bottom of the crucible must be polished periodically to produce discs of suitable quality for direct measurement). Loss-on-ignition (LOI) values were determined by heating a portion of the ground, dried sample to 950°C for 1 hour and measuring the weight loss.

Standards were prepared from blends of ultrapure, assayed compounds of the elements of interest, fused in a manner similar to that used for the samples. The standards were prepared in sets of 24, with 10 elements per standard. The sets were made to cover preselected concentration ranges. The instrument was calibrated in this manner because more accurate matrix corrections can be made when covering limited compositional ranges rather than using a single set to cover all samples. As little correlation as possible was maintained between any individual element distribution and that of any other element. This was done to obtain best results using the regression correction procedures described below.

Chemicals used for the standards were puratronic grade MgO, Al_2O_3, SiO_2, Al_2PO_4, $CaCO_3$ and Fe_2O_3 from Johnson Mattley Chemicals, 4-9's pure TiO_2 from Spex Industries, Inc., spectrographically pure $Li_2B_4O_7$ from Angstrom, Inc., and reagent grade $Na_2B_4O_7 \cdot 10H_2O$, $MnC_2H_4 \cdot 2H_2O$, $KHC_8H_4O_4$ and NH_4NO_3. All materials were analyzed to insure accurate concentration values and absence of trace contamination in the prepared standards. The materials were weighed into the fusion crucibles using a four place electronic balance for weights above 0.04 gram and with a micro balance for smaller weights.

The calibration method supplied by Siemens Corp.[3] is based upon either a linear or quadratic calibration curve with optional correction terms for the effects of accompanying elements. These effects include enhancement, absorption and direct line interference. The correction terms (calibration curve coefficients) are determined using a regression formula with a minimization condition on the difference between the calculated concentrations and the chemically determined values of the standards. The calibration curves thus determined from a set of standards are stored on floppy disk and then read into the computer prior to analysis. The day-to-day instrument drift is corrected by measurement of reference specimens.

As noted above, the most accurate analysis of samples with varying compositions requires multiple sets of standards. The calibration curves determined from each set of standards were stored on a single computer disk. The sample x-ray intensities were measured, then the set of standards giving the best analysis conditions was selected automatically by the operating program. The set of calibration curves is chosen which provides elemental answers on a sample which all fall within the calibrated ranges of that set of standards. If none of the sets of calibration curves meet this criterion, then the set giving the least total absolute value of out of range condition is chosen. We typically use from two to four sets of curves (10 curves per set) for a given mineralogical study.

Results and Discussion

The advantage of this method for major element analysis of geological samples is its speed coupled with a high degree of accuracy and precision. With two "Puff" fusion units in operation, 20 samples per day can be prepared. The 100 samples, prepared in one week, can be measured and the analytical reports completed in less than eight hours. Additional "Puff" units can increase the sample throughput (one technician can operate 4 units effectively). The accuracy and precision of the method are very good, resulting in an analytical tool well suited for geological exploration and production studies.

Applications of the method to well-documented geological reference materials show the typical accuracy obtained. Listed in Table 1 are the results on USGS reference rocks AGV-1, G-2, and GSP-1 compared to accepted values given by Abbey.[2] The results for each element coupled with the oxide totals show the XRF method has excellent absolute accuracy for each element and for the complete analysis of the sample as well.

Further evidence of the accuracy achieved with the method is given in Table 2. Listed are the average percent deviation for the major elements and oxide totals from the accepted values listed by Abbey[2] for 20 different geological reference materials. Included in the table are precision values: the standard deviations and 95% confidence levels determined from duplicate analysis of the reference materials. Both excellent absolute accuracy and relative precision are evident in the individual element analyses as well as in the complete analyses. The oxide totals include loss-on-ignition

Table 1. Analysis of USGS Geological Reference Materials; AGV-1, G-2 and GSP-1

Reference Material	SiO_2	AL_2O_3	Na_2O	K_2O	Fe_2O_3	CaO	MgO	TiO_2	MnO	P_2O_5	LOI[b]	Oxide Totals
AGV-1												
XRF	59.48	17.22	4.22	2.96	6.73	4.93	1.53	1.06	0.10	0.49		99.6
[a]Accepted	59.61	17.19	4.32	2.92	6.78	4.94	1.52	1.06	0.10	0.51	0.9	99.9
%Diff.	0.22	0.17	2.31	1.37	0.74	0.20	0.66	0	0	3.92		0.3
G-2												
XRF	69.18	15.49	4.09	4.42	2.70	1.90	0.82	0.49	0.03	0.13		99.9
[a]Accepted	69.22	15.40	4.06	4.46	2.69	1.96	0.75	0.48	0.03	0.13	0.6	99.8
%Diff.	0.06	0.58	0.74	0.90	0.37	3.06	9.33	2.08	0	0		0.1
GSP-1												
XRF	67.21	15.03	2.72	5.46	4.26	1.98	0.99	0.67	0.04	0.28		99.0
[a]Accepted	67.32	15.28	2.81	5.51	4.30	2.03	0.97	0.66	0.04	0.28	0.4	99.6
%Diff.	0.16	1.64	3.20	0.91	0.93	2.46	2.06	1.52	0	0		0.6

a) Accepted Values from Abbey[2].
b) LOI Values calculated from results listed in Abbey[2].

Table 2. Statistical Comparison to Geological Reference Materials

Oxide	Concentration Range (wt%)	"Accuracy" % Deviation From Accepted Values[a] %	"Precision" Std. Dev. Duplicates (wt%)	[b]95% Confid. Limits Duplicates (wt%)
SiO_2	33.9-75.7	.49	.213	.304
AL_2O_3	5.7-18.5	.97	.111	.159
Na_2O	0.4-9.0	3.38	.063	.091
K_2O	0.2-15.4	2.18	.003	.004
Fe_2O_3	1.4-17.8	2.17	.060	.086
CaO	0.7-20.7	2.13	.026	.037
MgO	0.1-13.5	4.08	.034	.049
TiO_2	0.04-3.7	2.85	.007	.010
MnO	0.01-0.8	1.98	.003	.004
P_2O_5	0.04-0.5	4.67	.003	.004
Oxide Total	97.2-100.6	.58	.314	.456

(a) Accepted values from Abbey[2].

(b) 95% Confidence Limits: Limits around the average of duplicates which will include the true value 95% of the time.

values but do not include analyses of other components such as fluorine, sulfur and trace elements.

The analytical method was then applied to 13 geological samples submitted by Phillips geologists. Precision data similar to those presented for the reference rocks are given for these samples in Table 3. The same level of precision was obtained for these samples as for the reference material; furthermore, oxide totals for these samples were within ±1% of 100.

The hardware system supplied with the MRS-400 spectrometer provides a powerful means of handling data and preparing analytical reports. Sample data are stored in floppy disk files and identified by date, customer initials, sample number, and analysis branch number by use of programs developed in-house. A summary report, which provides the data of most interest, and a more complete report, which provides detailed information on each sample analysis, are made for the customer. The data may also be transferred to the customer on floppy disk with the data formatted on the disk to suit the individual customer's requirements.

CONCLUSIONS

The analytical results obtained in this study illustrate that XRF analysis of geological samples using a simultaneous, multichannel x-ray spectrometer is accurate, precise and rapid. The method employed in calibrating the spectrometer, both preparation of standards and calculation of matrix corrections, as well as the method developed for preparing the samples are sufficient for the analysis. Although this work was performed with two fusion devices

Table 3. Precision Study: On 13 In-House Geological Samples

Duplicate Analysis of 13 In-House Samples

Oxide	Conc'n Range	Avg.	Std. Dev.	% Rel. Error
SiO_2	79.8-90.7	85.94	.28	.32
AL_2O_3	4.4-7.9	6.17	.029	.47
Na_2O	.8-1.4	1.06	.026	2.50
K_2O	1.9-2.5	2.19	.012	.54
Fe_2O_3	.4-1.9	1.15	.041	3.61
CaO	.2-1.4	.46	.0028	.60
MgO	.2-1.5	.57	.010	1.75
TiO_2	.1-.4	.31	.0039	1.26
MnO	.01-.03	.019	.00045	2.35
P_2O_5	.01-.10	.041	.00055	1.34
Total	99.3-100.3	99.85	.33	.33

in operation, the addition of more units has significantly increased the sample throughput at the same manpower levels.

Future applications for this method will include trace elements in the simultaneous analysis scheme and expand the calibration ranges to cover such samples as limestones, ores (i.e. iron ore) and coal or coal-ash type samples.

REFERENCES

1. J. T. Hutton and S. M. Elliott, Chemical Geology, 29:1-11 (1980).
2. S. Abbey, Geological Survey of Canada, Paper 80-14, Part 6. 1979 Edition of "Usable" Values, (1980).
3. R. Plesch, Siemens X-Ray Analytical Application Note No. 28, Siemens Corp.

CHEMICAL ANALYSIS OF COAL BY ENERGY DISPERSIVE X-RAY FLUORESCENCE UTILIZING ARTIFICIAL STANDARDS

Bradner D. Wheeler

EG&G ORTEC

Oak Ridge, Tennessee 37830 U.S.A.

ABSTRACT

Accurate determinations of the elemental composition of coal by classical methods can be quite difficult and are normally very time consuming. X-ray fluorescence utilizing the powder method, however, has the ability of providing accurate and rapid analyses. Unfortunately, well characterized standards, although available, are not plentiful. In addition, the durability or stability of ground and pelletized coal samples is poor resulting in deterioration with time. As a result, artificial coal standards were prepared from certified geological materials by fusing in lithium-tetra-borate in percentages approximating expected ash contents and compositions in coal. Since the lithium-tetra-borate comprises about the same percentage of the standard as does the carbon, hydrogen, and oxygen in coal, the ground and pelletized coal sample can be assayed against the fused calibration curves by compensating for the differences in the mass absorption coefficients of the two matrices.

INTRODUCTION

Background

As a result of the world's dramatic increase in energy demands coupled with increasing cost and decreasing availability of petroleum type fossil fuels, major interest in coal as a significant substitute for oil has developed due to the extensive reserves available. Although coal is quite abundant, its utilization as a fuel can create problems in design and operation of boilers such as fused slag deposits, ash corrosion and erosion, combustion ash release, slagging, fouling, and excess sulfur resulting in SO_2 emissions. The nature of the coal with the resulting characteristics of the ash are, therefore, of major concern both to the designer and the operator of the system. As a consequence, accurate and rapid analysis of the elemental composition becomes imperative.

Analysis of coal by conventional methods is a lengthy process usually involving several hours for the major constituents. Complete analysis including traces, even through atomic absorption or optical emission, can consume up to five to eight hours. Although these techniques are usually excellent for most trace element determinations, they are generally poor for the quantification of major elements. X-ray fluorescence, however, can be extremely accurate for the rapid determination of the major elements and excellent to good for the trace elements through atomic number 92.

ANALYTICAL PROCEDURES

Equipment and Operating Conditions

An EG&G ORTEC 6110 Energy Dispersive X-ray Spectrometer was utilized for this analysis and operated under the instrumental parameters as listed on Table 1.

TABLE 1: Operating Parameters for Coal Analysis

Elements Determined:	Na, Mg, Al, Si, P, S, K, Ca, Ti, Mn, Fe	Sr
Anode:	Rhodium	Rhodium
Filter:	none	Rhodium
Anode Voltage:	10 kV	35 kV
Anode Current:	75 μAmps	100 μAmps
Energy Scale:	0-10 keV	0-20 keV
Atmosphere:	Vacuum	Vacuum
Counting Rate:	200 seconds	20 seconds

Standards and Sample Preparation

Several sources of coal standards exist[1-3], but as stated previously, they are generally unstable with a period of time after preparation and also do not adequately cover the range of composition and ash content experienced by some of the consumers utilizing coal as a fuel. As a result, a system utilizing durable synthetic standards which could be compared directly to the quantitative analysis of whole coal samples appeared to offer a solution to the problem of standard availability and durability.

As mentioned previously, the availability of coal standards covering the expected range of compositions and ash contents are rather limited. As a consequence, a series of artificial standards were made by fusing varying amounts of well characterized materials in an amount of $Li_2B_4O_7$ to bring the total sample (on an ignited basis) plus the lithium-tetra-borate to a total sample weight up to grams to approximate the coal. Since the whole coal which will be analyzed against the artificial standards was made up with four grams of coal plus one gram of boric acid as a binder, one additional gram of $Li_2B_4O_7$ was added to the artificial standards to make a total weight of five grams. The standards and percentages used in making these standards are illustrated on Table 2.

TABLE 2. Compositions of Artificial Standards

STANDARD	COMPOSITION*	$Li_2B_4O_7$ (GRAMS)	NH_4NO_3** (GRAMS)
1	5% LIMESTONE 1c	4.8000	0.5000
2	10% LIMESTONE 1c	4.6000	0.5000
3	7% BAUXITE 697	4.7200	0.5000
4	7% SI BRICK 102 + 5% BAUXITE 697	4.5200	0.5000
5	2% SI BRICK 102 + 1% BUAXITE 697	4.8800	0.5000
6	10% OBSIDIAN 278	4.6000	0.5000
7	15% BASALT 688	4.4000	0.5000
8	15% CEMENT, BLUE	4.4000	0.5000
9	15% BAUXITE 697 + 20% K_2SO_4	4.0000	0.5000
10	10% CEMENT, BLUE	4.6000	0.5000
11	15% LIMESTONE 1c + 15% K_2SO_4	4.0000	0.5000
12	5% SI BRICK 102 + 2% DOLOMITE 88A	4.7200	0.5000
13	10% SI BRICK 102 + 10% BASALT 688	4.2000	0.5000
14	7% OBSIDIAN 278 + 6% LIMESTONE 1c	4.8000	0.5000
15	16% ALUMINA REFRACTORY 77A	4.3600	0.5000
16	8% SI BRICK 102 + 1% DOLOMITE 88A + 8% BAUXITE 697	4.3200	0.5000
17	11% ALUMINA REFRACTORY 77A + 8% SI BRICK 102	4.2900	0.5000
18	20% $CaSO_4$ + 5% SI BRICK 102	4.0000	0.5000

* ALL WEIGHTS ARE CALCULATED FOR A LOSS-FREE BASIS.

** OXIDIZING AGENT.

Each individual artificial standard as listed on Table 2 was prepared by fusing those percentages based on a one-gram sample (equated to loss free) with an appropriate amount of $Li_2B_4O_7$ to bring the total sample weight up to 5.0 grams. An oxidizing agent (NH_4NO_3) was also added in the amount of 0.5 grams. The entire sample was then fused and cast in a 95% Pt - 5% Au crucible at 1100°C for ten minutes.

Sample preparation techniques for the unknown coal samples utilized the powder technique. Since particle size effects can vastly affect the intensity relationships[4,5], one coal sample was ground in a tungsten carbide rotary swing mill (5 grams sample + 1 gram boric acid binder + 100 milligrams of sodium sterate grinding aid) for one to seven minutes. The resulting powders were pelletized with a boric acid backing at 15 tons per square inch and placed in the spectrometer for intensity measurement. In addition to establishing the optimum grinding time as illustrated on Figure 1, the optimum pelletizing pressure was also determined as displayed on Figure 2. As a result, all of the unknown coal samples were ground for five minutes and pelletized at 15 tons per square inch.

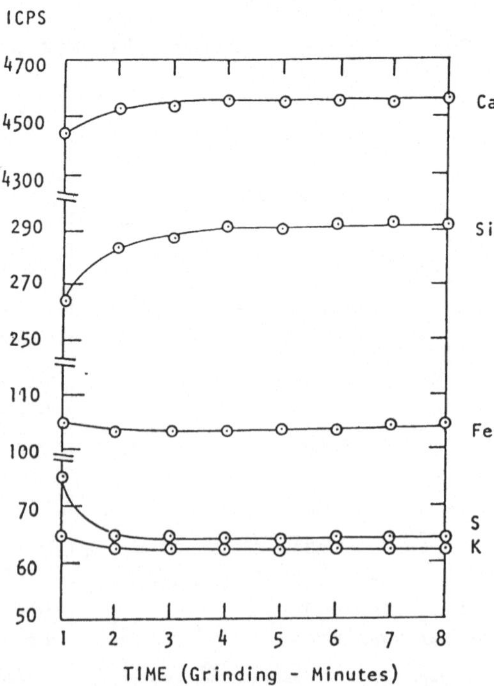

Fig. 1: Grinding Time vs Intensity

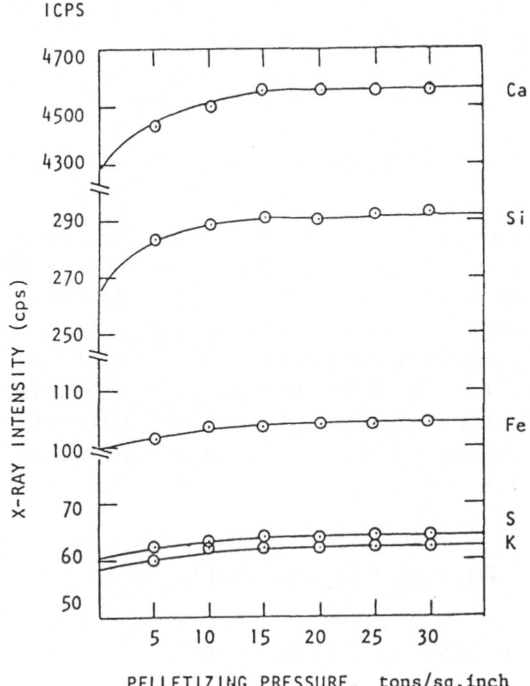

Fig. 2: Pelletizing Pressure vs Intensity

Analysis of Data

The intensity data was analyzed using the EG&G ORTEC matrix correction program, ATAC. This approach employs an exponential correction procedure for absorption and enhancement effects as a function of variations in the elemental intensities [6]. The basic equation utilized is as follows:

$$C_i = A_i + b_i I_i \exp\left(\sum_j m_{ij} I_j\right) \tag{1}$$

$$= A_i + B_i I_i \exp\left[\sum_j m_{ij} (I_j - \bar{I}_j)\right]$$

where $B_i = b_i \exp\left(\sum_j m_{ij} \bar{I}_j\right)$

I_i = intensity of assay element i

I_j = intensity of interferring element j

$\bar{I}_j$ = average intensity of element j of all standards in calibration m = proportionality constant derived from regression analysis of the standards

C_i = concentration of assayed element i

A_i = intercept of calibration curve

The interaction coefficients are determined by a non-linear multiple least squares fit of the standards concentrations/intensity data. This requires a minimum of n+6 standards where n is the number of interferring elements. Elemental concentrations were calculated with an iterative process using equation (1) with the interaction coefficients calculated from the standards. Analysis of this material required the utilization of the interelement correction equation (1) plus a stripping routine as stated in equation (2) in order to correct for overlapping lines.

$$C_i = a+b[I_iK\alpha+I_jL\alpha -(I_jK\alpha/I_jK\alpha)(I_jK\alpha_s)] \quad (2)$$

where C_i = concentration of element i

a = x intercept of calibration curve

b = slope of calibration curve

$I_iK\alpha+I_jK\alpha$ = measured intensity of the Kα of element i + Lα of interferring element j

$I_jK\alpha/I_jL\alpha$ = intensity ratio of Kα/Lα on pure element j

$I_jK\alpha_s$ = intensity of the Kα of element j in the sample

RESULTS AND DISCUSSION

The artifically lithium-tetra-borate glass standards were analyzed in an energy dispersive x-ray spectrometer. Interelement corrections were performed through the utilization of the equation (1). The interelement corrections are evident in low Z matrices with particular emphasis being placed at the low end of the periodic table. Examination of Figure 3 clearly illustrates the necessity of interelement correction particularly involving low Z elements which are periodic neighbors. A typical example would be in the analysis of relatively low concentrations of aluminum in the presence of high and variable concentrations of silica where the mass absorption coefficient of aluminum at the silica Kα energy is in excess of 4000.

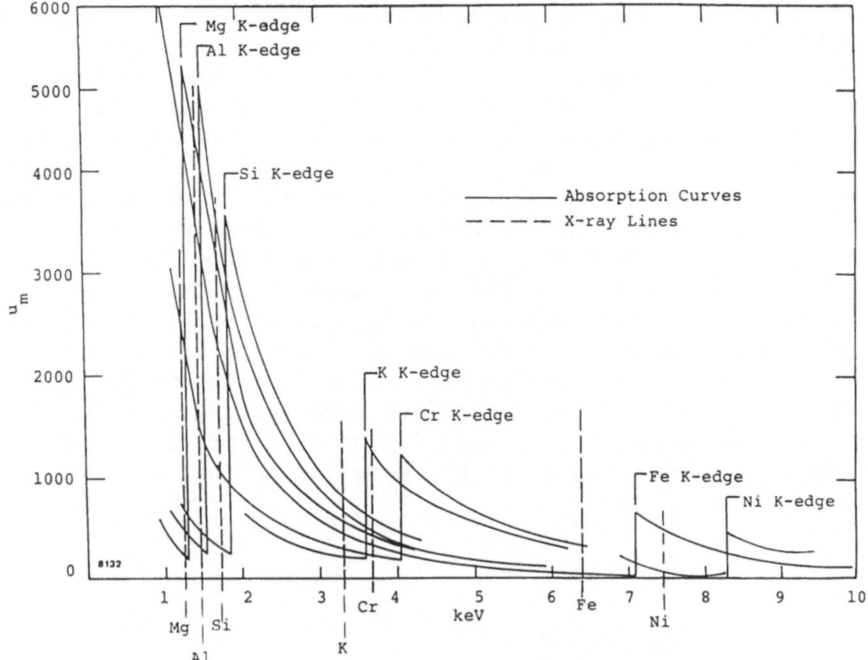

Fig. 3: Elemental Absorption Curves

Consequently, a correction must be calculated for the absorption of silica by aluminum in order to obtain reliable and meaningful results. Following this approach, the interelement corrections utilized for this analysis are illustrated on Table 3.

TABLE 3: Correction Protocol for Coal Analysis

Element	Interferring Element											
Assayed	Na	Mg	Al	Si	P	S	K	Ca	Ti	Mn	Fe	Sr
Na												
Mg			b	b		b						
Al		b		b		b						
Si		b	b			b						s
P				s		b						
S				b			b	b				
K						b		b				
Ca							b				b	
Ti								b			b	
Mn									b		b	
Fe*												

Notes: b = beer's exponential equation.
s = stripping equation.
* = linear equation.

The artificial standards are made from a lithium-boron-oxygen matrix while the whole coal samples are composed essentially of a carbon-hydrogen-oxygen matrix. Since the calibrations were performed utilizing the artificial standards, a correction equating the difference in the mass absorption coefficients of the two matrices becomes necessary. Examination of Figure 4 reveals that a direct correlation of mass absorption coefficients exists between a lithium-tetra-borate matrix to that of a carbon-hydrogen-oxygen. Although natural coals vary considerably in the concentrations of carbon-hydrogen-oxygen, the variation in mass absorption coefficients from a low carbon to one of high carbon content is rather small. As a result, an average mass absorption coefficient was calculated among the low to high quality coal as illustrated on Figure 4. Equating the real coal samples to the artificial standards involves substituting a proportionality constant into equations (1) and (2) which is merely a reflection in the ratio of the differences in the mass absorption coefficients of the two matrices.

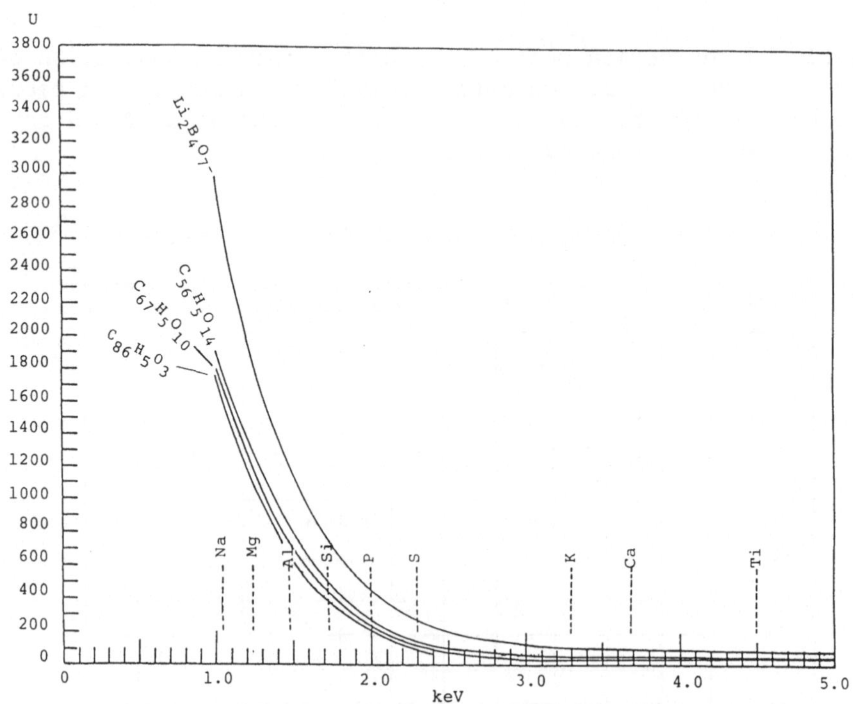

Fig. 4: Mass Absorption Coefficient of Coal

TABLE 4. Analysis of Coal (Summary)

Sample I.D.*	Composition (%)																					
	Na_2O		MgO		Al_2O_3		SiO_2		P_2O_5		S		K_2O		CaO		TiO_2		Fe_2O_3		Ash	
	List	Calc	List	Calc	List	Calc	List	Calc	List	Calc	List	Calc	List	Calc	List	Calc	List	Calc	List	Calc	List	Calc
100	-	<0.04	0.54	0.56	1.77	1.70	2.54	2.50	0.19	0.17	1.02	1.10	0.06	0.05	0.98	0.98	0.09	0.14	0.67	0.68	8.10	8.32
101	-	<0.04	0.52	0.48	4.22	4.06	11.60	11.60	0.12	0.09	0.57	0.52	0.37	0.36	1.55	1.42	0.13	0.10	0.65	0.62	20.67	20.50
102	-	0.09	0.29	0.26	4.16	3.99	11.49	11.62	0.045	0.051	4.19	4.10	0.51	0.41	1.06	0.99	0.19	0.12	3.40	3.20	22.58	22.70
103	-	0.09	0.15	0.14	2.42	2.43	6.11	6.05	0.025	0.019	3.94	3.79	0.26	0.21	0.64	0.56	0.12	0.10	2.41	2.34	12.87	12.65
104	-	0.10	0.11	0.08	1.90	1.96	4.48	4.54	0.015	0.019	4.50	4.31	0.24	0.19	0.36	0.32	0.09	0.07	3.30	2.94	10.74	10.95
105	-	0.13	0.11	0.10	2.32	2.29	4.19	4.20	0.045	0.052	1.25	1.36	0.25	0.30	0.06	0.06	0.10	0.11	0.67	0.82	7.83	8.00
106	-	<0.04	-	0.03	0.50	0.48	0.54	0.55	-	0.019	0.58	0.60	0.01	<0.01	0.04	0.02	0.02	0.05	0.12	0.15	1.29	1.15
107	-	<0.04	0.19	0.15	3.67	3.65	7.90	7.92	0.015	<0.015	0.49	0.53	0.36	0.42	0.08	0.08	0.17	0.19	0.52	0.61	13.05	13.20
108	-	<0.04	0.07	0.10	5.77	5.84	11.10	11.12	0.015	<0.015	0.70	0.75	0.58	0.62	0.03	0.03	0.37	0.37	0.25	0.18	18.48	18.67
109	-	<0.04	0.20	0.17	3.57	3.68	6.07	6.17	0.045	0.045	0.55	0.50	0.25	0.29	0.14	0.15	0.25	0.31	0.37	0.34	11.15	11.10
110	-	0.08	0.71	0.69	1.24	1.17	2.94	2.98	0.040	0.070	1.51	1.37	0.06	0.05	2.29	2.24	0.05	0.07	1.35	1.40	12.42	12.55
111	-	<0.04	0.42	0.47	2.03	1.92	3.58	3.80	0.14	0.14	0.91	1.04	0.01	0.01	2.26	2.23	0.14	0.14	0.45	0.39	10.60	10.50
120	0.09	0.10	0.14	0.15	2.26	2.35	6.47	6.36	0.030	0.031	3.52	3.58	0.25	0.23	0.73	0.73	0.10	0.09	2.06	2.06	13.06	13.05
121	0.06	0.08	0.16	0.18	2.60	2.67	7.26	7.22	0.036	0.040	3.47	3.33	0.36	0.34	0.60	0.67	0.13	0.10	3.31	3.44	15.01	15.15
122	0.03	0.04	0.34	0.34	1.72	1.75	4.79	4.60	0.052	0.075	0.50	0.59	0.12	0.15	1.96	2.11	0.08	0.07	0.73	0.83	11.00	10.91
123	0.05	0.08	0.10	0.12	2.05	2.10	4.78	4.83	0.027	0.022	3.45	3.55	0.24	0.22	0.35	0.39	0.10	0.08	2.79	2.84	10.99	11.06
124	0.07	0.05	0.18	0.21	2.85	2.91	7.61	7.51	0.016	0.052	1.04	1.14	0.35	0.42	0.52	0.66	0.17	0.18	1.40	1.58	13.61	13.56

*Samples plus listed values supplied by American Society Testing Materials, Committee D-5.

Note: List = certified value; Calc = value calculated by regression analysis.

The results of the analysis utilizing this procedure are summarized on Table 4 in which the calculated values are compared to the listed values[7]. Ash content can be directly calculated from the composition of the coal by summing the oxide concentrations and deducting the amount of sulfur which is in excess for combination with the available calcium and magnesium with correlation of the ash content illustrated on Figure 5.

CONCLUSIONS

As illustrated by the data, energy dispersive x-ray fluorescence is a viable technique for analyzing coal for elemental composition and ash content. Since coal standards are not particularly stable with time and some compositional ranges are not adequately covered, utilization of artificial standards and calculating the variations in the two types of matrices offers a convenient method of calibration.

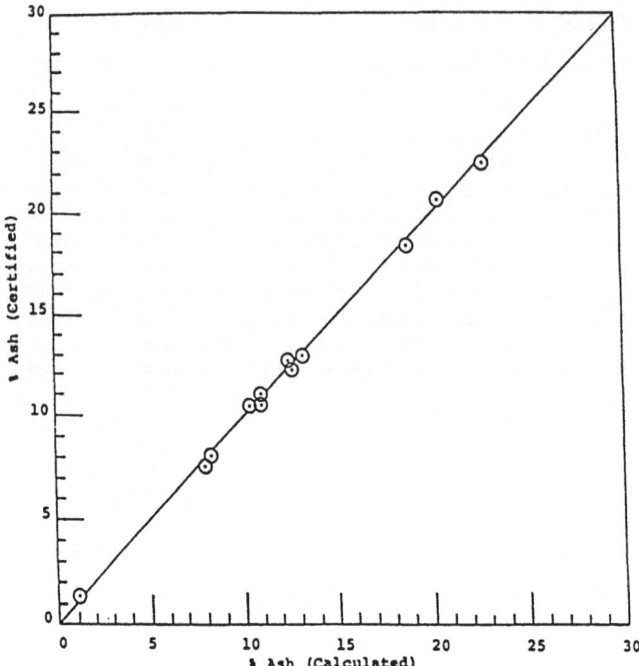

Fig. 5: Ash Content in Coal

REFERENCES

1. Alpha Resources, Inc. Stevensville, Michigan 49127.
2. Bramer Standard Company, Inc., Houston, Texas 77069.
3. National Bureau of Standards, Washington, D.C.
4. Bernstein, F., "Particle Size and Mineralogical Effects in Mining Applications", 11th Annual Conference for Application of X-ray Analysis, Denver Research Institute, University of Denver, 1962.
5. Wheeler, B.D., "Accuracy in X-ray Spectrochemical Analysis as Related to Sample Preparation", 1979 Symposium on X-ray Fluorescence Analysis in Agrochemistry, TZINO Institute, Moscow, U.S.S.R. September, 1979.
6. Hasler, M. F. and Kemp, J. W., "Suggested Practices for Spectrochemical Compositions", ASTM E-2 SM2-3, Philadelphia, PA, 1957.
7. American Society for Testing Materials, Committee D-5, Philadelphia, PA, 1982.

AUTHOR INDEX

Ames, L., 325
Arai, T., 423
Asada, E., 89

Ball, T. K., 409
Barral, M., 217
Baumgartner, F. C., 443
Beard, D. W., 99
Berry, W. S., 255
Boldrick, M. S., 275
Brown, A., 11, 53
Buman, A., 437
Byars, L. G., 355

Calvert, L. D., 105, 163
Carpenter, D. A., 307
Cherukuri, S. C., 99
Christensen, L. H., 377
Cline, J. P., 111
Cohn, S. H., 415
Conway, Jr., J. C., 299
Cross, J. B., 451

Dabrowski, A., 325
Das Gupta, K., 185, 341
DeGroot, P. B., 395
de Jesus, A. S. M., 391
DiMascio, P. S., 233
Drabaek, I., 377
Drummond, W., 325

Eaton, E. E., 137
Engler, P., 157

Fawcett, T. G., 171
Feather, C. E., 443
Foris, C. M., 53
Fujiwara, I., 89

Gainsford, G. J., 105, 163
Garbauskas, M. F., 81, 345
Garvey, R. G., 119
Gedcke, D. A., 355
Gehringer, R. C., 119
Gerron, R. A., 157
Gilfrich, J. V., 313
Goehner, R. P., 81, 345
Goldsmith, C., 259
Gurvich, Y. M., 437

Hare, T. M., 197
Hasegawa, K., 209
Hicho, G. E., 137
Hirose, Y., 291
Holomany, M., 87
Howard, S. A., 73
Huang, T. C., 35, 93
Hubbard, C. R., 45, 63, 87, 105, 149

Iwanczyk, J., 325

Jacobus, N. C., 355
Jenkins, R., 25, 141

Kamarchik, P., 129
Kato, T., 299
Kendall, D. S., 181
Kikkert, J., 401
Kirchhoff, P. M., 171
Kirkland, J. P., 313
Kodama, S., 283

LaBrecque, J. J., 337
Larchuk, T., 299
Lokshin, I., 437

Maeder, G., 245
Mantler, M., 351
Markho, P. H., 245
McCarthy, G. J., 119
McClune, W. F., 87
Melcher, D. M., 233
Misawa, H., 283
Mochiki, K., 209
Murray, J. J., 163

Nagel, D. J., 313
Newman, R. A., 171
Nichols, M. C., 189
Nielson, K. K., 385

Pangborn, R. N., 299
Parker, W. C., 337
Parrish, W., 35, 93
Perez-Mendez, V., 269, 275
Pfeiffer-Vollmar, H. W., 225
Post, B., 93
Prümmer, R. A., 225
Pyrros, N. P., 63

Qadri, S. B., 313

Ratliff, J., 129
Robbins, C. R., 149
Russ, J. C., 197
Ruud, C. O., 233

Sandborg, A., 431
Sanders, R. W., 385
Schreiner, W. N., 141
Sekita, Y., 283
Shen, R., 431
Sirianni, A. F., 105
Skelton, E. F., 313
Sleaford, B., 269
Smith, D. K., 119
Smith, G. S., 189
Smith, T. K., 409
Snyder, R. L., 1, 73, 99, 111, 149
Sohmura, T., 423
Sprauel, J. M., 217, 245

Tamenori, H., 423
Tanaka, K., 291
Taylor, J. B., 163
Torbaty, S., 217, 245
Toyohisa, S., 89

Ui, T., 89

van der Bank, D. J., 391
Vandermeer, R. A., 307
Vane, R. A., 369
Vartsky, D., 415

Wagner, C. N. J., 269, 275
Walker, G. A., 259
Webb, A. W., 313
Wesolinski, E. S., 391
Wheeler, B. D., 457
Wielopolski, L., 415
Wilson, L. V., 451
Wims, A. M., 205
Wong, E. R., 157
Wong-Ng, W., 87

Yajima, Z., 291
Yasumura, S., 415
Yeko, J., 157
Yoshioka, Y., 209

Zahrt, J. D., 331

SUBJECT INDEX

Absorption correction, EDXRF, 377
  with aerosols, 391
Accuracy in XRPD, 1
  figure of merit, 2
  peak finding, 6
  profile fitting, 6
  smoothing and stripping, 5
Accuracy in XRPD intensities,
  crystallite size, 35
  preferred orientation, 42
  profile fitting, 35
  Si, NBS powder, 36, 38, 41ff
Aerosols, correction of XRPD of, 391
Ag calibration for 2θ°, 46
Alignment of diffractometer, 25
  calibration curve, 30
  correction functions, 29
  systematic errors, 27, 31
α-Al_2O_3 (corundum), 49, 175
  particle size vs relative intensities, 114
  thermal expansion, 178
α-Fe_2O_3, SnO_2, 53
α-SiO_2 thermal expansion, 178
Alpha quartz XRPD standard, 30
Anatase, 172
Anisotropy, elastic, 219
Austenite, retained, 137
  vs Ni in Fe-Cr-Ni, 138
Automated diffractometer system, 197
Automated Huber-Guinier camera, 173
Automated search-match (see Search-match)

B_4C polarizer, 331
Background in XRPD, 66
Biological analysis, EDXRF, 385
Bone, Sr in, 415
Boron in glass, by XRF, 423
Bragg-Borrmann spectroscopy, 341
Broadening in diffraction of MgO, 301

Calibration, XRPD, 67
$CaNi_5$, 163ff
  hydrides, 163ff
Cast iron fractography, 291
Cell parameter refinement, 174
Cement, 439
CeO_2, 49
Chem-X, WDXRF spectrometer, 437
Coal, 388
  artificial standards, 457
  EDXRF analysis, 457
Coating, thickness by XRF, 431
Coherent/incoherent scattering, 380, 386
Correction functions, XRPD, 29
Crack propagation rates, 285
Cristobalite dust, 50
Cr_2O_3, 49
Cr_2O_3-Fe_2O_3, 120, 124, 126, 127
Crystallite size
  in XRPD, 38, 111
  standard, MgO, 50
Crystals, mosaic, 331

"d" values,
  basis of search-match, 89
  frequency distribution, 89

Data processing, JCPDS, 63
Deformation twinning, 309
Detector, mercuric iodide, 325 (see also Position sensitive)
Diffractometer,
  residual stress, 275
  use inside steel pipe, 233
Dölle-Cohen method for stresses, 245

Elastic constants,
  determination in Si wafers, 259
  macroscopic vs. X-ray, 220
Electronic circuits, PSD, 272
Empirical parameter equations, 351
Energy dispersive XRD,
  method for paint, 377
  search-match program, 345
  with fundamental parameters, 355, 369, 377
Enhancement correction, EDXRF, 377
EPRI pipe stress analyzer, 233
Errors in XRPD (see Accuracy in XRPD)
EXACT, fundamental parameters program, 369

Fatigue cracks, diffraction study of, 291
Fatigue, XRD study of, 283
$Fe(SO_4)_2(OH)_6$, 184
Figure of merit, XRPD, for search-match, 100
Filters, aerosol, 391
Fine-grained photo film used, 172
Flow and fracture from indentation, 299
Fluorophlogopite, for 2θ° calibration, 46
Flyash analysis, 153
Focussing XRPD camera
  convergent Soller slit, 185
  double monochromator, 187
Four-point bend tests, 235
Fractography, XRD, 283, 291
Fracture toughness vs. diffraction effects, 291
Fundamental parameters method,
  comparison of programs, 369
  programs, 355, 385

Gauss-Newton algorithm, 76, 79
Gaussian profiles, 75, 77, 173
Geochemical analysis, 401
Geological analysis, EDXRF, 385
Geological samples, XRF, 443, 451
Grain size effects, XRPD, phase identification with, 192
Guinier camera
  compared with diffractometer, 12, 17, 22
  errors and precision, 11ff
  for high temperatures, 171
Guinier powder patterns, 11, 53

Hanawalt search strategy, 99 (see also Search-match)
Hazardous wastes, 181
HgI_2, detector, 325
Huber-Guinier camera, 172
Hydriding, XRPD, in situ, 164

Intensity standards, XRPD, 49
  with solid solutions, 123
In vivo determination of Sr in bone, 415
IPT camera, Huber camera, 53
Iron,
  cast, fracture surfaces in, 291
  chromized, manganized, 225

Jarosite, 184
JCPDS
  data base,
    1982-1985 plans, 87
    evaluation, 88
    quality vs. search-match results, 101
  files and subfiles, 82, 86
  magnetic tape, simplifying the use of, 206
  round robin XRD test, 95
Johnson-Vand strategy, compared with Hanawalt's, 99

Limits of detectability, XRPD, 7

Liquid phase spherical agglomeration, 105
Long technique, for stress in thin film, 256
Lorentzian profiles, 75, 79, 142, 173

Marquardt algorithm, 76, 79, 143
Matrix, Compton scatter correction, 401
Mercuric iodide, detector, 325
MgO, 50
  flow, fracture, 299
  -NiO, 120, 124
Mica, 132
  synthetic for 2θ° calibration, 46
Microabsorption, 125
MoKα for steel XRPD, 270
Multichannel XRF, 437
Multiple wavelengths, in stress analysis, 217
Multi-scan XRPD for phase identification, 189

NBS
  standard reference materials, 45
  system for XRPD data, 63
NBS*AIDS80 program, 88
NBS*QUANT82 system, XRPD, 149
Ni vs. % retained austenite, 137
NIH elemental data base, 345
NiO-ZnO, 120
Nylon, 146

Oil shale standard COS-1, 389
Orchard leaves AGV-1, 388
Oxides, XRF analysis, 351

Paint pigments, EDXRF of, 129, 377
$Pb(NO_3)_2$, Mo_3Sb, As_2O_3, 53
Peak finding, XRPD, 6, 66, 69
  profile refinement, 69
Peak profile fitting, 173
Pearson VII profiles, 75, 77
Pellets vs. powders in EDXRF, 389
Phase transformation in U-Nb alloys,
  strain induced, 307
  thermoelectric martensitic, 309
Plastic strain distribution, XRD determination of, 296
Polarization, X-ray, 331
Position sensitive detector, 7, 8, 209
  curved, pressurized, 269
  electronic circuits, 271
  for liquids, glasses, 270
  for retained austenite, 270
  linear proportional, 281
  scintillation, 233
  using CrKα, 215
  using MnKα, 209, 215
Powder diffractometer (see also Position sensitive detector)
  errors and corrections, 13ff
  precision, 17ff
  slits, 12, 20
POWDER PATTERN data processing system, 63
Preferred orientation,
  aiding search-match, 190
  in mica samples, 48
  in XRPD, 2, 7, 49
  method for reducing, 105
  quantitative XRPD with, 129
  search-match with, 93, 97
Profile fitting, XRPD, 6, 46, 73, 141, 165
Profile refinement at NBS, 69
Profiles in diffraction from steels, 212
Pt-10% Rh thermal expansion, 177
Pt-Rh sample holder, 172

Qualitative analysis, EDXRF, 345
Qualitative XRPD programs, interactive, 81
Quantitative EDXRF, standard base, 355
Quantitative XRPD,
  NBS*QUANT82 phase analysis system, 149
  particle size effects, 111
  phase analysis with solid solutions, 119

profile fitting in XRPD, 74, 141
SCRIP program, 157
XRPD, effect of grain size, 160
Quartz,
CeO_2ThO_2, PtP_2, 53
quintuplet, profile fitted, 145
respirable, 50
XRPD analysis of, 153

Radial distribution function, apparatus for, 269
Radioisotope XRF excitation, 337
Reference intensity ratios, XRPD, 49
particle size effect, 111
Residual stress,
absorption calculation, 277
around indentation in MgO crystal, 299
averaging strain over depth, 277
Cr-Kβ used, 213
effect of absorption, 226
effect of gradients, 225, 277
effect of mechanical anisotropy, 219
fracture surface stress, 283
ground alloy steel, 214, 222
inside stainless steel pipe, 233
in Si wafers, 259
in stainless steel, 125, 209, 233
in steels, 225
integral ϕ method, 278ff
in thin film, 255
measurement accuracy, 237
MnKβ used, 215
ϕ-ψ diffractometer for, 275
polar distribution of, 252
principal stress near surface, inclined, 225, 245
PSD apparatus for, 209, 234
ψ-differential method, 278
$\sin^2\psi$ method, 214, 278
stress-free 2θ, 245
stress gradient effects, 217
stress tensor, 223
triaxial, with shear stresses, 245ff
vs. fatigue cracks, 284, 287, 288
Rock sample XRPD, 157
Rocking curve widths, 301
Round robin mixed phase samples, XRPD, 101
Rutile, 172

Scatter intensities, EDXRF, 385
SCRIP program for quantitative XRPD, 157
Search-match procedures,
automated with time-share, 205
grain size effects, 192
interactive programs, 81
modified Johnson-Vand program, 205
Siemens system, 99
using unique "d" values, 89
with 12 largest d's, 93
with Apple computer, 197
with complex organics mixture, 96
with multi-scans, 189
with preferred orientation, 190
with solid solutions, 119
without use of I's, 97
XRPD, computerized, 7, 81, 99, 197
Shape memory alloys, 307
Shear stresses from machining, 245
Si,
NBS powder, 36, 38, 41ff
particle size vs. relative intensities, 113
standard for 2θ°, 46
Silicon, XRPD relative intensities, 112, 153
Simplex algorithm, 76, 79
Simultaneous determinations of elements by XRF, 443, 451
Slip system activation, 303
Slit, θ compensating, 13, 20
Sodium ferrocyanide, 184
Soil5, 388
Solid solutions, XRPD intensities of, 119

Specimen preparation, XRPD, 36
Spectrometer, Borrmann transmission, 341
Spectrometer, wavelength dispersive Chem-X, 438
Spray drying XRPD samples, 105
$SrTiO_3$-$BaTiO_3$, 120, 124, 126
Standard addition method, XRF,
  sample preparation, 397
  with matrix correction, 395
Standard reference materials, XRPD, 45
  for retained austenite, 137
  preparation at NBS, 138
Standards, XRPD, for low angles, 30
Steel (see Residual stress)
Stress (see Residual stress)
Stress-free 2θ value, measured in steel, 245
Stress intensity factor, XRD study of, 283
Strontium, in bone, 415
Synchrotron radiation,
  for XRPD, 7
  XRF using, 313
SYNROC (multiphase ceramic), 198

Talc, 184
$TaSi_2$ film on Si, 259
Te deposit, 147
Thermal diffuse scattering, 146
Thermal expansion, by XRPD, 171
Thin films, stresses in, 255
TiC powder, relative XRPD intensities, 113ff
TiO_2, 49, 172
  anatase and rutile thermal expansions, 17, 18
Topograph for stress in thin film, 255
Tridymite dust, 50

U-Nb shape memory alloys, 307
Using backscatter, 385

Vegetation,
  XRF analysis, related to exploration, 409
Voigt profiles, 75, 77

W calibration for 2θ°, 46
WL spectrum photography, 342
WSi_2 film on Si, 259

XRF (see also Fundamental parameters method)
  as a prospecting tool, 443
  multichannel spectrometer, 443, 451
  of boron in glass, 423
  of light elements, 437
  spectrometer, WDXRF, 438
XRF11, 369
XRPD plus IR for identification, 181

Zinc yellow, 184
ZnO, 49